THE DARPA MODEL FOR TRANSFORMATIVE TECHNOLOGIES

The DARPA Model for Transformative Technologies

Perspectives on the U.S. Defense Advanced Research Projects Agency

Edited by William B. Bonvillian, Richard Van Atta, and Patrick Windham

OpenBook Publishers

https://www.openbookpublishers.com

All external links were active at the time of publication unless otherwise stated and have been archived via the Internet Archive Wayback Machine at https://archive.org/web

Updated digital material and resources associated with this volume are available at https://doi.org/10.11647/OBP.0184#resources

Every effort has been made to identify and contact copyright holders and any omission or error will be corrected if notification is made to the publisher.

ISBN Paperback: 978-1-78374-791-7
ISBN Hardback: 978-1-78374-792-4
ISBN Digital (PDF): 978-1-78374-793-1
ISBN Digital ebook (epub): 978-1-78374-794-8
ISBN Digital ebook (mobi): 978-1-78374-795-5
ISBN XML: 978-1-78374-796-2
DOI: 10.11647/OBP.0184

Cover design: Anna Gatti.

Contents

Notes on Contributors

William B. Bonvillian is a Lecturer at MIT, and Senior Director at MIT's Office of Digital Learning, leading a project on workforce education. From 2006 until 2017, he was Director of MIT's Washington Office, supporting MIT's historic role in science policy. He teaches courses on innovation systems at MIT and is coauthor of three books on innovation, *Advanced Manufacturing: The New American Innovation Policies* (2018), *Technological Innovation in Legacy Sectors* (2015), and *Structuring an Energy Technology Revolution* (2009), as well as numerous articles. Previously he worked for over fifteen years on innovation issues as a senior advisor in the U.S. Senate, and earlier was a Deputy Assistant Secretary of Transportation. He serves on the National Academies of Sciences' standing committee for its Innovation Policy Forum and chairs the Committee on Science and Engineering Policy at the American Association for the Advancement of Science (AAAS). He was elected a Fellow of the AAAS in 2011. He has a BA from Columbia, an MAR from Yale and a JD from Columbia.

Tamara Carleton, PhD, is the CEO and founder of Innovation Leadership Group LLC and lead author of the *Playbook for Strategic Foresight and Innovation* (2013), a hands-on guide that has been used by hundreds of the world's most innovative companies to make their teams more successful. She is the executive director of the Silicon Valley Innovation Academy at Stanford University and a visiting professor at the Osaka Institute of Technology in Japan. Previously she was an Innovation Fellow with the US Chamber of Commerce Foundation, a Fellow with the Bay Area Science and Innovation Consortium, and a Fellow at the Foundation for Enterprise Development. She has worked as a management consultant at Deloitte Consulting LLP, specializing in innovation, customer experience, marketing strategy, and enterprise applications. A multidisciplinary scholar, Dr. Carleton holds a doctorate in mechanical engineering from Stanford University, a master's of science from Syracuse University, and a bachelor's degree from The George Washington University.

David W. Cheney is a consultant and Managing Partner of Technology Policy International, a firm that provides analyses of science, technology, and innovation policy. He is the former Director of the Center for Science, Technology and Economic Development at SRI International, where his work focused on planning and evaluating science, technology, and innovation programs and institutions, primarily in the United States and Middle East. He is also a consultant to the World Bank and has been an adjunct professor at George Mason University. Before joining SRI in 1998, he was a senior executive in the U.S. Department of Energy, serving as director of the Secretary of Energy Advisory Board and advisor to the Deputy Secretary on industrial partnerships and national laboratories. He previously was a senior associate with the Council on Competitiveness, and an analyst with the Congressional Research Service. He has also held positions with the Internet Policy Institute, the Optoelectronics Industry Development Association, the Competitiveness Policy Council, and the Institute for Policy Science at Saitama University in Japan. He has a PhD in public policy from George Mason University, a MS in Technology and Policy from MIT and a BS in Geology & Biology from Brown University.

Phech Colatat is Assistant Professor of Strategy at the Olin Business School, Washington University in St Louis. He is a business school-trained sociologist with interests in healthcare, R&D, and strategic management. Motivated by alarming trends in the prevalence of autism spectrum disorder (ASD), his current research examines the way organizational and social network processes affects the diagnostic process. He completed his PhD at the MIT Sloan School of Management.

Robert Cook-Deegan, PhD, is a professor in the School for the Future of Innovation in Society, and with the Consortium for Science, Policy and Outcomes at Arizona State University. He founded and directed Duke's Center for Genome Ethics, Law & Policy (2002-12), and

Duke-in-Washington through June 2016. Prior to Duke, he was with the National Academies of Science, Engineering and Medicine (1991-2002); National Center for Human Genome Research (1989-90); and congressional Office of Technology Assessment (1982-88). His research interests include science policy, health policy, biomedical research, cancer, and intellectual property. He is the author of *The Gene Wars: Science, Politics, and the Human Genome* (1994) and more than 250 other publications.

Glenn R. Fong is an associate professor of global studies at Thunderbird School of Global Management. He is also the academic director of the school's Master of Arts in Global Affairs and Management. His areas of expertise include technology, global trade and industrial policies of the U.S., Japan and China, government and business relations and international political economy. Of Chinese-American ancestry, Fong has contributed commentaries and monographs to *Business and Politics, Issues in Science and Technology, International Security, International Studies Quarterly, Comparative Politics,* and the *Journal of Policy Analysis and Management*. In 1996, Fong authored *Export Dependence vs. the New Protectionism: Trade Policy in the Industrial World*. He has served as a consultant to the National Academy of Sciences, the U.S. Congressional Office of Technology Assessment, Japan's Ministry of International Trade and Industry (now METI), and IBM Corporation's e-Business Technology division. He earned his BA at the University of California, Berkeley, and MA and PhD degrees in government from Cornell University. Earlier in his career, he was an assistant professor at the University of Illinois-Chicago and a postdoctoral research fellow at Harvard University's Graduate School of Business Administration.

Erica R. H. Fuchs is a Professor in the Department of Engineering and Public Policy at Carnegie Mellon University, and a Research Associate with the National Bureau of Economic Research. Her research focuses on the development, commercialization and global manufacturing of emerging technologies, and national policy in that context. She was the founding Faculty Director of Carnegie Mellon University's Manufacturing Futures Initiative—an initiative across six schools aimed to revolutionize the commercialization and local production of advanced manufactured products. Over the past decade, Dr. Fuchs has played a growing role in national and international meetings on technology policy, including being one of twenty-three participants in the President's Council of Advisors on Science and Technology workshop that led to the creation of the Advanced Manufacturing Partnership, and serving on the expert group that supported the White House in the 2016 Innovation Dialogue between the U.S. and China. In 2012 she was selected a World Economic Forum "Young Scientist" (top 40 under 40 globally.) She currently serves on the National Academies' National Materials and Manufacturing Board; the Academic Advisory Board for MIT's Institute for Data, Systems, and Society, of which MIT's Technology Policy Program is a part; the World Economic Forum's Future of Production Global Futures Council; and the Advisory Editorial Board for *Research Policy*. Before coming to CMU, Dr. Fuchs completed her PhD in Engineering Systems at MIT in June 2006. She received her Masters and her Bachelor's degrees also from MIT in Technology Policy (2003) and Materials Science and Engineering (1999), respectively. Dr. Fuchs spent 1999-2000 as a fellow at the United Nations in Beijing, China. She grew up and attended K-12 in the Reading Public School District in Reading, PA. Her work has been published among other places in *Science,* the *Nature* journals, *Research Policy*, and *Management Science*; and has been covered on National Public Radio, by Bloomberg, and in the *New York Times*.

Larry Jackel is President of North-C Technologies, where he does professional consulting in robotics and machine learning. He also currently serves as a Learning Advisor to the Stanford Artificial Intelligence Lab (SAIL)-Toyota Center for AI Research at Stanford University. From 2003-07 he was a DARPA Program Manager in the IPTO and TTO offices. He conceived and managed programs in Autonomous Ground Robot navigation and Locomotion. For most of his scientific career Jackel was a manager and researcher in Bell Labs and then AT&T Labs. Members of Jackel's Adaptive Systems Department at Bell Labs laid the foundation for much of the machine learning that dominates AI today. He has also created and managed research groups in microscience and microfabrication, and in carrier-scale telecom services. Jackel was a founder of the Snowbird Workshop on Neural Networks for Computer and led the workshop many years. He was also a founder of the NIPS conferences. He has served as Program Chair for IJCNN. He has also been an organizer of the Frontiers in Distributed Information Systems (FDIS) workshop series. Jackel holds a PhD in Experimental Physics from Cornell University with a thesis in superconducting electronics. He is a Fellow of the American Physical Society and the IEEE. He has published over 150 papers and has over twenty patents.

Michael Piore has been on the faculty of the Department of Economics at MIT since 1966, and also currently holds a joint appointment with the Department of Political Science. He is also currently a Visiting Senior Fellow in International and Public Affairs at the Watson Institute for International and Public Affairs at Brown University. He earned his undergraduate and graduate degrees at Harvard University, where he wrote his doctoral dissertation under the direction of John T. Dunlop. He is the founding director of the MIT-Mexico Program and former associate director of the Center for Technology, Policy and Industrial Development. He has served as president of the Society for the Advancement of Socio-Economics (SASE) and as an elected member of the executive committee of the American Economic Association. He was a MacArthur Prize Fellow (1984-89), a member of the Executive Committee of the American Economic Association (1990-95), and a member of the Governing Board of the Institute for Labour Studies of the International Labour Organization (1990-96).

Jinendra Ranka, PhD, has over twenty-five years of experience in academic, commercial, and government research and development. He was a technical staff member at Bell Labs, Sycamore Networks, and MIT Lincoln Laboratory, and is currently the CEO at JASR Systems. Dr. Ranka served as a program manager in the Strategic Technology Office at DARPA and as a Deputy Office Director at IARPA. He has over 40 publications with over 5,000 citations and numerous patents. Dr. Ranka received his Doctoral degree in Applied & Engineering Physics from Cornell University in 1997 and his Bachelor of Science degree in Electrical Engineering from the California Institute of Technology in 1991. Dr. Ranka is a fellow of the Optical Society of America and is known for his discovery of supercontinuum generation in optical fibers.

Elisabeth B. Reynolds is the executive director of the MIT Industrial Performance Center and a lecturer in MIT's Department of Urban Studies and Planning. Reynolds works on issues related to systems of innovation, regional economic development, and industrial competitiveness. She is a member of the Massachusetts Advanced Manufacturing Collaborative as well as the Northeast Clean Energy Council. Her current research focuses on the pathways that U.S. entrepreneurial firms take in scaling production-related technologies, as well as advanced manufacturing, including the globalization of the biomanufacturing industry. Before coming to MIT for her PhD, Reynolds was the director of the City Advisory Practice at the Initiative for a Competitive Inner City (ICIC), a non-profit founded by Professor Michael Porter, focused on job and business growth in urban areas. Reynolds has an AB from Harvard in government and was a Fiske Scholar at Trinity College, Cambridge. She holds an MSc from the University of Montreal in economics and a PhD from MIT in urban and regional studies.

Richard Van Atta's career has focused on the national security policy, strategy, and technological capabilities of the United States for the Department of Defense, chiefly for the Office of the Secretary of Defense and the Defense Advanced Research Projects Agency, the Office of Science and Technology Policy (OSTP) of the White House, and the Intelligence Community. From 1983 to his retirement in 2018, he was a senior research staff member of the Strategy, Forces and Resources Division (SF&RD), the Science and Technology Policy Institute (STPI) and the Science and Technology Division at the Institute for Defense Analyses (IDA), with a focus on innovation for national security. His work at IDA included assessments of the programs and development strategies of the Defense Advanced Research Projects Agency (DARPA). From 1993 to 1998, he served as Assistant Deputy Under Secretary of Defense for Dual Use and Commercial Programs (on temporary assignment from IDA). He also was an adjunct faculty member in Georgetown University's Security Studies Program and the Science, Technology and International Affairs (STIA) program teaching courses on Emerging Technology and Security. Dr. Van Atta has a PhD in Political Science from Indiana University and a BA degree in Political Science from the University of California, Santa Barbara.

Patrick H. Windham is a Lecturer in the Public Policy Program at Stanford University and a Partner with Technology Policy International, a consulting firm. In the past he has taught at the University of California's Washington, DC, center and the University of Maryland. From 1984 until 1997 he served as a Senior Professional Staff Member for the Subcommittee on Science, Technology, and Space of the United States Senate's Committee on Commerce, Science, and Transportation. He helped Senators oversee and draft legislation for several major civilian science and technology agencies and focused particularly on issues of science, technology, and U.S. industrial competitiveness. He has served on five committees and roundtables of the U.S. National Academies. Mr. Windham received a BA from Stanford University and a Master of Public Policy degree from the University of California at Berkeley.

Acknowledgements

The editors are grateful to the authors who have contributed to this volume, and to the publications that originally published their articles and which have agreed to have them reprinted in this volume; each is acknowledged at the outset of each chapter.

Second, the editors wish to thank Peter L. Singer who provided great assistance in compiling this volume.

Third, the editors acknowledge and thank colleagues in Japan. This book is in part based on a briefing book prepared for a workshop at the National Graduate Institute for Policy Studies (GRIPS) in Tokyo on 25 February 2014. This workshop focused on lessons to be learned from the DARPA model as Japanese officials considered the structure of a DARPA variant. The authors particularly wish to thank Professor Atsushi Sunami, who organized that event. Several of the American authors of this book, including all three editors, participated in the workshop.

The editors also wish to thank the many colleagues and current and former DARPA officials who have provided valuable insights into how DARPA works. They also thank ARPA-E and IARPA former officials for their insights.

The editors thank their families for their support and patience as their various studies of DARPA were ongoing. Finally, Patrick Windham wants to particularly thank his spouse, Dr. Arati Prabhakar. In the interest of full disclosure, the editors want to note that Dr. Prabhakar served as the Director of DARPA from July 2012 to January 2017. However, Dr. Prabhakar was not involved in the creation of this book and the viewpoints expressed here should not be construed as representing her views or those of DARPA.

1. Introduction

DARPA—The Innovation Icon

Patrick Windham and Richard Van Atta

The Defense Advanced Research Projects Agency (DARPA) has become an "innovation icon," widely recognized for playing an important role in the creation and demonstration of many new breakthrough ("disruptive") technologies. Some of these technologies have strictly military applications, such as stealth and precision-guided munitions. Others are "dual-use technologies" that have benefited both the civilian world and the Department of Defense. Examples of these technologies include the Internet, Global Positioning System (GPS) receivers, voice recognition software, advanced semiconductor manufacturing processes, and un-manned aerial vehicles. It is a remarkable record.

This introductory chapter focuses on DARPA's key features—its mission, organization, linkages to other organizations, and "political design"—and how those features have contributed to its success. Later chapters and the book's Conclusion suggest some lessons that DARPA's experience offers for those interested in how this organization has worked over nearly sixty years and for those seeking to create similar technology agencies.

 https://doi.org/10.11647/OBP.0184.01

DARPA's Historical Mission and Organization

DARPA's Evolution

DARPA has existed for over sixty years and during that time it has evolved, changed, and, on a couple of occasions, come close to being dissolved. It has changed in its organizational structure and in some important operational mechanisms as well. There is no simple singular depiction of DARPA that is accurate because it has changed and adapted based on how the world around it has changed—especially on how the national security environment has changed, but also on what different Presidents and their Administrations have asked of it.

Importantly, even at a given point in time there are what might be termed several DARPAs, as different parts of the organization—as small as it is—have focused on very different things—both technologically and in terms of how they function. This is evident from its early history, as Richard Van Atta outlines:

> Indeed DARPA has morphed several times. DARPA has "re-grouped" iteratively—often after its greatest "successes". The first such occasion was soon after its establishment, with the spinning off of its space programs into NASA. This resulted in about half of the then ARPA personnel either leaving to form the new space agency, or returning to a military service organization to pursue military-specific space programs. A few years later, then DDR&E John S. Foster required ARPA to transition its second largest inaugural program—the DEFENDER missile defense program—to the Army, much to the consternation of some key managers within ARPA. Also early in its history ARPA was tasked to conduct a program of applied research in support of the military effort in Viet Nam.[1]

Thus, even by the early 1960s one could say there were three, perhaps four key DARPA thrusts—with the addition of its exploration of new, emerging technologies, such as materials, and the nascent information technologies. As the overview below shows, DARPA's history has been perturbed by political dynamics as well as the dynamics of the technologies it has pursued. Perhaps the most important hallmark of

---

1 Van Atta, R. (2008). "Fifty Years of Innovation and Discovery", in *DARPA, 50 Years of Bridging the Gap*, ed. C. Oldham, A. E. Lopez, R. Carpenter, I. Kalhikina, and M. J. Tully. Arlington, VA: DARPA. 20–29, at 25, https://issuu.com/faircountmedia/docs/darpa50 (Chapter 2 in this volume).

DARPA has been its adaptability and flexibility to respond to changing circumstances—often extremely rapidly.

DARPA's Origins: 1958–1970

In October 1957, the Soviet Union launched the first artificial satellite, Sputnik I, an accomplishment that shocked the United States. Many Americans worried that the country was losing technological leadership to its Cold War adversary.

After the launch of Sputnik, President Dwight Eisenhower followed the advice of Secretary of Defense Neil McElroy and leading scientists, including his science advisors, James Killian and then Dr. George Kistiakowsky, and proposed the creation of what became the Advanced Research Projects Agency (ARPA). ARPA was formed just four months after Sputnik on 7 February 1958 through DOD Directive 5105.15 by Secretary McElroy.[2] Herbert York, a Manhattan Project veteran and the first director of the Lawrence Livermore Laboratory, helped guide the early evolution of ARPA as its first Chief Scientist and then as the Defense Department's first Director of Research and Engineering.

Initially, the agency focused on three key assignments from the President: space, missile defense, and the detection of nuclear weapons tests. Eisenhower subsequently made it clear that space was to be the realm of a civilian agency, and later, in 1958, Congress and the President created the National Aeronautics and Space Administration (NASA), a civilian agency which took over the country's principal space programs, absorbing much of DARPA's Space Program. The two other Presidential assignments—missile defense and nuclear test detection—continued as the dominant foci for about fifteen years but eventually were moved to other parts of the Department of Defense (DOD).

Also, soon after its founding ARPA took on Project AGILE, as proposed by its Deputy Director, William Godel, which was a decade-long classified program supporting U.S. combat efforts in Vietnam and beyond. In retrospect, much of AGILE was naive, poorly managed and

2 Congress, through an amendment by Senator Mike Mansfield, renamed "ARPA" as "DARPA" in 1972, adding the word "Defense." Congress, through Senator Jeff Bingaman, renamed it "ARPA" again in February 1993, because of its "dual-use" role in creating technologies with commercial as well as military applications. The name reverted to "DARPA" in March 1996.

rife with amateurism. The ARPA Directors had little access or knowledge of what AGILE was doing as Godel "was running the AGILE office as his own covert operations shop".[3] There were important lessons learnt from AGILE (as a program run amok, with little oversight) on what not to do. It was hardly scientific and as an operational program it focused on near-term solutions. It became a key element in defining what DARPA would *not* be in the battle over competing visions for the agency's future.

With the quick transfer of the space program to NASA, ARPA spent the rest of the decade focused on missile defense, nuclear test detection and AGILE. However, in the early 1960s another role for ARPA emerged as it began to pursue a set of smaller, technically-focused programs under the general notion of "preventing technological surprise". Areas initially pursued were materials science, information technology, and behavioral science. In fact, one can argue that ARPA in essence "invented" these as areas of technological pursuit. These began in 1961 under Jack Ruina, the first scientist to direct ARPA, who hired J. C. R. Licklider as the first director of the Information Processing Techniques Office. That office played a vital role in the creation of personal computing and the ARPANET—the basis for the future Internet.

Resuscitation in the 1970s

It is important to note that ARPA in the late 1960s to early 1970s was a troubled agency—a victim of the Vietnam malaise and resource cutbacks that affected all of DOD, and with the additional issue that its post-space program thrusts (missile defense (DEFENDER) and nuclear

3 Weinberger, S. (2017). *The Imagineers of War: The Untold Story of DARPA, The Pentagon Agency That Changed the World.* New York, NY: Alfred A. Knopf, 81. Weinberger goes into considerable detail on Project AGILE and the role of Deputy Director Godel in shaping DARPA's involvement in tactical technologies related to not only U.S. combat in Southeast Asia, but also a much broader focus on counterinsurgency-related activities in other parts of the world. While there were some modestly successful early technology developments under AGILE, such as tactical remotely-piloted vehicles, much of this program was egregiously unsuccessful with harmful repercussions, including Agent Orange and other defoliation efforts, poorly conceived and methodologically suspect social science forays, and the "strategic hamlets" concept of population relocation. Perhaps most damning was the inclination of those running and overseeing these programs, including DARPA's director, to delude themselves that they were effective. Director Charles Herzfeld subsequently stated, "AGILE was an abysmal failure, a glorious failure" (Weinberger. (2017). *The Imagineers of War*, 185).

test detection (VELA)) had essentially run their course. Indeed, as early as 1965, Deputy Secretary of Defense Cyrus Vance, "came to advocate abolishing the Agency".[4] The 1965–1970 era was a crisis period. DARPA evolved both organizationally and programmatically from this crisis largely due to John S. Foster, who became Director of the Defense Research and Engineering (DDR&E)[5] in 1965 and remained for eight years. By the mid-1970s DARPA had jettisoned the AGILE program and transitioned DEFENDER to the Army. DARPA was explicitly looking for new directions first under Director Eberhardt Rechtin, who created a Strategic Technologies Office, and then his successor Steven Lukasik, who saw AGILE as "an embarrassment" and closed it down, transitioning parts of it into a new Tactical Technology Office. Thus, by the mid-1970s DARPA had substantially refocused on technology offices and moved away from the original mission-focused assignments. Crucial to this rejuvenation was DARPA taking on a broad new focus aimed at finding technological alternatives to the use of nuclear weapons to respond to the Soviet Union. This was a key imperative stemming from the concerns of President Richard Nixon and his National Security Advisor, Henry Kissinger, and which continued with Secretary of Defense James R. Schlesinger as a leading proponent under President Gerald Ford. DARPA identified and developed new tactical capabilities based on then emerging technologies through programs on stealth, standoff precision strike, and tactical surveillance via unmanned aerial vehicles (UAVs).

DARPA in the 1980s: Transformative Technology Development and Transition

With this refocusing DARPA survived the axe. Through years of persistent efforts, working with the DDR&E in the Office of the Secretary of Defense (OSD), DARPA transitioned these capabilities to the military,

4 Barber Associates, R. (1975). *The Advanced Research Projects Agency, 1958–1974*. Report Prepared for the Advanced Projects Research Agency, vii-3. Springfield, VA: Defense Technical Information Center.

5 The DDR&E was created in 1958 as the third ranking position in the Pentagon, below only the Secretary and Deputy Secretary of Defense, as essentially the Chief Technology Officer. DARPA reported to the DDR&E. Subsequently this position became the Undersecretary of Defense for Acquisition, Technology and Logistics (USD(AT&L)).

creating what Under Secretary of Defense William Perry and Secretary of Defense Harold Brown (under President Jimmy Carter) would call the "offset strategy"—ways to offset the Soviet Union's conventional war capabilities and lowering the corresponding risk of nuclear war. These key DARPA programs are among the most important programs in terms of the agency's impact on defense capabilities and are often touted as DARPA's impact in ushering in a "revolution in military affairs" evidencing how DARPA helped to transform tactical warfare.

Parallel to DARPA's transformational programs in military technologies in the 1970s-80s were its programs revolutionizing information technology, stemming from the early 1960s focus of IPTO (Information Processing Technology Office) Director Licklider. ARPA/DARPA fundamentally affected what was to become computer science. President John F. Kennedy and Secretary of Defense Robert McNamara became very concerned about a "command and control" communication crisis during the Cuban Missile Crisis; ARPA Director Jack Ruina brought in Licklider to work on it, who saw the problem in a context of evolving computing systems. While one element of this was the ARPANET, this was part of a much broader and increasingly coherent program of research begun under Licklider. His concept of "man-computer symbiosis" provided a multi-pronged development of the technologies underlying the transformation of information processing from clunky, room-filling, inaccessible mainframe machines to the ubiquitous network of interactive and personal computing capabilities.[6] This transformation continues today in DARPA's pursuit of cognitive computing, artificial intelligence and robotics—key DARPA thrusts.

DARPA in the 1990s: End of the Cold War

Early in the 1990s, DARPA, as well as the rest of DOD, had to adapt to the fact that the main adversary, the USSR, had collapsed. Thus, the focus of its weapons research had disappeared. Moreover, the U.S. was in a budget crisis partly due to the vast defense spending of the 1980s. The Clinton Administration entered office with the rubric

6 This transformation is detailed in Waldrop, M. M. (2001). *The Dream Machine: J. C. R. Licklider and the Revolution that Made Computing Personal*. New York, NY: Viking Press.

"dual-use"—technologies that would have both defense and civilian economy payoffs—as one way to make the economy more competitive. Under this approach, DOD could leverage off the civilian sector in cutting costs to develop new technologies. This era of dual-use programs was a major redirection of DARPA and it became highly contentious with elements in Congress. The Technology Reinvestment Project (TRP) was created to partner defense technology developers with commercial firms and universities.

OSD and DARPA worked with the White House to develop this program to continue DARPA's exploration and development of "breakthrough" technologies in the mode of the information revolution, despite the lack of a peer security adversary. Secretary of Defense William Perry emphasized the dual-use concept. During this period emphasis was heavily on fostering new technologies in information and electronics including advanced sensing, while programs in unmanned systems and precision strike continued. Also, programs in biotechnology were started. At the end of the 1990s, DARPA took on a program in partnership with the Army seeking a radical approach for using robotics for ground combat—the Future Combat System—which ultimately proved to be hugely unsuccessful. It was overly ambitious and rushed into acquisition by the Army, and, after the expenditure of about $20 billion, was eventually cancelled by the Secretary of Defense.

DARPA in the 2000s: War on Terror

The 2000s is the period of DARPA Director Anthony Tether—the longest tenured DARPA Director. Within months of taking the role, the terror attacks of September 11 occurred and DARPA became enmeshed in the "War on Terror". The Total Information Awareness (TIA) program became the most notable DARPA response. This became a controversial program as the use of information technologies to identify possible terrorists and terror attacks raised issues of privacy. Tether's tendency to supervise program managers (PMs) also raised questions about whether DARPA should be inherently bottom-up, PM-driven or more director driven. DARPA also developed programs in sensors and sensor systems to support combat needs in Iraq and Afghanistan. During this period DARPA also developed programs in cognitive computing (artificial intelligence) and autonomous systems with the "DARPA Challenge" contests for self-driving cars as highly visible examples initiating the implementation of these technologies. These Challenges were successful in creating interest and incentivizing teams of researchers to demonstrate integrated autonomous capabilities.

DARPA in the 2010s: Technology for Security in a Globalized World

Through the current decade DARPA has continued on a primarily technology focused agenda in which the emphasis is on pursuing technologies that can create technological surprise. However, it recognized that the world of technology has changed considerably with the advent of globalization. Where the U.S. and DOD led in technology development in the past, now there are global competitors pursuing many of the technologies that DARPA had pioneered. At the same time, there is a growing peer competition in the security arena while terrorism is an ongoing concern. Thus, DARPA's mission of avoiding technological surprise and also creating technological surprise for our adversaries is even more daunting. Under Barack Obama Defense Secretaries Chuck Hagel and Ash Carter, DOD announced a new "Offsets" strategy to attempt to build a new U.S. technological lead as new peer competitors developed capabilities in areas DARPA had created in the previous offset strategy.

DARPA also responded to the era of major advances in life sciences, most visibly, the Human Genome initiatives led by the National Institutes of Health (NIH) and their private sector competitor, J. Craig Venter. DARPA had long been conducting some biotechnology research but in 2013 created a new Biological Technologies Office to focus on this area. Fields like synthetic biology created new kinds of threats that needed counters, and DOD's own massive health care system and injured soldiers from two Middle Eastern wars required new medical responses. While NIH's research remained largely focused on biology, DARPA's flexibility enabled it to pursue a "convergence" approach, creating unified research efforts combining engineering, physical and computational sciences with biology for a new research model pursuing new kinds of therapies. In the information domain, DARPA is focusing on artificial intelligence, cognitive computing, and approaches for advancing microelectronics to advance quantum computing and neuro-synaptic processors based on how the brain processes information. With a foundation on previous research in aeronautics and propulsion, DARPA is embarking on a major thrust in hypersonic systems. Meanwhile, growing cyber threats spurred several ambitious DARPA programs in cybersecurity.

Thus, the agency's technical and security foci have changed with the times, although its mission has remained largely the same:

> DARPA's original mission, established in 1958, was to prevent technological surprise like the launch of Sputnik, which signaled that the Soviets had beaten the U.S. into space. The mission statement has evolved over time. Today, DARPA's mission is still to prevent technological surprise to the U.S., but also to create technological surprise for our enemies.[7]

However, to carry out this mission today the agency must focus on creating and demonstrating breakthrough technologies for national security, in which there are many more highly capable players and where technologies quickly disseminate globally.

DARPA's Organization and Budget

To achieve its mission of technology leadership DARPA has evolved a highly adaptive and responsive organization. The hallmark of DARPA is agility. At the heart of DARPA are its "technology offices"—the offices where program managers fund the development of new technologies. The agency also has a series of "support offices", which provide services in areas such as contracting, human resources, legal matters, and accelerating the transition of new technologies to the military services. The number of technology offices and their specific roles change over time. Below are the DARPA's current technical offices:

- Biological Technologies Office (BTO)
- Defense Sciences Office (DSO)
- Information Innovation Office (I2O)

7 DARPA. (2005). *DARPA—Bridging the Gap, Powered by Ideas,* 1. Arlington, VA: Defense Advanced Research Projects Agency, http://www.dtic.mil/cgi-bin/GetTRDoc?Location=U2&doc=GetTRDoc.pdf&AD=ADA433949.

- Microsystems Technology Office (MTO)
- Strategic Technology Office (STO)
- Tactical Technology Office (TTO)

Sometimes DARPA officials and outside observers informally refer to some of these technology offices as "systems offices". In the list above, the two systems offices are the Strategic Technology Office (STO) and the Tactical Technology Office (TTO). These offices create new "proof-of-concept" engineering systems for DOD, such as new unmanned aerial vehicles or small GPS receivers. The goals here are to develop and demonstrate significantly new or improved capabilities and, DARPA hopes, to change people's minds about what is technically possible. The work sponsored by these systems offices is often inspired by long-term national security challenges, needs, or opportunities.

The "systems offices" and the other technology offices typically fund different types of R&D (Research and Development) performers. In the non-systems offices, many of the R&D performers are in universities or component manufacturers. The systems offices usually fund engineering teams that may include defense companies and government laboratories. However, at times the systems offices encounter technical challenges that lead them to also support fundamental research, and the other technology offices sometimes work on military systems. In practice, the line between non-systems technology offices and systems offices is not rigid.

Each DARPA office has multiple "programs" (the term used to refer to R&D funding activities in specific areas of technology). Program managers propose these programs, get approval and funding from senior DARPA officials, write the funding solicitations, select the R&D performers (sometimes with help from other technical reviewers), and supervise and assist the performers. A program manager may supervise several programs. Typically, a program will have specific technical objectives, a budget of tens of millions of dollars, and will last for three to five years. In many cases, an individual program will fund multiple R&D projects run by different performers, so as to test different technical ideas. Having a good set of diverse technical approaches early on in a program is helpful.

Each DARPA technology office also can fund small "seed" programs, which provide a way for program managers to generate and test new ideas. In recent years, each office also has run an annual "open" competition in which applicants can propose work in areas of technology not covered in the office's programs. These "open" competitions help generate additional new ideas from the technical community.

DARPA therefore uses a "portfolio" approach: it funds a wide range of R&D programs and also often funds multiple projects within a single program. Its program managers are experts who make thoughtful decisions, but since the R&D focus is high-risk to achieve "high payoff" results, the outcomes are unpredictable and the agency and its program managers invest in a range of promising technologies. Some programs and projects will work while others will not. However, by investing in a number of options, the agency seeks to increase the chances of success while accepting the inherent risk that some research may not succeed.

DARPA itself does not build actual operational prototypes of new systems; it turns over "proof-of-concept" prototypes to other parts of the defense and commercial worlds—a process that DARPA calls "technology transition".[8]

At the heart of DARPA are approximately one hundred program managers ("PMs") and the office directors, deputy office directors, agency director and deputy director who supervise them. While these are all government employees, most are hired using special hiring authorities on a term basis—usually of three to five years. Importantly, none of these are permanent staff—all are in essence temporary, although some individuals' tenure may get extended by becoming an office director or deputy director. The agency also has approximately one hundred other government employees who provide important services, such as contracting, legal services, human resources, and security, and at any one time it also has several military liaisons. Additionally, contractors support these government employees. Some of these contractors are highly-trained PhD scientists and engineers who provide valuable technical assistance to program managers, and others are support staff.

The agency's budget for 2019 is $3.427 billion a year. DARPA has no laboratories of its own. It is a funding agency.

8 There was one significant exception. DARPA did develop operational technology for seismic detection of Soviet underground nuclear tests. DARPA was only able to transition this seismic detection network to the Air Force after running it for approximately twenty years.

Important Features of the DARPA Model

DARPA's Focus on Ambitious Goals

Ambitious goals

DARPA focuses on ambitious technological goals, not on incremental improvements.

First, DARPA is a *technology* agency. It funds *advanced research* to develop or create new technologies, not just to explore science. Its mission is to create valuable new technologies. It can support basic scientific research, but as means toward new technology.

Second, DARPA focuses on ambitious, difficult ("DARPA Hard"), and potentially revolutionary projects. It does not focus on immediate or incremental improvements in technology.[9] It focuses on trying to achieve significant changes or shifts in technical capabilities.

Third, DARPA seeks to create "breakthrough", "transformative" or "disruptive technologies"—all terms that are popular today. This means something different than just the creation of novel new devices or tools. Rather, the objective is to create new possibilities and capabilities and particularly seek "change-state" technologies—that is, technologies that significantly change existing capabilities. As a result, the focus is more on outcomes and results rather than the specific character of the technologies that they nurture. So, for example, sometimes an entirely new technology may dramatically improve capabilities. One could argue that the ARPANET was such an example and was a "breakthrough" or "transformative" technology. But at other times integrating existing technologies in new ways may significantly transform capabilities, perhaps by dramatically reducing costs or reducing the time it takes to perform tasks. For example, a new system that significantly reduces the cost and time involved in launching small satellites into orbit may not involve radically new "breakthrough" technologies but rather combine and upgrade existing technologies to create dramatically better capabilities. This

9 There have been times, usually to meet a wartime need, when DARPA has focused on short term technologies, notably under project AGILE during the Vietnam conflict, but these have become exceptions. Under such circumstances, it is important to ask whether DARPA is the best place to pursue such near-term technology developments.

type of improvement is also valuable. Moreover, projects that integrate existing technologies in new ways may carry as much technical risk and offer as much potential benefit as projects to create individual new technologies.

A Challenged-Based R&D Model

DARPA's goals are not only ambitious; they are also focused on specific challenges and opportunities rather than on general discovery or invention. One of this book's editors (William B. Bonvillian) has noted two important aspects of this model: it is "challenge-based", and it is a "connected model" that connects scientific research to these technical challenges.[10]

By "challenge-based", we mean that DARPA program managers identify specific technical capabilities that they think would be both valuable and achievable. Again, DARPA focuses on trying to reach ambitious technical goals but also it tries to demonstrate those capabilities as quickly as possible. It seeks to accelerate the creation of valuable new technologies.

It also uses a "connected model" of R&D—a deliberate process of connecting basic science and engineering to specific technical goals and challenges. This makes DARPA significantly different from some other U.S. R&D agencies. The National Science Foundation (NSF), for example, supports intellectually interesting basic research in universities that is often unconnected to any specific technical goals. NSF funds "pure" research. Practical results may eventually come out of that research, but NSF does not set ambitious technical goals and then create programs designed to achieve those goals. This is not a bad thing. NSF's mission is to advance general knowledge, by drawing upon the talents and curiosity of brilliant researchers. While DARPA draws upon that new knowledge, as well as the skilled researchers that universities train, it nonetheless remains an agency focused on achieving specific technical goals.

10 Bonvillian, W. B. (2009). "The Connected Science Model for Innovation—The DARPA Model", in *21st Century Innovation Systems for the U.S. and Japan*, ed. S. Nagaoka, M. Kondo, K. Flamm, and C. Wessner. Washington, DC: National Academies Press. 206–37, https://doi.org/10.17226/12194, http://books.nap.edu/openbook.php?record_id=12194&page=206 (Chapter 4 in this volume).

DARPA also sometimes funds basic scientific research itself, if that research is connected to important technical goals. The agency's Defense Sciences Office, for example, funds research in fundamental physics, materials, and mathematics, but mainly for the purpose of helping to advance important capabilities. In this sense, DARPA connects science with technical challenges in ways that it hopes will lead to valuable new technical capabilities.

High-Risk/High-Payoff Projects

DARPA focuses on "high-risk/high-payoff" projects and has developed a philosophy and set of procedures for managing this type of research.

First, the agency is willing to take big technical risks in order to try to get "change-state" results. DARPA is not interested in incremental improvements in technologies or weapons systems. While these improvements are important, especially to the military, they are the province of other R&D agencies. DARPA's specific mission is to develop significant new or better technologies; to do so, it focuses on projects that involve high risk and the possibility of failure but that also will create high payoffs, if successful.

Second, however, there is nothing haphazard or nonchalant about the way in which DARPA takes risks. In fact, one could call its approach one of "thoughtful" or "rigorous" risk-taking. New program managers and office directors are encouraged and expected to fund programs that offer the possibility of significant advances. But they must also think rigorously about whether ambitious goals are achievable and what technical approaches are most promising. Agency leaders expect their program managers to consult widely with relevant technical communities, test and retest their ideas, and constantly learn.

This two-part emphasis on both ambitious goals and rigorous thinking is best seen in a set of questions originally written down by George Heilmeier, a noted inventor and DARPA director from 1975 to 1977. These are questions ("The Heilmeier Catechism") that program managers should ask themselves when designing new programs, and these are the questions that DARPA office directors and the agency director will ask when those program managers propose new initiatives, and when they review these programs:

- What are you trying to do? Articulate your objectives using absolutely no jargon.
- How is it done today, and what are the limits of current practice?
- What's new in your approach and why do you think it will be successful?
- Who cares?
- If you're successful, what difference will it make?
- What are the risks and the payoffs?
- How much will it cost?
- How long will it take?
- What are the midterm and final "exams" to check for success?

Third, in addition to this overall philosophy, the agency has evolved ways that can help optimize results in this high-risk environment. Here, again, the agency's "portfolio" approach is important. The agency makes thoughtful decisions—which are possible because it recruits world-class experts—in full knowledge that R&D is unpredictable and some programs and projects will fail. Indeed, if none failed, the agency's culture asserts that it would not be doing its job; it would not be bold enough. Investing in a wide range of programs and in a range of projects and technical approaches within those programs increases the chances that the agency's investments will lead to some significant successes as well as some failures.

In addition, DARPA expects that programs and R&D projects within those programs often will not go as planned. These are research projects tackling unknowns and thus it is likely that promising R&D ideas will fail, that new opportunities will be discovered, and therefore that R&D plans need to be adjusted. So, DARPA program managers constantly evaluate projects and work with performers to identify obstacles and opportunities and to make adjustments; DARPA contracts allow them to do this. DARPA does not force its program managers or R&D performers to adhere to unrealistic or ineffective plans or milestones. Projects certainly have technical objectives, but it is expected that R&D projects will change as R&D performers learn what works and what does not. Program managers and R&D performers themselves continuously evaluate and adapt.

Thus, at DARPA technical failures are expected, since these are high-risk projects and not all will succeed. DARPA and the overall technical community will learn from these dead ends, and the agency will terminate unsuccessful programs and shift funding to more promising ideas. Because the agency has no laboratories or researchers that it must fund year in and year out, it has the freedom to move away from unsuccessful projects to focus on promising ones. Some DARPA leaders state that the only "true failures" occur when R&D performers are unwilling or unable to be candid about the technical problems they are encountering, and therefore the learning process breaks down.[11]

DARPA's Organization and Management

Several of the articles in this compendium identify organizational and management features that have contributed to DARPA's success. These include:

Independence

While DARPA is a DOD agency, under the Secretary of Defense, it has usually had a great deal of independence in determining its overall programs.

However, this does not mean that DARPA does not respond to the national security priorities and strategic directions set by the Secretary of Defense and the President. Recall that ARPA was initially focused on a set of three Presidential issues—areas of national security priority that were identified as being given insufficient focus by the military services. Importantly, these were broad overall research thrusts and ARPA was given wide latitude on how to conduct the research. Generally, this has been the case ever since. This is crucial to DARPA's focus on change-state, revolutionary capabilities: unless a DARPA-type organization is truly independent, then that organization will feel pressure to work on short-term, incremental projects rather than long-term, potentially

11 We are grateful to Dr. Jane Alexander, a former deputy director of DARPA, for making this point.

breakthrough technologies. A related point is that this type of organization can only maintain its independence and budget if it has support and protection from high-level officials.

A Flat, Non-Hierarchical Organization, with Empowered Program Managers

Hiring technically-accomplished program managers and letting them propose and then run R&D programs is a central feature of the DARPA model.[12] Program managers have the authority and responsibility to prepare all the details of a new proposed program: its scope, its rationale (why should we fund it?), the science and engineering behind it, the specific technical objectives, the metrics for measuring technical progress, and the proposed budget and schedule.

Program managers need to be recruited and supervised. DARPA is able to do so using only two layers of management: office directors and their deputies and then the agency director and deputy director. Since these managers are themselves technically very well trained, they can make informed decisions quickly and competently—including which experts to hire as program managers, when to approve or not approve a proposed R&D program, and how to ensure that program managers operate their programs in a technically effective way.

A unique aspect of DARPA's management is that it brings in its key assets—the program managers—on a temporary, short-term basis, usually for three to five years each. Thus, there is roughly a 25 percent turnover every year. Hiring new program managers allows for new ideas and capabilities. But hiring talented program managers can be a challenge, given that private-sector salaries are higher, that the DARPA job only lasts three to five years, and that program managers must move to the Washington, DC, area. However, DARPA also offers exciting opportunities to create new technology, so many people are interested in the possibility of working at the agency. The agency has been able

12 Bonvillian, W. B., and Van Atta, R. (2011). "ARPA-E and DARPA: Applying the DARPA Model to Energy Innovation", *The Journal of Technology Transfer* 36: 469–513 (Chapter 13 in this volume); and Bonvillian, W. B., and Van Atta, R. (2012). *ARPA-E and DARPA: Applying the DARPA Model to Energy Innovation.* Presentation at the Information Technology and Innovation Foundation, Washington, DC, February, https://www.itif.org/files/2012-darpa-arpae-bonvillian-vanatta.pdf

to attract highly capable people who want to work on important and exciting ideas.

Outside Performers and Temporary Project Teams

Research and development are performed entirely by outside performers. DARPA has no internal research laboratory that it must maintain and fund every year and the agency is free to hire whomever it thinks are the best people for specific projects. This emphasizes several key points about the DARPA model: it relies on technically-capable program managers, R&D teams include world-class experts, and the projects DARPA funds are limited in time and focused on specific scientific and technical objectives.

Multi-Generational Technology Investments

If a particular DARPA program is successful, then the agency may fund additional "generations" of three- to five-year programs in this technical area.[13] By working on important technical ideas over longer periods of time, DARPA can create enduring new technologies (technology "motifs") that truly change the technology landscape over time. Each generation of R&D may have different specific objectives and metrics but can be based on a common technical area. Usually each generation learns from prior experience. This may even include supporting a radically different approach to those tried previously, especially if the objective is seen as an enduring national security challenge.

This point about multi-generational investments is important and not always well understood. The fact that DARPA programs typically run from three to five years suggests that the agency funds relatively short-term engineering experiments. It is true that the agency funds many different technical ideas for limited periods of time, but when agency leaders find a new technology that they think offers significant new capabilities for the Defense Department and the country as a whole, they will make sustained investments over many years.

13 For a fuller discussion of DARPA technology thrust areas, see Van Atta, R., Deitchman, S., and Reed, S. (1991). *DARPA Technical Accomplishments. Volume III.* Alexandria, VA: Institute for Defense Analyses (chapter 4), https://apps.dtic.mil/dtic/tr/fulltext/u2/a241680.pdf

This is usually with a new program manager focused on achieving even more ambitious outcomes, or an entirely new approach, perhaps integrating prior results into a promising new technical idea and creating working prototypes. Technology examples include computing and networking investments, which led to the Internet, iterative advances in artificial intelligence, new concepts for quantum computing and spintronics. On military systems DARPA sponsored many years of investments in stealth, precision-guided munitions, and unmanned aerial vehicles.

Investments in Complementary Strategic Technologies

DARPA sometimes will fund work in additional technical areas relating to major new technology. These related ("complementary") areas are important for the overall success of the new technology, and developing them also builds political support for commercialization and implementation by showing Defense Department leaders and others that the entire system around that new technology will work. For example, DARPA not only invested in early computer routers and the software to run them (the ARPANET) but also in applications of computer networking (file transfers, e-mail, etc.) and later in new computer communications protocols (TCP/IP) that would allow different computer networks to talk to each other. In short, DARPA and its R&D performers created and demonstrated a complete system.

Flexible Hiring and Contracting Authority

The work of DARPA managers and their administrative staff is helped by special laws that apply to DARPA hiring and contracting. For example, DARPA has legal authority to hire program managers very quickly. In the case of program managers from universities or other government agencies, DARPA may use what is called "Intergovernmental Personnel Agreements" (IPAs). Under an IPA, the individual stays an employee of his or her university or laboratory, but he or she is temporarily assigned to DARPA and becomes a temporary government employee under a contract with DARPA. The National Science Foundation and other government R&D agencies also have this authority. The IPA process allows DARPA to hire quickly and to pay the same salary people earned earlier.

In the case of people from industry, another provision of law (Section 1101 of the Strom Thurmond National Defense Authorization Act for Fiscal Year 1999) allows DARPA to hire experts quickly, although people from industry must leave their companies while they are at DARPA. Congress provided these laws about hiring in part because DARPA program managers are temporary, not permanent federal employees.

All program managers and all senior DARPA managers must follow rules to prevent conflicts of interest—that is, to prevent them from making decisions about whether to award contracts to their current or former employers or to companies in which they own stock. But DARPA has a clear process in which other government employees can make these contract decisions, if the need arises.

In addition to flexible hiring authority, DARPA has legal permission to use a wide range of flexible contracting procedures, including "other transactions authority" (OTA).[14] This OTA power releases DARPA from highly restrictive government procurement requirements. DARPA also has "prize authority". For example, in the robotics field DARPA has sometimes used its legal authority to organize contests and provide prizes, in order to draw in groups that do not usually work with the government.

Creating New Technical Communities

By funding multi-disciplinary teams that both compete and cooperate with each other, DARPA often stimulated new technical communities and new academic fields. Examples over the years include materials science and engineering, computer science, and now synthetic biology/engineering biology. In fact, one can argue that DARPA actually makes two very

14 DOD offers this explanation: "For DOD, 'other transactions' is a term commonly used to refer to the 10 U.S.C. 2371 [Title 10, United States Code, section 2371] authority to enter into transactions other than contracts, grants or cooperative agreements. OTA provides tremendous flexibility since instruments for prototype projects, awarded pursuant to this authority, generally are not subject to federal laws and regulations limited in applicability to procurement contracts." This description is from Office of the Under Secretary of Defense for Acquisition, Technology, and Logistics. (2001). *"Other Transactions" (OT) Guide for Prototype Projects*, www.acq.osd.mil/dpap/docs/otguide.doc.

important contributions: it not only helps create and demonstrate new technologies but also helps create important new technical communities.

These researchers then can perform additional R&D, teach students, and contribute further ideas to DARPA. In addition, DARPA-funded communities are a primary means for transitioning the newly-developed technologies to the military and to commercial companies

How DARPA Transfers Its Technologies

DARPA succeeds in large part because other organizations in government and the corporate world further develop and then commercialize and buy the new technologies. In other words, since DARPA itself does not usually build full prototypes or early operational systems, it must rely on other parts of the U.S. national innovation system to perform those tasks. What features of the DARPA model and the overall national innovation system help technology transfer (what DARPA calls "technology transition")?[15]

DARPA's Willingness to Challenge Incumbent Technologies

DARPA is willing to challenge existing technologies and the organizations that produce and use them. Again, the agency sees its job as changing people's minds about what is possible. So, for example, it showed that a computer network using open standards could replace proprietary networking systems. It created and then, with support from the Office of the Secretary of Defense, pushed for the adoption of stealth, unmanned aerial systems, precision strike, and night vision. It uses conferences, prize competitions, "technology insertion projects" (demonstrations of new technology in actual military systems), and other techniques to demonstrate and publicize new technical capabilities.

A Community of Technology Advocates

As discussed earlier, DARPA and its performers create new technical communities. Besides helping DARPA undertake new research,

15 This section draws largely from Bonvillian and Van Atta. (2011). "ARPA-E and DARPA".

researchers in these new communities also often become knowledgeable, enthusiastic advocates for new technologies.

Some of these experts work in the government, some in universities, some for large firms, and some start new entrepreneurial companies. They share an overall vision of what can be done, and they often become what Bonvillian and Van Atta call "communities of change-state advocates"—people who are willing and able to change the technology world. This is a very important reason why DARPA has been so influential.

Close Ties to DOD Leaders

The agency's close ties to Secretaries of Defense and other senior officials not only help DARPA maintain its independence; these ties also mean that these officials become "champions" who want to further develop and then use DARPA-created technologies. Their support is very important for technology transfer. For example, senior DOD officials pushed the U.S. Air Force to adopt both stealth aircraft and unmanned aerial vehicles. Bonvillian and Van Atta see DARPA and DOD using an "island/bridge" model of organization: DARPA is a type of organizational island, with a high degree of autonomy, but it also has a close link ("bridge") to senior DOD officials, helping it to transfer its new technologies to the wider defense world.[16]

Connection to Technically-Sophisticated, Well-Funded Customers

The process of turning a radical new technology into actual products is usually risky, difficult, and expensive. DARPA and the overall Defense Department deal with this difficulty in two ways.

First, DARPA is fortunate that the Defense Department can be both willing and able to turn new prototype technologies into actual products. Its senior leaders may want advanced technologies, and its other laboratories, contractors, and large procurement system can enable the Department to refine and buy these new products. Even so, the "transition" of new technologies from DARPA to the military services is often difficult because DARPA-developed capabilities usually

16 Bonvillian and Van Atta. (2011). "ARPA-E and DARPA", 486.

challenge the current way of doing operations. Thus, DARPA spends considerable time and effort on the transition process, recognizing that it is often difficult.

Second, the agency also works directly with private sector companies that are interested in commercializing new DARPA-demonstrated technologies. One example is DARPA's long work with the semiconductor industry on advanced chip-making technologies which has led to better and less expensive computer chips for both military and civilian customers. Examples includes silicon-on-insulator technology and MMIC signal processing chips. The new commercial frontier of self-driving vehicles is another example of an industry adopting and building upon DARPA-funded research. Many firms and venture capitalists in the commercial world avidly follow DARPA programs.

U.S. intellectual property law helps facilitate this transfer of DARPA-funded technology to the corporate world. Under the Patent and Trademarks Act Amendments of 1980 (popularly known as the "Bayh-Dole Act"), universities and small companies may keep legal title to inventions developed with federal money. When DARPA projects create new technologies, universities may license inventions to companies and small firms can easily use their inventions to help create new products.

A Good Political Design

In addition to the points made above about the way DARPA is organized and how it operates to succeed it must also have a good "political design".[17] Senior government officials, members of the national legislature, and the larger technical community must support the agency or at least not fight its operations and budget. DARPA succeeds because its mission (national defense) is important, because it has a reputation for producing valuable and high-quality technology, and because it does not threaten the budgets of other agencies.

17 Bonvillian, W. B. (2013). *Evolution of U.S. Government Innovation Organization: From the Pipeline Model, to the Connected Model, to the Problem of 'Political Design.* Presentation at the National Graduate Institute for Policy Studies (GRIPS) GRIPS Innovation, Science, and Technology Seminar, Tokyo, April.

The Remainder of this Book

The rest of this book is divided into four parts: Part I, "Perspectives on DARPA"; Part II, "The Roles of DARPA Program Managers"; Part III, "Applying the DARPA Model in Other Situations"; Part IV, "Conclusions".

Part I, "Perspectives on DARPA", has seven chapters. Chapter 2, by Richard Van Atta, is a history of DARPA's first fifty years. Chapter 3, by Michael Piore, Phech Colatat and Elisabeth Beck Reynolds, compares DARPA's culture with more traditional federal R&D agencies, including NSF. Chapter 4, by William B. Bonvillian, discusses the "DARPA Model", and particularly how it follows an approach developed during World War II that connects cutting-edge science with the solution of specific technical challenges. Chapter 5, by Tamara Carleton, discusses the central role of technical vision in DARPA's operations and results. Chapter 6, by Glenn R. Fong, is a history of how DARPA-funded inventions placed a central role in the development of personal computers and their software. Chapter 7, by Erica R. H. Fuchs, discusses DARPA's governance approach as embodying an imbedded network. Chapter 8, by David W. Cheney and Richard Van Atta, explores the processes through which DARPA creates new programs, looking at the origins of several past DARPA programs. Chapter 9, by Patrick Windham, addresses a set of questions that have been raised concerning the DARPA model.

Part II, "The Roles of DARPA Program Managers", contains Chapters 10 and 11, written by Jinendra Ranka and Larry Jackel, two former DARPA program managers.

Part III, "Applying the DARPA Model in Other Situations", contains two chapters. Chapter 12, by William B. Bonvillian, examines the lessons that DARPA's model of creating innovation provides for other, older, "legacy sector" parts of the Department of Defense. Chapter 13, by William B. Bonvillian and Richard Van Atta, discusses how leaders might effectively apply the DARPA model to the (then) relatively new Advanced Research Project Agency-Energy (ARPA-E) as well as organizational lessons from ARPA-E itself. Chapter 14, by William B. Bonvillian, discusses IARPA, another DARPA clone. Chapter 15, by Robert Cook-Deegan, explores the possible application of the DARPA model to the National Institutes of Health (NIH).

Part IV, "Conclusions", consists of Chapter 16, by Richard Van Atta, Patrick Windham and William B. Bonvillian, summarizing key lessons from DARPA's experience on how to structure an organization to successfully create new, innovative technologies.

These various chapters overlap to some degree. However, the editors of this book hope that together they will provide readers with a comprehensive set of insights on how this remarkable government agency works and why it has succeeded as well as it has.

References

Barber Associates, R. (1975). *The Advanced Research Projects Agency, 1958–74*. Report prepared for the Advanced Projects Research Agency. Springfield, VA: Defense Technical Information Center.

Bonvillian, W. B. (2013). *Evolution of U.S. Government Innovation Organization: From the Pipeline Model, to the Connected Model, to the Problem of Political Design*. Presentation at the National Graduate Institute for Policy Studies (GRIPS) GRIPS Innovation, Science, and Technology Seminar, Tokyo, April.

Bonvillian, W. B. (2009). "The Connected Science Model for Innovation—The DARPA Model", in *21st Century Innovation Systems for the U.S. and Japan*, ed. S. Nagaoka, M. Kondo, K. Flamm, and C. Wessner. Washington, DC: National Academies Press. 206–37, https://doi.org/10.17226/12194, http://books.nap.edu/openbook.php?record_id=12194&page=206 (Chapter 4 in this volume).

Bonvillian, W. B., and Van Atta, R. (2012). *ARPA-E and DARPA: Applying the DARPA Model to Energy Innovation*. Presentation at the Information Technology and Innovation Foundation, Washington, DC, February, https://www.itif.org/files/2012-darpa-arpae-bonvillian-vanatta.pdf

Bonvillian, W. B., and Van Atta, R. (2011). "ARPA-E and DARPA: Applying the DARPA Model to Energy Innovation", *The Journal of Technology Transfer*, 36: 469–513, https://doi.org/10.1007/s10961-011-9223-x (Chapter 13 in this volume).

Chesbrough, H. (2003). *Open Innovation: The New Imperative for Creating and Profiting from Technology*. Boston, MA: Harvard Business School Press.

Christensen, C. M. (1997). *The Innovator's Dilemma: When New Technologies Cause Great Firms to Fail*. Boston, MA: Harvard Business School Press.

DARPA. (2005). *DARPA—Bridging the Gap, Powered by Ideas*. Arlington, VA: Defense Advanced Research Projects Agency, http://www.dtic.mil/cgi-bin/GetTRDoc?Loca- tion=U2&doc=GetTRDoc.pdf&AD=ADA433949

Dugan, R. E., and Gabriel, K. J. (2013). "'Special Forces' Innovation: How DARPA Attacks Problems", *Harvard Business Review* 91/10: 74–84.

Heilmeier, G. (1992). "Some Reflections on Innovation and Invention", Founders Award Lecture, National Academy of Engineering, Washington, DC.

National Research Council. (2013). *21st Century Manufacturing: The Role of the Manufacturing Extension Partnership Program*. Washington, DC: The National Academies Press, https://doi.org/10.17226/18448, https://www.nap.edu/catalog/18448/21st-century-manufacturing-the-role-of-the-manufacturing-extension-partnership

National Research Council. (2012). *Rising to the Challenge: U.S. Innovation Policy in the Global Economy*. Washington, DC: The National Academies Press, https://doi.org/10.17226/13386, https://www.nap.edu/catalog/13386/rising-to-the-challenge-us-innovation-policy-for-the-global

Office of the Under Secretary of Defense for Acquisition, Technology, and Logistics. (2001). *"Other Transactions" (OT) Guide for Prototype Projects*. Washington, DC: Department of Defense, www.acq.osd.mil/dpap/docs/otguide.doc

Shinohara, K. (2014), "High-Risk & High-Impact Program in Japan: ImPACT", in *Weekly Wire News from East Asia and Pacific*, National Science Foundation Tokyo Regional Office, July 4, 2014.

Singer, P. L. (2014). *Federally Supported Innovations: 22 Examples of Major Technology Advances That Stem from Federal Research Support*. Washington, DC: Information Technology and Innovation Foundation, http://www2.itif.org/2014-federally-supported-innovations.pdf

Van Atta, R. (2013). *Innovation and the DARPA Model in a World of Globalized Technology*. Presentation at the National Institute of Science and Technology Policy and the Center for Research and Development Strategy, Tokyo, July.

Van Atta, R. (2008). "Fifty Years of Innovation and Discovery", in *DARPA, 50 Years of Bridging the Gap*, ed. C. Oldham, A. E. Lopez, R. Carpenter, I. Kalhikina, and M. J. Tully. Arlington, VA: DARPA. 20–29, https://issuu.com/faircountmedia/docs/darpa50 (Chapter 2 in this volume).

PART I

PERSPECTIVES ON DARPA

2. Fifty Years of Innovation and Discovery[1]

Richard Van Atta

The Advanced Research Projects Agency (ARPA)—which came to be known as DARPA in 1972 when its name changed to the Defense Advanced Research Projects Agency—emerged in 1958 as part of a broad reaction to a singular event: the launching by the Soviet Union of the Sputnik satellite on 4 October 1957. While in retrospect, Sputnik itself does not seem to be a particularly significant technological achievement, it had massive psychological and political impact. As recounted in Roger D. Launius' "Sputnik and the Origins of the Space Age", found on the website for NASA's Office of History, "The only appropriate characterization that begins to capture the mood on 5 October involves the use of the word hysteria".[2] Launius wrote in the same document that then Senate Majority Leader Lyndon B. Johnson, recollected, "Now, somehow, in some new way, the sky seemed almost alien. I also remember the profound shock of realizing that it might be possible for another nation to achieve technological superiority over this great country of ours."

1 This contribution originally appeared as a chapter entitled "Fifty Years of Innovation and Discovery", in *DARPA, 50 Years of Bridging the Gap*, ed. C. Oldham, A. E. Lopez, R. Carpenter, I. Kalhikina, and M. J. Tully. Arlington, VA: DARPA. 20–29, https://issuu.com/faircountmedia/docs/darpa50. This book was published in 2008 to commemorate the agency's fiftieth anniversary.

2 Launius, R. D. "Sputnik and the Origins of the Space Age", *NASA History*, http://history.nasa.gov/sputnik/sputorig.html.

 https://doi.org/10.11647/OBP.0184.02

For the United States to find itself behind the Soviet Union in entering space signified that something was seriously wrong not only with America's space program but with its organization and management of advanced science and technology for national security. Sputnik evidenced that something was substantially wrong with U.S. defense science and technology and that a fundamental change was needed. Out of this ferment—in fact one of the first actions to emerge from it—was a bold new concept for organizing defense advanced research: the Advanced Research Projects Agency. This agency—renamed the Defense Advanced Research Projects Agency (DARPA) in 1972—refocused and rejuvenated America's defense technological capabilities. Moreover, DARPA has also instigated technological innovations that have fundamentally reshaped much of the technological landscape not only in defense capabilities but much more broadly with breakthrough advances in information technologies, sensors, and materials that have pervasive economic and societal benefits.

The "DARPA Model"

DARPA's primary mission is to foster advanced technologies and systems that create "revolutionary" advantages for the U.S. military. Consistent with this mission, DARPA is independent from the military Services and pursues higher-risk research and development (R&D) projects with the aim of achieving higher-payoff results than those obtained from more incremental R&D. Thus, DARPA program managers are encouraged to challenge existing approaches and to seek results rather than just explore ideas. Hence, in addition to supporting technology and component development, DARPA has on funded the integration of large-scale "systems of systems" in order to demonstrate what we call today "disruptive capabilities".

Underlying this "high-risk—high-payoff" motif of DARPA is a set of operational and organizational characteristics including: relatively small size; a lean, non-bureaucratic structure; a focus on potentially change-state technologies; a highly flexible and adaptive research program. We will return to these characteristics later. What is important to understand at the outset is that in contrast to the then existing Defense research environment, ARPA was designed to be manifestly different. It did not have labs. It did not focus on existing military requirements. It was separate from any other operational or organizational elements. It was explicitly chartered to be different, so it could do fundamentally different things than had been done by the Military Service R&D organizations.

The reason for this dramatic departure, as elaborated below, was that President Dwight D. Eisenhower and his key advisors had determined—as evidenced by the Sputnik debacle—that the existing R&D system had failed to respond to the realities of the emerging national security threat embodied by the Soviet Union.

DARPA's Origins: Strategic Challenges ~1958

Sputnik itself demonstrated that the USSR not only had ambitions in space, but also had developed the wherewithal to launch missiles with nuclear capabilities to strike the continental United States. Therefore, at the outset ARPA was focused initially on three key areas as Presidential Issues: space, missile defense and nuclear test detection.

The first issue, achieving a space presence, was a large element of the initial ARPA, but was spun off to become NASA, based on President Eisenhower's determination that space research should not be directly under the Department of Defense (DOD). According to Herbert York's book, *Making Weapons, Talking Peace: A Physicist's Odyssey from Hiroshima to Geneva*, it was well understood in ARPA that its role in space programs was temporary and that the creation of NASA was already in the works both in the White House and in Congress.[3]

To address ballistic missile defense (BMD), ARPA established the DEFENDER program, which lasted until 1967, performing advanced research relating to BMD and offensive ballistic missile penetration. This program was ARPA's largest over the decade and included pioneering research into large ground-based phased array radar, Over the Horizon (OTH) high-frequency radar, high-energy lasers, and a very high acceleration anti-ballistic missile interceptor, as well as extensive research into atmospheric phenomenology, measurement and imaging, and missile penetration aids.

---

3 York, H. (1987). *Making Weapons, Talking Peace: A Physicist's Odyssey from Hiroshima to Geneva*. New York, NY: Basic Books, 143.

ARPA's nuclear test detection program, VELA, focused on sensing technologies and their implementation to detect Soviet weapons testing. VELA Hotel satellites successfully developed sensing technology and global background data to detect nuclear explosions taking place in space and the atmosphere, providing monitoring capability supporting the Limited Test Ban Treaty in 1963. VELA also included seismic detection of under-ground explosions and ground-based methods to detect nuclear explosions in the atmosphere and in space.

By 1960, a counter-insurgency project (AGILE) was started as the Vietnam War heated up. This included diverse tactical systems ranging from field-testing experiments leading to the M-16 rifle to foliage-penetrating radar capable of automatically detecting intruders, an acoustically stealthy aircraft for night surveillance, and initial work in night vision.

In 1962 ARPA initiated the Office of Information Processing Techniques and Behavioral Sciences to address information processing "techniques" with a focus on possible relevance to command and control. As is elaborated below, under the expansive vision of its first director, J. C. R. Licklider, this office went on to effect a fundamental revolution in computer technologies, of which the now-famous ARPANET was only one element.

What is DARPA?

DARPA was first established as a research and development organization immediately under the Secretary of Defense with the mission to assure that the U.S. maintains a lead in applying state-of-the-art technology for military capabilities and prevent technological surprise from her adversaries.

ARPA was created to fill a *unique* role, a role which by definition and in its inception put it into contention and competition with the existing Defense R&D establishment. As the Advanced Research Projects Agency, ARPA was differentiated from other organizations by an explicit emphasis on "advanced" research, generally implying a degree of risk greater than more usual research endeavors. As former ARPA Director Dr. Eberhardt Rechtin emphasized, research, as opposed to development, implies unknowns, which in turn imply the possibility of failure, in the sense that the advanced concept or idea that is being researched may not be achievable. Were the concept achievable with little or no risk of failure, the project would not be a *research* effort, but a *development* effort.

It is clear from DARPA's history that within the scope of this mission the emphasis and interpretation of *advanced* research have varied, particularly in terms of the degree and type of risk and how far to go toward demonstration of application. Risk has several dimensions: (1) lack of knowledge regarding the phenomena or concept itself; (2) lack of knowledge about the applications that might result if the phenomena or concept were understood; (3) inability to gauge the cost of arriving at answers regarding either of these; and (4) difficulty of determining broader operational and cost impacts of adopting the concept. As answers about (1) become clearer through basic research, ideas regarding applications begin to proliferate, as do questions of whether and how to explore their prospects. DARPA is at the forefront of this question and has the difficult job of determining whether enough is known to move toward an application and, if so, how to do so. At times this can be very controversial, as researchers may feel they do not know enough to guarantee success and are concerned that "premature" efforts may in fact create doubts about the utility and feasibility of the area of research, resulting in less funding and (from their perspective) less progress. DARPA, however, has a different imperative than the researcher to strive to see what can be done with the concepts or knowledge, even if it risks exposing what is not known and what its flaws are. This tension is endemic in DARPA's mission and at times has put it at odds with the very research communities that it sponsors.

During times of changing circumstances, the agency has had to reassess its project mix and emphasis due to determinations both internally and within the Office of the Secretary of Defense regarding the appropriate level of risk and the need to demonstrate application potential. In a sense, these somewhat contradictory imperatives serve as the extreme points on a pendulum's swing. As DARPA is pulled toward one of the extremes, often by forces beyond itself, including Congressional pressures, there are countervailing pressures stressing DARPA's unique characteristics to do *militarily relevant advanced research.*

At the other end of the spectrum, as projects demonstrate application potential, DARPA runs into another set of tensions, not with the researcher, but with the potential recipient of the research product. Given that the ideas pursued are innovative, perhaps revolutionary, they imply unknowns to the user in terms of how they will be implemented and how this implementation will affect the implementer's overall operations. To this end, the potential military users seek to reduce their uncertainty in what is a highly risk-intolerant environment by encouraging DARPA, or some other development agency, to carry forward the concept until these risks are minimized, or by simply ignoring, delaying or stretching out its pursuit. While achieving transition can be increased by additional risk reducing research, this also entails substantial additional cost and raises the issue of mission boundaries. Perhaps one of the most critical and difficult aspects of the DARPA Director's job is to decide that DARPA has concluded its part of a particular technology effort and while there is surely more work to be done, it is not DARPA's job to do it.

There have been several occasions in DARPA's history when its management has determined that it has done enough in an area to demonstrate the potential of a specific concept—such as Unmanned Air Vehicles (UAVs)—and that it is thus time for others to fund development of its application and acquisition. These decisions have at times meant that a potential concept becomes a victim of the "valley of death", with the application either failing to be realized, or, as in the case of UAVs, taking over a decade with special high-level attention from the Office of the Secretary of Defense (OSD) to come to fruition

Over the years DARPA has made considerable effort to develop mechanisms to engage potential "customers" in an emerging concept. Working with prospective developers and users as the ideas mature is a key aspect of DARPA project management. However, DARPA has to remain aware that over-extending its involvement in a particular technology development has costs as well—specifically, it means that resources and capabilities are not available to explore other potentially revolutionary ideas. Indeed, this lesson goes back to the very beginnings of DARPA, when it transferred the incipient space program to the newly created NASA. Herbert York, ARPA's first Chief Scientist recalls, that the civilian space program being moved to NASA (and remainder back to the Services) was "what left room for all the other things that ARPA has subsequently done... including the Internet. If ARPA had been left completely tied up with all these space programs, all kinds of other good things would never have happened".[4]

DARPA's Key Characteristics

It was recognized from the outset that DARPA's unique mission required an organization with unique characteristics. Among the most salient of these are:

- It is independent from Service R&D organizations

DARPA neither supports a Service directly nor does it seek to implement solutions to identified Service requirements. Its purpose is to focus on capabilities that have not been identified in Service R&D and on meeting defense needs that are not defined explicitly as Service requirements. This does not mean that DARPA does not work with the Services, but it does mean that it does not work the requirements that drive Service R&D.

- It is a lean, agile organization with risk-taking culture

DARPA's charter to focus on "high-risk/high-payoff" research requires that it *be tolerant of failure and open to learning*. It has had to learn to manage risk, not avoid it. Because of its charter, it has adopted organizational, management and personnel policies that encourage individual responsibility and initiative, and a high degree of flexibility in program definition. This is one reason that DARPA does not maintain any of its own labs.

A primary aspect of DARPA's lean structure is that it centers on and facilitates the initiative of its program managers. **The DARPA program manager is the technical champion who conceives and owns the program.** As the program manager is the guiding intelligence behind the program, the most important decisions of DARPA's few Office Directors are the selection of and support of risk-taking, idea-driven program managers dedicated to making the technology work.[5]

- It is idea-driven and outcome-oriented

4 York, H. (2007). *Interview*, 5 January.

5 Currently DARPA has Directors for six Offices: Defense Sciences; Information Processing Technology; Information Exploitation; Microsystems Technology; Strategic Technology; and Tactical Technology.

The coin of the realm at DARPA is promising ideas. The Project Manager succeeds by convincing others—the Office Director and the DARPA Director—that he or she has identified a high potential new concept. The gating notion isn't that the idea is well-proven, but that it has high prospects of making a difference. The DARPA program manager will seek out and fund researchers within U.S. defense contractors, private companies, and universities to bring the incipient concept into fruition. Thus, the research is outcome-driven to achieve results toward identified goals, not to pursue science per se. The goals may vary from demonstrating that an idea is technically feasible to providing proof-of-concept for an operational capability. To achieve these results the program manager needs to be open to competing approaches, and be adroit and tough-minded in selecting among these.

Which DARPA?

While the concept of DARPA as a "high-risk—high pay-off" organization has been maintained, it also has been an intrinsically malleable and adaptive organization. Indeed, DARPA has morphed several times.

DARPA has "re-grouped" iteratively—often after its greatest "successes". The first such occasion was soon after its establishment with the spinning off of its space programs into NASA. This resulted in about half of the then ARPA personnel either leaving to form the new space agency, or returning to a military service organization to pursue military-specific space programs. A few years later then DDR&E (Director of Defense Research and Engineering) John S. Foster required ARPA to transition its second largest inaugural program—the DEFENDER missile defense program—to the Army, much to the consternation of some key managers within ARPA. Also, early in its history ARPA was tasked to conduct a program of applied research in support of the military effort in Vietnam.

More important than the variety of the programs is that they demonstrate the speed with which DARPA took on a new initiative and also how rapidly its programs can move—sometimes more rapidly than its supporters within DARPA may desire. However, particular programs or technologies have not become the identifier of what DARPA is. Rather, DARPA's identity is defined by its ability to rapidly take on and assess new ideas and concepts directed at daunting military challenges or overarching application prospects. While the dwell time on new ideas may vary and DARPA may return to the concept iteratively over its history—most notably with its return to missile defense in the 1970s leading to the Strategic Defense Initiative (SDI) in the 1980s—its hallmark is to explore and create new opportunities, not perfect the ideas that it has fostered. A crucial element of what has made DARPA a special, unique institution is its ability to re-invent itself, to adapt, and to avoid becoming wedded to the last problem it tried to solve.

DARPA Roles

Emphasizing DARPA's adaptability is not to say that there are not some underlying elements to what DARPA does. While there have been some additional ad hoc activities thrown in over time, DARPA has had significant roles in the following:

- Turning basic science into emerging technologies
- Exploring "disruptive" capabilities (military and more generic)
- Developing technology strategy into a Defense strategy
- Foster revolution or fundamental transformation in a domain of technology application (e.g., the Internet or standoff precision strike)

Key Elements of DARPA's Success

There are several key elements in DARPA's succeeding in its unique role as an instigator of radical innovation.

- Create surprise; don't just seek to avoid it

DARPA mission is to investigate new emerging technological capabilities that have prospects to create disruptive capabilities. It is differentiated from other R&D organizations by a charter that explicitly emphasizes "high-risk, high payoff" research.

- Build communities of "change-state advocates"

DARPA program managers may often themselves foster a specific concept or technological approach that they seek to explore and develop. But almost never are they the main, let alone sole, investigator of the concept/approach. Rather it is DARPA's motif to instigate cooperation among a group of forward-looking researchers and operational experts. In this sense, DARPA's success depends on it being a leader and catalyst in developing this community of interest.

- Define challenges, develop solution concepts, and demonstrate them

One aspect of DARPA's success has been efforts to define strategic challenges in detail. Since its inaugural Presidential Issues, DARPA has been problem focused, seeking breakthrough, change-state approaches to overcome daunting issues. This has been true in the military realm from the outset. DARPA-sponsored researchers under Project DEFENDER conducted detailed assessments of intercontinental missile phenomena for both defense and offense. For example, in the 1960s and 1970s, DARPA funded studies at the then new Institute for Defense Analyses on missile offense and defense first under the STRAT-X project on ICBM offense-defense followed by then PEN-X study which assessed both U.S. and Soviet capabilities to penetrate missile defense systems. Subsequently, in the late 1970s, DARPA funded studies to understand how the Warsaw Pact was postured against Western Europe in order to determine how technology could provide a means to offset the Warsaw Pact's numerical and geographic advantages. According to *Transformation and Transition: DARPA's Role in Fostering an Emerging Revolution in Military Affairs*, a paper by the Institute for Defense Analyses, this planning led to DARPA research in both stealth and standoff precision strike, which provided the basis for Secretary of Defense Harold Brown's and Director of Defense Research and Engineering William Perry's "offset strategy".[6]

Such detailed conceptual work also facilitated DARPA's non-military research—explicitly that in information technology. J. C.

6 Van Atta, R., Lippitz, M., et al. (2003). *Transformation and Transition, DARPA's Role in Fostering a Revolution in Military Affairs. Volume 1.* Alexandria, VA: Institute for Defense Analyses, https://doi.org/10.21236/ada422835, https://fas.org/irp/agency/dod/idarma.pdf

R. Licklider came to DARPA as head of the Information Processing Techniques Office with a vision on man-computer symbiosis that grew in specificity as he collaborated with others, especially Robert Taylor, to present a perspective of internetted computers providing capabilities for collaboration and data interchange amongst researchers.[7] Some of this work is described in Licklider's article, "Man-Computer Symbiosis", and Licklider and Taylor's, "The Computer as a Communications Device".

Tension Between DARPA Roles

DARPA has been a pursuer of new breakthrough technologies *independent of defined needs*. It also has been a developer of concept prototypes and demonstrations that *address needs* (but not defined requirements). While complementary, these are substantially different roles requiring different management approaches and different types of researchers. The first type of endeavor requires an exploratory, somewhat unstructured approach seeking out alternatives amongst competing ideas. The latter focuses on taking a specific set of emerging capabilities and combining them into a demonstration of proof-of-concept. Such demonstrations are generally larger in scale and more resource intensive than exploratory research. Moreover, rather than exploratory, they are aimed at assessing the merit of a specific concept. Indeed, demonstration prototype efforts can be "resource sumps", as they are both uncertain and costly. Therefore, the DARPA Director has needs to attentively oversee these while maintaining and protecting the more exploratory research efforts.

DARPA's Successes

Over the fifty years since its inception DARPA has had several major accomplishments that distinguish it as an innovative organization.

7 Licklider, J. C. R. (1960). "Man-Computer Symbiosis", *IRE Transactions on Human Factors in Electronics* 1: 4–11, https://doi.org/10.1109/thfe2.1960.4503259; Licklider, J. C. R., and Taylor, R. (1968). "The Computer as a Communications Device", *Science and Technology* 76: 21–31. See Waldrop, M. M. (2001). *The Dream Machine: J. C. R. Licklider and the Revolution that Made Computing Personal*. New York, NY: Viking Press.

Third Generation Info Tech—the Creation Interactive Information

The singularly most notable technology accomplishment that DARPA is known for is the development of what is now known as modern computing, as embodied in the personal computer and the Internet. While this achievement had its origins in the remarkable vision of one man, J. C. R. Licklider, its coming to fruition speaks volumes for the nature of DARPA as an organization and the willingness of its management to support and nurture the pursuit of such an extraordinary perspective.[8]

The vision that Licklider brought to DARPA was one of a totally revolutionary concept of computers and how they could be used. He foresaw that rather than being fundamentally highly automated calculating ma- chines, computers could be employed as tools in supporting humans in creative processes which he discussed in the article "Man-Computer Symbiosis" in March 1960's *IRE Transactions on Human Factors in Electronics*, volume HFE-1. However, to do so would require entirely new, yet non-existent computer capabilities that included interactive computers, internetted computing, virtual reality, and intelligent systems.

Licklider's extraordinary notion of "man-computer symbiosis" was a fundamental vision that foresaw using new types of computational capabilities to first achieve augmented human capabilities, and then possibly artificial intelligence. Licklider brought these inchoate notions to DARPA when he was named Director of its Information Processing Techniques Office (IPTO). He brought a powerful vision of what could be and used this as the basis for sustained investment in the underlying technologies to achieve the vision. This concept became the gestation of a concerted effort that culminated in the ARPANET, as well as a number of technological innovations in the underlying computer graphics, computer processing, and other capabilities that led to DARPA's fundamental impact on "making computers personal": a truly change-state vision which had fundamental impact in fostering a transformational concept and the creation of an entire industry.

---

8 Waldrop, M. M. (2001). *The Dream Machine: J. C. R. Licklider and the Revolution that Made Computing Personal*. New York, NY: Viking Press, provides considerable detail on DARPA's fundamental role in advancing computer technology.

DARPA's Role in Creating a Revolution in Military Affairs[9]

DARPA has been instrumental in developing a number of technologies, systems and concepts critical to what some have termed the Revolution in Military Affairs (RMA) that DOD implemented in the 1990s based on R&D conducted by DARPA over the prior fifteen years, according to the Institute for Defense Analyses paper *Transformation and Transition: DARPA's Role in Fostering an Emerging Revolution in Military Affairs*. It did so by serving as a virtual DOD corporate laboratory: a central research activity, reporting to the top of the organization, with the flexibility to move rapidly into new areas and explore opportunities that held the potential of "changing the business". DARPA acted as a catalyst for innovation by articulating thrust areas linked to overall DOD strategic needs, seeding and coordinating external research communities, and funding large-scale demonstrations of disruptive concepts. In doing so, the DARPA programs presented senior DOD leadership with opportunities to develop disruptive capabilities. When these programs received consistent senior leadership support, typically from the highest levels of the Office of the Secretary of Defense, they transitioned into acquisition and deployment. At other times, without this backing from the highest reaches of the department, only the less disruptive, less joint elements moved forward.

An example of one of the most successful DARPA programs is its championing of stealth. A radical and controversial concept, DARPA's stealth R&D harnessed industry ideas. Low-observable aircraft had been built before, for reconnaissance and intelligence purposes, but not pursued for combat applications. The Air Force had little interest in a slow, not very maneuverable plane that could only fly at night. After considerable engineering work, the Have Blue proof-of-concept system enabled top OSD and Service leadership to proceed with confidence to fund and support a full-scale acquisition program. OSD leadership kept the subsequent F-117A program focused on a limited set of high priority missions that existing aircraft could not perform well. For example, the program focused on overcoming Soviet integrated air defenses, and

9 This section draws upon Van Atta, et al. (2003). *Transformation and Transition.*

worked with Congress to protect its budget, with a target completion date within the same administration. The result was a "secret weapon" capability—exactly what DARPA and top DOD leadership had envisioned.

Sustaining the DARPA Vision

DARPA's higher-risk, longer-term R&D agenda distinguishes it from other sources of defense R&D funding. Perhaps the most important effect of DARPA's work is to change people's minds as to what is possible.

DARPA's fifty-year history reveals a constant mission to create novel, high-payoff capabilities by aggressively pushing the frontiers of knowledge—indeed demanding that the frontiers be pushed back in order to explore the prospects of new capabilities. As an entity DARPA has many of the same features as its research.

DARPA began as a bold experiment aimed at overcoming the usual incremental, tried and true processes of technology development. Like the research it is chartered to develop, DARPA has consistently been purposively "disruptive" and "transformational" over its fifty years.

Sustaining this unique ethos has not always been easy. There have been several efforts over the years to "tone DARPA down;" make its research more compatible and integrated into the rest of DOD R&D; have it focus more heavily on nearer term, more incremental applications—in other words make it behave like a normal R&D organization. There have been efforts to broaden its charter into system prototyping well beyond the proof-of-concept demonstrations it has constructed on several breakthrough systems. However, with strong internal leadership, both within DARPA and in the OSD, as well as support from Congress, DARPA has been able to perform a truly unique role—it has been and continues to be DOD's "Chief Innovation Agency", pushing the frontiers of what is possible for the benefit of national security and the nation as a whole.

References

Launius, R. D. "Sputnik and the Origins of the Space Age", *NASA History*, http://history.nasa.gov/sputnik/sputorig.html

Licklider, J. C. R. (1960). "Man-Computer Symbiosis", *IRE Transactions on Human Factors in Electronics* 1: 4–11, https://doi.org/10.1109/thfe2.1960.4503259

Licklider, J. C. R., and Taylor, R. (1968). "The Computer as a Communications Device", *Science and Technology* 76: 21–31.

Van Atta, R., Lippitz, M., et al. (2003). *Transformation and Transition, DARPA's Role in Fostering a Revolution in Military Affairs. Volume 1.* Alexandria, VA: Institute for Defense Analyses, https://doi.org/10.21236/ada422835, https://fas.org/irp/agency/dod/idarma.pdf

Waldrop, M. M. (2001). *The Dream Machine: J. C. R. Licklider and the Revolution that Made Computing Personal.* New York, NY: Viking Press.

York, H. (1987). *Making Weapons, Talking Peace: A Physicist's Odyssey from Hiroshima to Geneva.* New York, NY: Basic Books.

3. NSF and DARPA as Models for Research Funding

An Institutional Analysis[1]

Michael J. Piore, Phech Colatat, and Elisabeth Beck Reynolds

The Federal government expends roughly $33 billion annually on scientific research and development in academic institutions, or 60 percent of total academic R&D funding. The former figure represents roughly one percent of U.S. GDP. These funds are allocated through a number of different government agencies and organizations, each operating in a somewhat different way. This study is designed to identify different organizational models of the way in which these funds are allocated to academic research and make a very preliminary assessment of the impact of these different models on the way in which researchers behave and the products their work produces. This has important implications for national science policy and the emergent field of "the science of science policy".

The study grew out of a much narrower project focused on the attempt to create an agency within the Department of Energy designed to foster radical innovation in energy technologies. The new agency, Advanced Research Projects Agency-Energy (ARPA-E), was modeled

1 This article was originally released as an MIT Industrial Performance Center Working Paper in July 2015.

 https://doi.org/10.11647/OBP.0184.03

on the Defense Advanced Research Projects Agency (DARPA), an agency in the Department of Defense (DOD) that was credited with having generated a variety of new, discontinuous technologies and was generally contrasted with other agencies in the DOD, but more particularly, the National Science Foundation (NSF) and the National Institutes of Health (NIH), which were considered more cautious and conservative, and which fostered more continuous or incremental technological developments.

It rapidly became apparent, however, that the critical characteristics of the DARPA model—if indeed there was such a model—were not obvious. The project was consequently restructured to focus on DARPA as an organization, and, subsequently, on the attempt to identify what was peculiar about DARPA, relative to NSF. Material on NIH and other funding provided by the Defense Department was also collected but it is more limited in scope.

From the very start, the project has been conceived in the context of the broader debate about the effectiveness of government, i.e., public sector, initiatives. DARPA attracted our attention in no small measure because of the reputation of the agency as a great success in a period when government has been generally disparaged and government initiatives, especially in the promotion of particular industries, enterprises or technologies, have been viewed with great skepticism. In recent years, there has been a revival of interest in active government. The NSF and DARPA have garnered new interest as countries—particularly developing countries—look to the United States for models for the promotion of economic growth via what has become the new mantra of economic development: "innovation and entrepreneurship in the knowledge economy".

DARPA attracted our attention for a third reason too: the central role the program managers play in its organization and operation and the power and discretion which is lodged in the hands of these agents at the base of the organizational pyramid. In this respect, it constitutes a "street-level" bureaucracy, a class of governmental organizations that we have been studying in other contexts and which appear to offer a model for public sector management that is alternative to both the classic Weberian bureaucracy, widely viewed as rule-bound and rigid, on the one hand, and the new public management, which uses the profit maximizing firm in a competitive market as a template to construct a more flexible alternative, on the other hand.[2]

This chapter is divided into sections as follows: the first section discusses the methodology and research approach. The second section presents the basic findings. It is divided into three subsections, focusing first on DARPA, then on the National Science Foundation (including some background material on NIH), and finally on the origination and motivation of the faculty researchers whose work these Federal organizations fund. The third section of the chapter then turns to an interpretation of the results. I conclude with a discussion of some of the broader implications of the study and the further research toward which they point.

I. Methodology and Research Approach

Our study is centered on MIT. It is based primarily upon data gathered at MIT itself and from outsiders with whom our contacts at MIT had worked directly or whom they recommended as particularly good informants The MIT focus creates a relatively well-defined universe, but obviously limits the generalizability of the results. We discuss those limits in the body of the text.

The focus was dictated by challenges of access. We talked early on with some of the top officials at DARPA, but the agency would not provide us with the data or the names of personnel that would have been required to draw a random sample of researchers or Agency personnel or even to select our informants in a more systematic way.

The study has both a quantitative and a qualitative dimension. The qualitative dimension is based on interviews with key informants. We sought out MIT faculty members who had previously worked on DARPA projects and were knowledgeable about the agency. All of them had also received funding from other sources as well, and hence were able to compare their experiences across Federal agencies, and to a limited extent, with non-Federal funding sources. Virtually all of

2 Piore, M. (2011). "Beyond Markets: Sociology, Street-Level Bureaucracy, and the Management of the Public Sector", *Regulation & Governance Special Issue: Sociological Citizens: Practicing Pragmatic, Relational Regulation* 5/1: 145–64, https://doi.org/10.1111/j.1748-5991.2010.01098.x; Lipsky, M. (1980). *Street-Level Bureaucracy: Dilemmas of the Individual in Public Services.* New York, NY: Russell Sage Foundation.

them had experience with the NSF. Some had also received funding, or considered applying for funding, directly from one or more of the military services, from NIH, and from private organizations (e.g., companies, foundations, and the like). We tried to interview the DARPA program managers of the projects on which our MIT respondents had worked, but we were limited to program managers who had left the agency. In total, we held formal, but open-ended, interviews with twenty-two MIT faculty members, and twelve current or former program managers and agency officials. Fourteen of these came from DARPA, eight from NSF, and five from NIH.

For the quantitative dimension of the study, we started with a data set of all research projects which received outside funding at MIT in the years 1997–2008. We then linked this data to data on patents, licenses, commercial ventures (startups) and citations in scholarly journals. The bulk of this data was provided directly by various offices at MIT, to whom we are greatly indebted for their cooperation. The citations, however, we collected ourselves with the help of a team of MIT undergraduate research assistants.

We focus here on the qualitative dimension of the study, but report preliminary results of the quantitative dimensions as background in the next section below.

II. Basic Findings

DARPA

Background

To appreciate the nature of this Agency and its role in the debates surrounding Federal research policy, it is important to understand its history, and the nature of its success, particularly in the period of widespread skepticism and general depreciation and disparagement of government and its ability to create and maintain dynamic, innovative programs.

DARPA was created in 1958 in reaction to the launching of the Soviet space satellite Sputnik, and the universal surprise with which it was greeted by the U.S. military, the country's scientific establishment and the political class. That surprise was widely attributed to the conservative bias of scientific and engineering research, particularly the National Science Foundation that provided the major component of Federal research support and was the principle vector of research policy. The conservative bias was in turn attributed to the peer review process through which funding was allocated and the research effort more generally evaluated. A second component of military research was financed by the Offices of Research of the various branches of the armed services through grants but also through their own laboratories. The obligation of these offices to support the existing infrastructure was a second conservative force in the existing structure. A new agency was then conceived in large measure in reaction to these other organizations. Thus, DARPA was effectively given carte blanche to develop its research projects on its own, unconstrained by the existing research establishment. The institution that we set out to study was the result. It is partly the result of a mission and ethos defined in opposition to these other agencies and partly of organizational characteristics created to escape the constraints under which they operated. In this study, we use the NSF as a foil against which to define and understand the DARPA model, since for academic research it is by far the most important of the various institutions against which DARPA was conceived.

Evaluation of Success

The organization that has emerged over time is, as we shall see, distinctive and poses a challenge to the principles of organization that guide these other agencies. But it has proven to be very resistant to systematic evaluation. The resistance is in part conceptual—it is hard to know how the agency ought to be evaluated. But it is also institutional: DARPA has refused quite explicitly to help support an effort at evaluation, at least in connection with the present study. It rejected our request for data which would have enabled us to define a list of projects, trace the participants drawn into the agency's orbit, and assess the impact upon conventional measures of scientific output such as patents and citations in scholarly journals. Their claim is that the agency has to be evaluated in terms of its contribution to the mission of the armed forces, a mission that is notoriously difficult to define.

The most extensive evaluation effort of which we are aware is a three-volume study by Van Atta, et al.[33] The study reviews approximately forty projects and develops a narrative account both of DARPA's contribution to the projects and the contribution of the technology which emerged in the process to the military mission and to civilian uses. A great strength of the study is that it includes most of the projects upon which the agency's reputation in the general public or the science policy community rests, and in that sense it both reflects and sustains the esteem in which the agency is held. But the projects were selected largely on the basis of the data available to evaluate them in this way, and there is no effort to map them onto the larger universe of projects in which DARPA has been engaged, or might have been engaged in the period. Indeed, in the sense that the study purports to evaluate the agency's success, the projects studied are selected on the dependent variable. The study does not include projects that were considered and never undertaken, or undertaken but abandoned or, as apparently is frequently the practice, folded into other very different projects. It is, moreover, difficult on the basis of this study to compare DARPA to other funding agencies with a different organizational structure and approach.

On the other hand, it is not clear how one would evaluate an agency of this kind. Conventionally, programs are evaluated in terms of benefits and costs. But in the case of research on new technologies the costs are the opportunity costs of research in domains whose pay-offs, since they were never actually undertaken, are impossible to know and the benefits of these projects accrue not only in military preparedness, which even when it is not classified is ill-defined, and some of the projects—the World Wide Web, for example—have so fundamentally altered the texture of everyday existence and have such widespread commercial ramifications that the benefits seem virtually infinite. The Agency is certainly right: Its mission cannot be reduced to the patents and citations in terms of which research results are conventionally measured in academic studies.

Nonetheless, in order to make any systematic comparison, it would be helpful to have some of these conventional measures of success. And

3 Van Atta, R., Deitchman, S., and Reed, S. (1990–1). *DARPA Technical Accomplishments*. 3 Volumes. Alexandria, VA: Institute for Defense Analyses.

for this study, we have constructed such measures starting from data provided by our own institution: MIT maintains a roster of grants and contracts obtained by its faculty and researcher staff. We have linked that individual contract data to several outcomes which are conventionally used as indicators of success. The granting agencies include DARPA, NSF, and NIH as well as the various military research offices, and a number of nongovernmental funding sources (private companies, foundations).

The outcomes which we looked at are threefold: patents, citations, and technology licenses. In addition, we linked the technological licenses to data on new business ventures. The results of this project will be reported in a separate paper. Preliminary findings with respect to patents, technology licenses and new business ventures, are contained in Tables 3-1 and 3-2. As can be seen there, DARPA performs better than any of the other agencies on all of these measures, notwithstanding the fact that the agency explicitly rejects them as measures of its performance.

Table 3-1 Patents supported by sponsored research at MIT, 1997–2008. (Table prepared by the authors)

	# patents	# awards	# awards leading to patents	Total funding ($ mil)	Funding per patent	P (award has patent)	# patents per award (award has patent)
Agency	[a]	[b]	[c]	[d]	[d/a]	[c/b]	[a/c]
NSF	258	2988	90	1671	6.48	3.0%	2.87
NIH	181	2645	82	3955	21.85	3.1%	2.21
DARPA	153	519	67	1090	7.12	12.9%	2.28
Navy	94	1037	44	569	6.05	4.2%	2.14
Consortium	78	205	16	1518	19.46	7.8%	4.88
Army	52	471	22	692	13.31	4.7%	2.36
DOE	46	787	23	3683	80.07	2.9%	2.00
Air Force	38	856	28	470	12.37	3.3%	1.36
NASA	25	1586	18	1071	42.84	1.1%	1.39
MIT — Internal	24	128	4	1491	62.13	3.1%	6.00

Table 3-2 Startups supported by sponsored research at MIT, 1997–2008. (Table prepared by the authors)

Agency	Number of associated startups (awards)	Total awards	Total funding ($ mil)	P (award supported startup)	Funding per startups
	[a]	[b]	[c]	[b/a]	[c/a]
DARPA	20 (21)	519	1090	4.0%	54.5
NSF	20 (25)	2988	1671	0.8%	83.6
NIH	14 (23)	2645	3955	0.9%	282.5
Navy	6 (9)	471	692	1.9%	115.3
Army	6 (6)	1037	569	0.6%	94.8
DOE	5 (6)	787	3683	0.8%	736.6
Air Force	3 (4)	856	470	0.5%	156.7

Finally, our own work has been particularly influenced by the research of our colleague Erica R. H. Fuchs, who originally called our attention to the significance of DARPA as a possible model of government organization. Fuchs focuses specifically on the role of DARPA in one particular technology, the technology of computing, and places emphasis on the role of the program manager in creating and maintaining networks of researchers or research communities. We follow Fuchs in this last respect, but the broader range of projects which we examine (albeit much more superficially) and the contrast with the NSF complicates this picture.[44]

Qualitative Findings

Our findings are best understood against the backdrop of a standard peer-review model, which our respondents seemed to carry in the backs of their heads. Central to this model is an academic or scholarly discipline. The financing agency issues a call for proposals from such a discipline. Researchers from that discipline are invited to submit proposals. A panel from within the discipline is then recruited to

44 Fuchs, E. R. H. (2010). "Rethinking the Role of the State in Technology Development: DARPA and the Case for Embedded Network Governance", *Research Policy* 39/9: 1133–47, https://doi.org/10.1016/j.respol.2010.07.003 (Chapter 7 in this volume).

review these submissions. The panel ranks the proposal, and the agency awards its funds in order of rank, progressing from the highest ranked proposals down the list until the funds are exhausted. The funds are typically awarded in the form of a grant, generally with reporting requirements but with minimal reviews of the research results and no effort to ensure adherence to the original proposal. The model is actually very close to the way in which research funding is organized at the NSF and NIH, albeit, as we shall see, with important qualifications. But the DARPA model is very different. Which of the differences is important for the research outcomes is, of course, an open question, and given the number of dimensions along which practice departs from the standard model, not an easy question to answer.

The DARPA Model

The central figure in the DARPA model is the program manager (PM). The PMs typically comes into the agency with a very specific technological idea which they want to develop. They then spend some period of time—often a year or more—researching that technology and the domain (or domains) in which it lies through their own reading, visiting and talking to key figures who are thought to have something to contribute to the technology or to its development, and colloquia, conferences, small group meetings and other encounters, which he or she typically organizes, in which the technology is discussed and various approaches to its development are debated. After this initial exploratory period, the PM works up a plan for development of the technology and writes and issues RFP's soliciting proposals for the various components of that plan. At DARPA, these are known as Broad Agency Announcements (BAA). The proposals are sent out for review to experts whom the PM selects, within the government (particularly the military) and outside. But the ultimate decision as to which proposals to fund rests with the PM alone. Proposals that are accepted then serve as the fulcrum for a research contract which is negotiated with would-be contractors. Contracts typically include specific performance requirements. Contractors are required to submit frequent progress reports and progress is continually monitored through these reports and through site visits. Contracts are subject to revision or cancellation in the light of research experience. In addition to the review process, the organization holds regular seminars and conferences, comparable to those out of which the project initially emerged: contractors (who at DARPA are called performers) are required to attend these meetings, where they are expected to report their own progress and to listen and comment on the reports of others.

Given the central role of the PMs, the way the organization operates depends a lot on the way in which the PMs are recruited and managed. Hence key to the organizational model is the fact that the PMs come from the research community outside the organization, have relatively short tenure in the agency itself (an average of four to five years), and then leave the organization to pursue their careers elsewhere. We have not been able to follow these careers systematically, but it is significant that no obvious pattern emerged in the interviews. Most of the PMs whom we interviewed came from an academic or military background, and afterwards returned to their home institutions, often as a research administrator, but sometimes as rank-and-file professors and researchers, or, alternatively, joined the supporting consulting firms which surround DARPA (to which we will return shortly). Significantly, all of the PMs to whom we talked thought of their DARPA experience as a high point in their careers, one of the most exciting and stimulating periods in their professional lives (this point is stressed particularly by Fuchs).

The Agency operates outside of the civil service recruitment, hiring regulations and salary structure; and although it seems unable to pay exactly what the PMs would earn in the private sector, it is able to negotiate pay scales and contract terms significantly better than those that other government agencies can offer.

Emphasis was placed in virtually all of our interviews upon the fact that the PMs come to the agency with their own project, an idea which they essentially originate and to which they have a personal commitment (respondents talked of that commitment in fact as if it were an obsession—although that was not the term they actually used). In turn, it is obvious that the environment in which the agency operates and its structure determine who brings proposals to the agency and which of those proposals, i.e., which potential PMs, are actually recruited and hired.

DARPA is a flat organization, a hierarchy with three levels: a director, a series of office managers, and the program managers. The director has an associate director who works with him or her but not as a separate level in the hierarchy. The director sets the broad outlines of the research agenda. The research itself is grouped into program areas, largely on the basis of technology and mission, and the office managers flesh out the agenda in their own areas. The PMs coming to the agency with their own ideas present them to the director and/or the office managers. DARPA cultivates a reputation for being open to new, radical ideas originating outside the organization (indeed, listening to people talk, one is led to believe that the ideas *always* originate from outside the organization) whether or not they fit the defined program. But the office managers and the director play an active role in recruiting ideas that fit into the program and in screening proposals to ensure that the program has some coherence and direction.

While the program itself originates with the director and is fleshed out by the office managers and the PMs whom they hire, it is conceived in consultation with the military services, with Congress and with the Administration. And it is clear in discussions with the agency that careful attention is paid to cultivating support within the political and administrative environment in which it operates. Particular emphasis is placed in virtually all discussions with people about the program upon the military mission of the agency and the way in which that operates to shape the programs.

Another significant factor shaping the programs is the agency's mission in supporting radical, discontinuous technological change. That mission, as we have already mentioned, is rooted in DARPA's origins in 1958 as a response to the Russian launching of Sputnik and the way in which Sputnik caught the U.S. military and scientific establishments by surprise.

These two factors—the military mission, and the focus on discontinuous technological development—surface repeatedly in interviews. The Agency is always looking at whether, on the one hand, the research would be undertaken elsewhere in the government or the society, or, on the other hand, whether there is a constituency—already existing or one which could be cultivated—in the military services which would adopt the new technologies and actually deploy them. To the outside observer, the role of the military mission in the operation of the agency—and particularly in the ability of the organizational model to operate in other contexts—is difficult to understand. This is because the technologies under development are often so distant from actual military application that it is hard to imagine a technology for which no military application could be found, and much of what the agency does seems to have no obvious constituency within the military establishment. Nonetheless, reference to the critical role played by the military missions in the success of DARPA was stressed so repeatedly and by so many different informants, especially in discussions of transferring the DARPA model to the Department of Energy in the form of ARPA-E, that one had to believe it is indeed central to the organizational model.

In sum, the characteristics which distinguish DARPA as a funding organization are:[5]

1) The discretion and authority lodged in the PMs;
2) Awards in the form of contracts with specific deliverables and specified performance measures periodically monitored for specific performance. Typically, performance measures specified in contracts are set unrealistically high—targets which stimulate and focus debate about the characteristics of the technology;
3) PMs recruited and compensated outside of the regular civil service regulations;
4) Flat organization consisting of only three levels—PMs, the office managers, and the Director with an assistant director;
5) The tenure of the direct employees of the organization is very short—three to five years for the PMs, even less for many of Director (with the major exception of Tony Tether, who held the position for the full eight years of the Bush Administration 2001–2009).

In addition, two characteristics, which have received little attention in the literature and which we have not discussed so far, stand out:

5 Bonvillian, W. B. (2006). "Power Play", *The American Interest* 2/2, November/December, 39–48, at 48.

6) The very extensive use of support personnel hired from outside subcontractors, typically consulting firms, not independent contractors. These consulting firms—but often the particular personnel assigned by the firm to work with DARPA as well—have a long-term relationship with the agency. The tasks which they assume and the roles they play range from clerical and administrative support to high level professional functions. The latter include scientific and engineering research, but also key administrative, training and supervisory tasks. Contractors are used, for example, to "orient" (and in effect to train) new PMs and also to advise them in the development and execution of their programs throughout their careers in the agency. Given the short tenure of DARPA's own personnel, the contractors provide the organizational continuity. And many of the subcontractors who work with DARPA have a long history with the agency, some having actually served as PMs or as performers.

This role of the outside contractors, and particularly the consulting firms, is a complete reversal of the usual relationship between temporary and permanent employees and, from the point of view of organizational studies, is probably the most interesting aspect of DARPA as an institution. Temporary employees typically have short tenure with the organization and are used to smooth out the variation in personnel requirements, a buffer against flux and uncertainty. The role of these outsiders suggests that a great deal of the much-vaunted flexibility (or malleability) of the organization, and the adaptability which it is supposed to confer on the agency's program relative to other federal research agencies such as the National Laboratories or NSF, is illusory.

Parallel to the use of consultants, but somewhat different, is the way the agency draws on outsiders to audit and police its contracts with researchers. The outsiders in this case, however, are experienced government employees who are certified to perform this function. The Agency looks for the most qualified auditors within the military services, people who are able to use government contracting regulations in a creative way to accommodate the needs of the performers the PMs want to recruit—although the specific examples which were cited in the interviews related to the requirements of private industry, not academics.

The academics, however, reported that the auditors were surprisingly knowledgeable about the technical dimensions of the projects and helpful as the researchers tried to provide explanations for why they were unable to meet contract requirements—explanations that could then be used by the PM in defending his or her program within the agency.

7) The interaction which occurs in the process of contract administration should be understood as part of a final characteristic of the DARPA organizational model: the continual review and discussion which surrounds a program from its very inception until it is completed or phased out. That discussion takes place through a variety of vehicles, including small group meetings; larger and more formal seminars and conferences; formal meetings when seeking funding for new program proposals and on continuing or expanded funding for ongoing programs in meetings between the PMs, the office managers and the DARPA director; and reviews and auditing of contracts with outside auditors and with the PM. It involves continual questioning both of the ends of the program (why do we want to have this research in the first place? Why is DARPA, and not the private sector or some other government agency, financing it? How do you assess its success in doing so? What are the proper metrics? Etc.). We will come back to the significance of this review process shortly.

The NSF

The central thrust of NSF research support—and the focus in the present study—is its grants awards for discipline-based scientific research and education. The Agency also has a series of ancillary programs and activities which are organized around specific scientific and policy problems, and/or are explicitly interdisciplinary in character (among which is the program which supports our own research project). Other special programs support research institutions as opposed to individuals and sponsor special conferences.

In its disciplinary programs, NSF presents a sharp contrast to DARPA. Its organization and mode of operation resembles the model which faculty members carry in the back of their mind, as we noted initially. It is basically organized around scholarly disciplines and is designed to support and sustain them. Funds are awarded in the form of grants through a competitive process organized and administered by a program manager. Competitions take place on a regular basis in a schedule announced and publicized in advance. The NSF does not actively solicit proposals. Applicants select the division to which they wish to apply, almost invariably the division corresponding to the discipline in which they were trained. Submissions are evaluated in a peer review process by a panel drawn from members of the discipline. The panel ranks the proposals relative to each other. Funds are allocated to the various divisions at higher levels of the organization (through a process which we did not investigate for the study). Within each division, funds are then generally awarded to proposals in the order in which they have been ranked by the review panel until they have been exhausted.

The role of the PM is, however, not as limited as this conventional picture seems to suggest. program managers at the NSF certainly do not have the wide latitude to define their program and to pick out the investigators who will participate in it that their analogues do at DARPA. However, they are not completely bound by the peer review process. They actually have the power and responsibility to fund proposals out of the order established in the peer review process if, for one reason or another, they believe it is desirable to do so. Furthermore, the attention devoted to the procedures for funding proposals out of rank order in the training and orientation of the PMs implies that this is not an incidental part of their job; that they are expected to continually review and evaluate the panels' rankings, although they may not often actually act to contravene it. When they do fund a proposal out of order, the decision is usually justified by its importance to the health and progress of the discipline. In this, they do not act alone; they must first obtain the approval of their supervisor in the division. The procedures for obtaining that approval apparently vary somewhat across the agency, but, as it was described to us in interviews, it typically entails a written memorandum which is then discussed and evaluated by the division director. In at least some divisions, these "out of line" proposals are discussed formally and informally among the PMs as a group. Those discussions are part of an ongoing discussion within the division about the direction of the discipline and the kind of research that would be required to sustain it and maintain a balance among its different components. These discussions, we will argue, play a role analogous to the continual discussion and debate which surrounds the research support process at DARPA.

The NSF has a reputation for being extremely conservative with an overwhelming bias in favor of proposals which hover very close to the center of the discipline, in terms of the hypotheses which they entertain and the methodology which they employ. As we have already noted, the surprise launching of the Russian Sputnik in 1958 was attributed to this conservative bias and DARPA was explicitly and deliberately designed to counter-balance it. NSF continues to have that reputation. It was reflected in comments of MIT faculty in virtually every interview we conducted, often spontaneously, but always when respondents were asked to compare NSF and DARPA funding. Many commented that so much emphasis was placed on feasibility at NSF that you actually had to have done the research (or a good part of it) before you submitted the proposal for funds to finance it. Several faculty members said their strategy was to submit proposals to fund research already underway and use the funds to initiate new projects, which then became the foundation for their next grant proposal.

The conservative bias is widely attributed to the peer review process through which funds are awarded. But it appears that the bias is not inherent in the process itself but rather in the way it is organized and administered. That in turn reflects the way in which the agency conceives of its mission, which is to sustain the country's scientific capability through education and research, a capability which is in turn embedded in the academic disciplines. The PMs have an incentive to emphasize the awards as the outcome of the peer review process to avoid having to justify the outcome to rejected applicants. Their responsibilities, in contrast to those of DARPA managers, leave them very little time to give detailed feedback, a point which our faculty respondents emphasized repeatedly. But more fundamentally, if the PMs fail to intervene in the process it is because they share the biases of the review panels. They are very much a part of the scientific community which the discipline defines. Their backgrounds make it natural that they would think in these terms. Indeed, they are selected for that reason. In contrast to DARPA

PMs, the PMs at NSF are drawn from the disciplines whose research proposals they manage. About half of the PMs are career civil servants, the other half are on short-term contracts of one to three years, on leave from university research positions and are often actually paid through their universities at the levels they were receiving as faculty members.

This is not to say that the PMs add nothing to the process. The role of the NSF in reviewing a wide variety of research proposals and the PMs own position within that process gives them a broader vision than any particular review panel is likely to have. But it is still very much a vision of what Thomas Kuhn would call "normal science",[6] a vision in which progress occurs within the boundaries of the discipline, through adherence to the standards of the community that develops within those boundaries, and which the community promulgates and enforces through the control which it exercises over the careers of its members. The way in which the PMs represent the community was driven home in one of our interviews by one of the respondents who, when confronted with the criticism that the most important criteria in judging a research proposal at NSF was feasibility, gave us a long defense of feasibility as a cannon of "good science".

One can see this as well in another area where the PMs act with power and discretion helping researchers whom they do not fund themselves find support through other government agencies, acting essentially as brokers and at times even putting together packages of funds from several different agencies. These efforts are facilitated by the extensive contacts which career PMs develop with the Federal research establishment. But they do not seem to see this activity as part of their regular responsibilities to oversee the health of the disciplines for which they are responsible, and they talk about it in very different terms, terms which make a sharp distinction between the discipline approach of NSF and other criteria which might justify a given research project (potential contribution to social welfare or to economic progress, for example).

A final piece of evidence suggesting that it is not the peer review process per se but the orientation of the organization which uses it is provided by the comment of one faculty member who had participated in NSF panels: he argued that the conservative bias in the research which

6 Kuhn, T. (1962). *The Structure of Scientific Revolutions*. Chicago, IL: University of Chicago Press.

the panels funded reflected the instructions which the panel members received. He and his colleagues, he insisted, were perfectly capable of evaluating and ranking the kind of high risk, original research which DARPA sought out and funded, if they were instructed to do so. It is to be noted that this comment calls into question the central role of the PM at DARPA as much as that of the peer review process at NSF.

We emphasize the dichotomy between the way in which the NSF actually operates and the way in which MIT faculty members perceive its operation, because in terms of the impact of the organization upon the research community, it is not clear which is more important. It is after all the faculty who must actually conceive the research program and carry it through. To appreciate how their perceptions influence the research process, it is important to understand how they think about their work and how they design their research programs. A second set of findings that emerged from this study relate directly to this question.

The NIH

It is perhaps worth adding at this point a few limited observations about what we learned about the NIH. It is virtually impossible to make broad generalizations about the NIH, given its $30 billion annual budget (fully half of all civilian R&D expenditures)[7] across twenty-seven Institutes and Centers. But several interviews with MIT faculty and Program Officers (Pos, as opposed to PMs) at institutes within NIH provide some context for thinking about the role of the Program Officer at NIH relative to NSF and DARPA.

Program Officers have relatively little discretion is selecting proposals to receive funding. Proposals across the NIH first go to the Center for Scientific Review (CSR) that then categorizes the proposals and assigns them to the relevant institute. The proposals are reviewed by "study sections" (equivalent to a review panel) which score the proposals. The final scores and reports are sent to the Pos who then gather within each institute for a "Paylist" meeting within their division (one level below Institute level) to discuss the awards and decide which programs to fund at what level.

7 Cook-Deegan, R. (2015). "Has NIH Lost Its Halo?", *Issues in Science and Technology* 31/2: 36–47. (Chapter 15 in this volume).

Like PMs in the NSF, Pos can challenge the scoring of a particular proposal, but instead of approaching their supervisor in their division like in the NSF, Pos approach the "Advisory Council", a body that reviews the study section process, and ask for a special review of a proposal that they consider a "high program priority". However, this seems to happen infrequently and internal research at the NIH shows that there is a fairly smooth curve demonstrating that as the scores get higher, the percentage of awards at that level gets lower. Going outside the payline doesn't happen that often. As one PO stated, as much as they like to think they are finding the diamonds in the rough, they are not as aggressive in going beyond the payline as they like to think they are.

Where Pos seem to have more influence is in supporting the overall direction of the Institute's agenda and new areas of science where they see a lack of investment. For areas of research that are new and where "you would never get something like that approved in a regular study section", Pos can make the case within their Institute that there should be more attention and investment. This could come through "funding opportunity announcements" (FOAs) which indicate the Institute's interest in a new area. The NIH may also encourage more research through the creation of new program areas that receive formal set-asides for funding. This currently represents approximately 15–20 percent of all NIH funding. Pos talked about the impact they felt they have had on the development of their field in important new areas of research. This might be in the form of a new program or through a process of "coaching and coaxing" applicants on their proposals for funding in these new areas of research.

As with the NSF, Pos have relatively limited contact with their grantees, usually connecting once a year when progress reports are due. They are also less engaged today in sponsoring conferences than in the past due to budgetary constraints. However, they seem to play an active role in supporting and encouraging next generation Pos to apply for NIH grants and help them navigate the system. This aligns with the NIH's efforts to lower the average age of grant recipients (the average age is forty-two, with a median of fifty-two).[8]

8 Harris, A. (2014). "Young, Brilliant and Underfunded", *New York Times*, 2 October, https://www.nytimes.com/2014/10/03/opinion/young-brilliant-and-underfunded.html?_r=0

MIT Faculty

The funding agencies are only one side of the research equation. On the other side are the scientists and engineers whom the agencies need to attract if the work they want to support is actually to be carried out. At DARPA, these researchers are aptly referred to as *performers*. In this study, they are represented by those faculty whom we interviewed at MIT. The interviews suggested that they have a dual motivation. On the one hand they have a profound intellectual commitment to science and engineering, although not necessarily a well-fleshed out research agenda. On the other hand, their position at MIT requires them to raise substantial funds from agencies and organizations on the outside. These funds are not required to support their family. The wide range of opportunities open to the faculty at an elite school like MIT ensures that they will always be able to earn a comfortable living. But the Institute is only committed to paying the academic portion of their salary support. An additional two to three months is viewed as "summer support" and must be raised through research grants and contracts on the outside. In addition, faculty are expected to support a mini-research establishment consisting of overhead on lab space, equipment and administration and a team of graduate students who work with them over the course of three or four years on projects related to the faculty member's own research. In many respects the research establishment is like a small business and the terms in which faculty members discuss it makes them sound like independent entrepreneurs.[9]

Evaluation of Experiences with Funding Agencies

All of the faculty members with whom we talked were very enthusiastic about the intellectual experience of working with DARPA. This is perhaps not surprising given the fact that we were talking primarily to faculty members who had received DARPA funding, although the unanimity of opinion on this score was striking. There were a number

9 For a somewhat different view of the relationship between economic and intellectual motivation see Freeman, R. B. (2011). "The Economics of Science and Technology Policy", in *The Science of Science Policy: A Handbook*, ed. K. Fealing, J. Lane, J. Marburger III, and S. Shipp. Stanford, CA: Stanford University Press. 85–103, https://doi.org/10.1111/j.1541-1338.2011.00523.x

of components to this experience. These included the opportunity to interact with other researchers in the various meetings and conferences which DARPA PMs organized in the process of putting together and then executing their programs.

Often these involved encounters with researchers from other disciplines or from outside the university, in private industry and/or in government labs. Several respondents reported that they had developed relationships in this way that fundamentally altered their research trajectories and/or created the foundations for long-term research collaborations. It is to be noted that several of the PMs suggested that this is exactly what they were trying to do in developing their program—although the MIT faculty did not seem to be simply echoing the comments they had picked up at DARPA.

Faculty members also emphasized their interactions with the PMs themselves whom they tended to talk about as colleagues and collaborators rather than merely as research funders or supervisors. These intellectual interactions with the PMs ranged from the initial discussions when the PM was preparing his or her research program to the extensive feedback which the DARPA PMs provided when a proposal was turned down. But they also mentioned the interaction with colleagues working on similar projects in seminars where they were required to present their research in progress as stimulating intellectually and important in the research process.

As noted earlier even the interactions with contract auditors were viewed as part of the intellectual experience, a feature of the way DARPA operates which is not accidental. The auditors are typically seconded from the military and recruited because of their ability to understand the substance of the research and its relevance for the agency's mission. Since performance standards specified in the DARPA contracts are often deliberately set at levels that are virtually impossible to achieve, auditors spend considerable time trying to understand the obstacles to attaining the specified standards and identifying more realistic targets. Indeed, it is precisely to stimulate this type of discussion that targets are set above realistic expectations.

In addition to the intellectual experience of working with DARPA, two other features were mentioned in interviews. One is the size of the awards, which were, by and large, much larger than could be obtained through the NSF or NIH. The second was the ability to buy expensive lab equipment which could then be used for other projects.

On the downside was the threat that the agency would cut off funding in the middle of a project. Because funds are awarded in the form of contracts rather than grants, and because, as just noted, specified performance requirements were often unrealistic, the agency is in a position to cut off funding not just because of the research performance itself, but actually for any reason. This was a major threat under the administration of Tony Tether; he was believed by our MIT respondents to have cut contracts when budget cuts forced him to reorder the agency's priorities in ways that were unrelated to the research which the contract initially covered. Funds were also cut when the research suggested that the project itself was not viable and the goals could not be achieved, or when a competing approach to the problem proved to be more successful. Whatever the actual reason, the sudden loss of funding was a particular problem for faculty members who are using the funds to finance graduate students working on doctoral dissertations, and several respondents reported that as a result of their DARPA experience, they had moved to a portfolio strategy for financing, in which they were careful to avoid excessive dependence on a single agency.

The other downside of DARPA funding is the frequent reporting requirements, in many cases every three months. This was particularly a problem for faculty doing basic science (as opposed to applied work), since they often did not have results at these reporting intervals.

The NSF

In contrast to DARPA, the intellectual experience of working with the NSF was universally characterized as dull, indeed pedestrian. It certainly involved none of the excitement or intellectual stimulation associated with DARPA. Proposal writing was seen as a chore. There was no thought of showcasing the intellectual excitement associated with the work. The widely expressed view that you had to have done much if not all of the work in advance of proposing it eliminated the element of surprise and discovery which the researcher might originally have felt and gave the process a slightly dishonest flavor (although the respondents did not put it in precisely those terms). Our respondents generally view the NSF's program managers as competent; they talked of them as colleagues and, although they were not asked to compare them directly to DARPA PMs, the comparison was not unfavorable to NSF. But there was little opportunity to interact with them in the way that they interacted with DARPA PMs; they provided little help in preparing proposals and little feedback when the proposals were rejected. NIH project managers incidentally were not respected as colleagues in the way that PMs at NSF and DARPA were; they also do not have the capacity to fund proposals outside of the rank order established by the peer review panels.

Most of our respondents who had received NSF grants had also participated in review panels, but this participation was seen as a chore: people felt obligated to participate to support the discipline and in return for funding they had received, but it was not viewed as a rewarding experience. One could imagine the discussions in the review panel meetings as comparable to the small group meeting which DARPA organized, but they were never discussed in those terms. The range of proposals that the panel members were required to read could have been seen as an opportunity to get an overview of the field but it was never discussed in these terms either.

In sum, the advantages of the NSF were on the "business side". Here, the main advantage of NSF funding was that once a grant was awarded, the funding was secure, and one could count on it, especially in supporting graduate students. This contrasts with DARPA, where there was always the possibility that funding would be cut off in the midst of a thesis project. Also, NSF grants involved minimal reporting requirements; the major incentive to perform was to gather material to support the next grant proposal.

III. Interpretation

Economic and Sociological Perspective

The DARPA material lends itself to two quite different interpretative lenses. From the point of view of standard economic theory, with its preference for market mechanisms and individual incentives and its distrust of government bureaucracy, the salient feature of the DARPA organizational structure is the way in which it suspends the rules and regulations which normally constrain government officials. The mechanisms here include the freedom from the regulations governing hiring and salary scales, the use of contracts with requirements for specific performance (as opposed to grants), the way in which program managers are hired from outside the organization, their short and very limited tenure, and the very extensive use of outside contractors who can be replaced easily and at will. On the other hand, the standard theory which would emphasize the rules which normally constrain government actors rests upon a rational choice theory of individual behavior in which the actors are presumed to make a sharp separation between means and ends, and the technical relationships that determine the way in which the former affects the latter, and then to maximize the ends given the means at their disposal. The characteristic of the problems which DARPA, and NSF as well, are designed to address is that the ends are ill-defined and unclear, and the causal relationships between the means and the ends are exactly what the organization is supposed to be investigating. This entails what economists call "Knightian uncertainty", i.e. uncertainty about what the possible outcomes actually are let alone what the probability of realizing any one of them.[10] Neither the competitive market nor the rational choice model has much to say about how this should be addressed.

The standard rational choice theory has a second problem too. The theory attempts to understand and explain behavior in terms of individual self-interest. It has very little to say about the agent's behavior when he or she has no particular interest in the choice among the alternatives we are attempting to understand. The choices of the faculty researchers are, up to a point at least, understandable in those terms, but the role of the PMs is not; or, at least, they do not yield an obvious interpretation of our findings. At both NSF and DARPA, the PMs seem to be motivated primarily by the intellectual interest and excitement of the work in which they are engaged. They seem to believe in the mission of the organization and see little difference between their own interests and that of the organization for which they worked. This was of course no accident. The agencies consciously recruited them with

10 Knight, F. H. (1921). *Risk, Uncertainty, and Profit*. Boston, MA: Hart, Schaffner & Marx.

this in mind. However, it called not for a theory of individual choice but rather a theory of how the agency's mission was conveyed to the agents, and how it was understood by them.

The second interpretative lens through which the material gathered for this study might be addressed is *organizational theory*. We use this term very loosely here to refer to a range of theoretical ideas drawn from sociology, cognitive theory, language theory, and social psychology, all of which, however, suggest that human behavior must be understood in terms of the social context in which it occurs. Behavior in this view cannot be reduced to individual actions, coordinated indirectly and impersonally by a market (or market-like) mechanism, but rather must be understood in terms of the way in which people interact with each other. Applied to science studies, the basic idea is that scientific inquiry takes place within a community and is governed by a set of rules, habits and customs, partly explicit but with a substantial tacit or implicit component, which the community generates. These rules have both a social and an intellectual dimension. The funding agencies are then understood in terms of their impact upon such communities. The same basic conceptual apparatus can be applied to understanding the internal operation of the funding agencies themselves, for they are also communities of practice which arise and evolve over time.[11] This is true of both DARPA and NSF. The major difference between them is that DARPA is creating new communities and NSF is managing scientific disciplines which are communities that already exist.

Our own understanding of this perspective derives from a series of case studies conducted by the Industrial Performance Center at MIT on the organization of product design and development in the private sector.[12] Related understandings can be found in Fuchs and Phech Colatat,[13] which are however not independent of the current project, and also in Donald Schön, and Kuhn.[14] In the IPC study, we conceptualized

11 Schön, D. (1983). *The Reflective Practitioner: How Professionals Think in Action*. New York, NY: Basic Books

12 Lester, R., and Piore, M. (2004). *Innovation—the Missing Dimension*. Cambridge, MA: Harvard University Press.

13 Fuchs. (2010). "Rethinking the Role of the State"; Colatat, P. (2015). "An Organizational Perspective to Funding Science: Collaborator Novelty at DARPA", *Research Policy* 44/4: 874–87.

14 Schön. (1983). *The Reflective Practitioner*; Kuhn. (1962). *The Structure of Scientific Revolutions*.

a research community as like a language community. Like language, it emerges and evolves through conversation, discussion and debate. We termed that conversational process *interpretation*. Particular product ideas, or in the present case, research projects, are drawn out of this conversation and pursued through a second process, *analysis*. Analysis proceeds very much as it does in engineering (and economics) textbooks: there is a clear statement of the end or ends which the product is designed to achieve, and one then organizes alternative resources, or means, so as to optimize (or maximize) the ends. But the interpretative process is under-theorized and requires some amplification. It is, we argued in the IPC study, like a conversation, a discussion or debate. It depends on who participates in that conversation, what they actually talk about, how the conversation proceeds from one subject to the next. The role of the manager in this process is to foster the conversation and to guide it. In this, he or she is like a host at a cocktail party, inviting the guests, introducing them to each other, suggesting topics of discussion that might be of common interest, introducing new topics or new people to the conversation group when the discussion flags and the participants begin to lose interest, breaking up groups when the discussion becomes too intense and threatens to collapse in mistrust and acrimony. Ultimately this discussion and debate leads not to agreement but to a common understanding that serves as the basis for further discourse. We think of that common understanding as like a language.

The interpretative process then essentially divides into two phases. In the first, or initial phase, the community is in formation. The participants are building a common understanding, generating a new language so to speak. In the second, or mature, phase they are using that language to discuss the technology in which they are interested and the products or research projects to which it might lead. In so doing, they do not make the clear distinction between means and ends that is central to analysis; indeed, they move back and forth between means and ends, revising (or reinterpreting) the ends in the light of the means and vice versa. Importantly, the common understanding that sustains the community, and, in a sense, defines it, continues to evolve through discussion and debate even in this mature phase.

Understood in these terms, what is distinctive about DARPA is that the PMs are essentially creating an interpretative community and then driving it toward the generation of novel products. They bring together around a technological problem people who would not be in contact with each other without the PMs intercession, guiding them through a variety of different encounters, meetings, discussions and seminars to talk to each other, to enter into a conversation in a way that effectively develops a language of community and then sustaining that conversation and encouraging them to draw out of it specific research projects that they then "analyze". But what is striking to the outside observer listening to the participants describe this experience is the priority accorded to interpretation even in the later stages of project development. This is most apparent in the administration of contracts when the performers fail to meet the specific goals. The failure triggers a discussion in which the first question is whether the goals were correctly specified and how they might be redefined in the light of the research that has already taken place. It is not, as it would normally be in the analytical phase of product development, focused solely on what means would be required to achieve these goals. The Agency refuses to estimate the success rate of the projects it undertakes precisely because rather than kill a project outright, it is redefined.

In contrast to DARPA, NIH and NSF are entering into research communities that already exist and seek to support and perpetuate them rather than either create them or direct them. These communities too are sustained by an internal conversation that evolves over time. The discussions that occur among the PMs at NIH and NSF or among the members of the review panels as they evaluate different proposals are a part of that conversation. However, the conversation is largely autonomous of the funding agencies and those conversations that occur in the funding process are more the expression of a set of values and criteria of judgment that have been developed elsewhere than a direct determinant of those values. In sharp contrast to DARPA, the project proposals cannot be revised in the light of the discussion within the agency, and in that sense the panel's judgment tends to involve the analytical application of criteria that the panelists bring with them, rather than an interpretative conversation about those criteria.

On the other hand, the PMs are engaged in a discussion within the agency about the direction of the discipline. The discussion is largely undirected although the division director must exert some influence over it. Unfortunately, we did not explore the nature of that discussion in our interviews. It is an area left for further research. That research could focus on the documents that are generated when the PMs intervene to fund a proposal that would not have received money on the basis of the peer review ranking. An understanding of this process is, in certain respects, more important than understanding DARPA, since a number of developing countries look to NSF as a model of how to support their own education and research establishments.

Conclusions

This study is part of an attempt to understand the structure and operation of Federal agencies supporting academic research in science and engineering. It centered on the contrast between DARPA and NSF, drawing on the experience of faculty members of MIT who have received funding from both organizations. The focus has been on the role of the program (or project) managers in the two agencies. In both agencies the program managers have substantial discretion in the selection of projects to fund and in the management of the funding process. That discretion was anticipated in the case of DARPA, and was one of the major reasons for selecting that agency for study. The degree of discretion at NSF, on the other hand, was surprising. It is much greater than the faculty whom we interviewed generally believed, and is one of the major findings of the study.

The program managers stand at the base of the organizational pyramid in both agencies, and given the discretion that is lodged there, both organizations are in effect street-level—as opposed to classic Weberian—bureaucracies. But the two agencies operate very differently.

At the NSF, proposals are evaluated and ranked through a peer review process, and the discretion of the program manager consists of his or her ability to fund proposals out of the order of peer review ranking. The process for doing so is carefully supervised and reviewed by higher levels of the organization. The procedures for the exercise of discretion are carefully laid out for new PMs in their initial orientation, along with the basic criteria upon which these decisions are supposed to be made. Written reports are required along the way.

Moreover, there is an ongoing discussion among the PMs within the organization about the way in which the academic discipline they are funding is evolving, and about possible biases in the review process. It is in the context of that discussion that funding decisions by the staff are made. There is, however, very little direct interchange between the PMs and the researchers whom the agency funds. The process here is totally consistent with the literature on the management of discretion within street-level bureaucracies.

DARPA is managed very differently. The program managers receive very little orientation or training. While there is extensive interaction between the PMs and the research community they are seeking to draw into their project, there is very little interaction among the PMs themselves (the quip is that the only thing they share is a travel agent). There is a strong organizational culture and a high degree of organizational continuity, but, given the very high turnover and the short tenure of the PMs and, with a few exceptions (like that of Tony Tether) the agency's directors as well, it is very hard to understand how that continuity is maintained and the strong organization culture is created and sustained. It appears that a critical factor here (possibly *the* critical factor) is the network of consultants and consulting firms that support the organization; many of these consultants have worked with DARPA over a long period of time and some of them have actually been PMs within the organization.

The existence of that network and the role it seems to play is the second major finding of this study. DARPA has a reputation for flexibility and is often contrasted to classic bureaucratic organizations. However, given the role of outside consultants in maintaining organizational continuity, it would appear that a good deal of the flexibility of the organization is illusory, and that to the extent that it exists, the flexibility must reside in the role assigned to the PMs and not the way they perform that role.

The findings of the study are incomplete. In focusing on the role of the PMs, we have neglected other aspects of the organizational models, and especially those levels of the organization where the basic budgetary decisions are made, allocating funds among competing disciplines in the case of NSF and broad project areas, in the case of DARPA. Moreover, while the contrast between the two organizations helps us to identify and highlight key aspects of each, it leaves the impression that they are competitive with each other and that the choice between them is a key to national science policy, whereas in fact at the national level at least they are complementary. The NSF is responsible for maintaining the country's basic scientific establishment, ensuring the supply of technical manpower and maintaining its basic research capabilities; DARPA is dependent upon that establishment for the raw material from which its projects are created.

But the most important implications of this project are not its substantive findings but the implications for how one thinks about science policy and the conceptual issues in the emergent field of "the science of science policy".[15] While the field is ostensibly interdisciplinary, it has been heavily influenced by the discipline of economics and what might be termed the conceptual biases of that discipline as a lens for understanding public policy. The influence is pervasive, and it would take a true outsider coming from some other discipline (which we are not) to identify what these are. But one perspective that seems particularly important is a view of government policy in which government intervention consists of imposing restrictions upon, and creating incentives for, action and that its impact can be understood in terms of the self-interest of individuals whose behavior is a response to the price incentive in the market. In science policy, this seems to imply that the budgetary allocations made in our cases at the peak of the organizational hierarchy are the critical policy decisions. But what this study emphasizes is that, in the United States at least, government institutions intervene at the very micro level in the way the projects are conceived and executed. The way that these interventions are conducted is the product of an active debate and discussion within the organization, and also between the organization and the scientific community. We have drawn here upon our own research to understand the nature of that debate, and how the way it is conducted and managed influences the outcome. But the more general point is that that understanding is critical to science policy, and that one has to reach far beyond the conceptual framework of economics to analyze it

15 Fealing, K. H., Lane, J., Marbuger, J. III, and Shipp, S., eds. (2011). *The Science of Science Policy: A Handbook*. Stanford, CA: Stanford University Press, https://doi.org/10.1111/j.1541-1338.2011.00523.x

References

Bonvillian, W. B. (2006). "Power Play, The DARPA Model and U.S. Energy Policy", *The American Interest* 2/2, November/December, 39–48, https://www.the-american-interest.com/2006/11/01/power-play/

Christensen, C. M. (1997). *The Innovator's Dilemma: When New Technologies Cause Great Firms to Fail*. Boston, MA: Harvard Business School Press.

Colatat, P. (2015). "An Organizational Perspective to Funding Science: Collaborator Novelty at DARPA", *Research Policy* 44/4: 874–87.

Cook-Deegan, R. (2015). "Has NIH Lost Its Halo?", *Issues in Science and Technology* 31/2: 36–47. (Chapter 15 in this volume).

Fealing, K. H., Lane, J., Marbuger, J. III, and Shipp, S., eds. (2011). *The Science of Science Policy: A Handbook*. Stanford, CA: Stanford University Press, https://doi.org/10.1111/j.1541-1338.2011.00523.x

Freeman, R. B. (2011). "The Economics of Science and Technology Policy", in *The Science of Science Policy: A Handbook*, ed. K. Fealing, J. Lane, J. Marburger III, and S. Shipp. Stanford, CA: Stanford University Press. 85–103, https://doi.org/10.1111/j.1541-1338.2011.00523.x

Fuchs, E. R. H. (2010). "Rethinking the Role of the State in Technology Development: DARPA and the Case for Embedded Network Governance", *Research Policy* 39/9: 1133–47, https://doi.org/10.1016/j.respol.2010.07.003 (Chapter 7 in this volume).

Harris, A. (2014). "Young, Brilliant and Underfunded", *New York Times*, 2 October, https://www.nytimes.com/2014/10/03/opinion/young-brilliant-and-underfunded.html?_r=0

Knight, F. H. (1921). *Risk, Uncertainty, and Profit*. Boston, MA: Hart, Schaffner & Marx.

Kuhn, T. (1962). *The Structure of Scientific Revolutions*. Chicago, IL: University of Chicago Press.

Lester, R., and M. Piore. (2004). *Innovation—the Missing Dimension*. Cambridge, MA: Harvard University Press.

Lipsky, M. (1980). *Street-Level Bureaucracy: Dilemmas of the Individual in Public Services*. New York, NY: Russell Sage Foundation.

Piore, M. (2011). "Beyond Markets: Sociology, Street-Level Bureaucracy, and the Management of the Public Sector", *Regulation & Governance Special Issue: Sociological Citizens: Practicing Pragmatic, Relational Regulation* 5/1: 145–64, https://doi.org/10.1111/j.1748-5991.2010.01098.x

Van Atta, R., Deitchman, S., and Reed, S. (1990–1991). *DARPA Technical Accomplishments*. 3 Volumes. Alexandria, VA: Institute for Defense Analyses.

Schön, D. (1983). *The Reflective Practitioner: How Professionals Think in Action*. New York, NY: Basic Books.

4. The Connected Science Model for Innovation—The DARPA Model[1]

William B. Bonvillian[2]

Introduction: Fundamentals of Defense Technology Development[3]

The rise of the U.S. innovation system in the second half of the twentieth century was profoundly tied to U.S. World War II and Cold War defense science and technology investment.[4] However, this late twentieth-century military technology evolution is only part of a much bigger picture of innovation transformation. Growth economist Carlotta Perez argues that an industrial—and therefore societal—transformation has occurred roughly every half century, starting with the beginning of the industrial

1 This contribution originally appeared as a chapter in *21st Century Innovation Systems for the U.S. and Japan*, ed. S. Nagaoki, M. Kondo, K. Flamm and C. Wessner. (2009). Washington, DC: National Academies Press. 206–37, https://doi.org/10.17226/12194, http://books.nap.edu/openbook.php?record_id=12194&page=206

2 This chapter was written in 2006 with updates added in May 2008, reflecting developments through that time.

3 Major portions of this chapter appeared in Bonvillian, W. B. (2006). "Power Play, The DARPA Model and U.S. Energy Policy", *The American Interest* 2/2, November/December, 39–48, https://www.the-american-interest.com/2006/11/01/power-play/, and appear here by permission of that journal.

4 Ruttan, V. W. (2006). *Is War Necessary for Economic Growth? Military Procurement and Technology Development*. New York, NY: Oxford University Press.

 https://doi.org/10.11647/OBP.0184.04

revolution in Britain in 1770.[5] These technology-based innovation cycles flow in long multi-decade waves. Arguably, not only do these waves transform economies and the way we organize societies around them, they transform military power as well; U.S. military leadership has paralleled its technological innovation leadership. Perez found that the U.S. led the last three innovation waves—the information technology revolution represents the latest. Will this leadership continue? At stake is not only economic leadership, but U.S. military leadership.

In other words, for the U.S. there has been a deep interaction between war and technology—war has greatly influenced technology evolution, and the converse is also true. While this has been the case for centuries, this interaction has been accelerating. Defense technology cannot be discussed as though it were separate from the technology that is driving the expansion of the economy—they are both part of the same technology paradigms. Military historian John Chambers has argued that few of the critical weapons that transformed twentieth century warfare came from a specific doctrinal need or request of the military;[6] instead, the availability of technology advances has driven doctrine. If technology innovation is a driving force in both U.S. economic progress and military superiority, and these elements have interacted, we need to understand the causal factors behind this innovation.

One factor involves critical institutions, which represent the space where research and talent combine, where the meeting between science and technology is best organized. Arguably, there are critical science and technology institutions that can introduce not simply inventions and applications, but significant elements of entire innovations systems. We will focus on aspects of the U.S. innovation system supported by the defense sector—particularly the Defense Advanced Research Projects Agency (DARPA). An Eisenhower creation, DARPA was the primary inheritor of the World War II connected science model embodied in Los Alamos National Laboratory and MIT's Radiation Laboratory (Rad Lab).

---

5 Perez, C. (2002). *Technological Revolutions and Financial Capital: The Dynamics of Bubbles and Golden Ages*. Northampton, MA: Edward Elgar. See also Atkinson, R. D. (2004). *The Past and Future of America's Economy—Long Waves of Innovation that Power Cycles of Growth.* Northampton, MA: Edward Elgar.

6 Chambers, J., ed. (1999). *The Oxford Companion to American Military History*. Oxford: Oxford University Press, 7.

DARPA came to play a larger role than other U.S. R&D mission agencies in both the Cold War's defense technology and the private sector economy that interacted with it.[7] DARPA will be used as a tool to explore the deep interaction between U.S. military leadership and technology leadership. As we attempt to understand where DARPA came from, we will also ask where it goes next, particularly in IT, as a way of focusing on the continuing strength of the defense innovation system.

Role of Technology Innovation and Talent in Growth

Defense and civilian sector innovation in the U.S. are part of one economic system; that system includes not only sharing the same technology paradigms but sharing the societal wealth—economic growth—thrown off by that economic system, which funds both the military and the technology it increasingly depends on for leadership. Therefore, we need to understand the nature of innovation in economic transformation. Keeping in mind the argument that economic growth has dramatically affected military transformation, what are the causal factors in economic growth?

To briefly summarize more than three decades of work in growth economics: Robert Solow, a Professor of Economics at MIT, won the Nobel Prize in 1987. Solow was profoundly dissatisfied with the growth model of classical economics, where growth was understood in a static model of the interaction between capital supply and labor supply. Solow posited a dynamic model, arguing that while capital and labor supply remained significant, there was a much bigger factor. Studying five decades of U.S. economic growth he found that more than half of this growth flowed from technological and related innovation.[8] He argued that growth rates are not in an equilibrium but can be altered through innovation advance, with societal well-being expanding

---

7 Van Atta, R., et al. (1991). *DARPA Technological Accomplishments, An Historical Review of DARPA Projects*. Alexandria, VA: Institute for Defense Analyses; Goodwin, J. C., et al. (1999). *DARPA, Technology Transition*. Arlington, VA: Defense Advanced Research Projects Agency.

8 Solow, R. M. (2000). *Growth Theory, An Exposition*. 2nd ed. Oxford: Oxford University Press, ix–xxvi, http://nobelprize.org/nobel_prizes/economics/laureates/1987/solow-lecture.html

correspondingly. The key factor behind his growth through innovation thesis, his work suggests, was the research and development system. However, because technology development is complex and not easy to measure, he treated it as "exogenous" to the economy. Economist Paul Romer of Stamford University (and later NYU) articulated what I will call a second direct growth factor.[9] If the first is Solow's technological innovation founded on R&D, Romer argued that technical knowledge drives economic growth, and that it is an "endogenous" element in the economy. The key factor standing behind this knowledge is science and technological talent, the "human capital engaged in research". He suggested a prospector theory of innovation—the nation or region that fields the largest number of well-trained prospectors will find the most gold, i.e., the most innovative advances.[10]

These two direct factors—in shorthand, talent and R&D—don't stand in isolation from each other, but rather are interacting parts of an intricate ecosystem of innovation. There are many other factors that are important parts of this system, elements that are more indirect, implicit, and peripheral to innovation advance than the two direct factors essential to economic growth posited above, but these indirect factors are nonetheless ones that a society must also get right for innovation advance.

The list of indirect innovation factors is long and, because growth economics is relatively new to the economics scene, the metrics for understanding the interaction of these factors are largely unexplored. On the government side they include fiscal, tax, and monetary policy; trade policy; technology standards; technology transfer policies; government procurement; intellectual property protection; the legal and liability systems; regulatory controls; accounting standards; and export controls. On the private sector side, which in a capitalist enterprise must dominate innovation, they include investment capital, including angel, venture, IPO's, equity, and lending; markets; management principles and organization; talent compensation and reward; and quality of plant and equipment.

Keep in mind that that these direct and indirect innovation factors all interact, and that it is the interaction that is most important. Therefore,

9 Romer, P. (1990). "Endogenous Technological Change", *Journal of Political Economy* 98: 72–102, http://pages.stern.nyu.edu/~promer/Endogenous.pdf

10 See discussion of Solow and Romer in Warsh, D. (2006). *Knowledge and the Wealth of Nations.* New York, NY: W. W. Norton.

they represent a common system for both economic and defense sector advance.[11]

Is There a Third Direct Innovation Factor?

What does innovation organization look like? This factor must be seen and understood at least at two levels, the institutional level and the personal, face-to-face level. We will explore these in succession.

U.S. Innovation Organization at the Institutional Level

In addition to the two direct and the numerous indirect innovation factors suggested above, arguably there is a third direct factor: the way that R&D and talent, in particular, come together to form an innovation system. In other words, if R&D is factor A, and talent is factor B, they form an interacting combination, AB, which in itself is a third factor: the meeting space for science and technology and the talent behind it.

11 We have been discussing innovation in the context of economics. However, growth economics—because it is founded on a dynamic model of innovation—has begun to break down the focus of economics, since the late 1940's (neoclassical economics), on the mathematical modeling suited to analysis of limited numbers of variables in a closed equilibrium. Instead, as growth economist Brian Arthur has argued, innovation can create increasing returns, not just diminishing returns, leading to transformational phase shifts in an economy. Growth economics requires not only the neo-classical economics of physics-like fundamental principles subject to formulaic proof, but an economics of complexity, where a rich array of interacting elements must be accounted for in systems that are not static but evolve. For example, if innovation organization is a key factor in innovation and therefore economic growth, this element pushes economics towards its original roots in the social sciences and away from neo-classical economic modeling, which cannot fully capture organizational elements. This concept puts an orange in what economics has viewed as a mix of apples. In other words, growth economics is gradually broadening economics' explanatory depth and toolset to reach and understand complex systems, and the third innovation factor discussed below, innovation organization, arguably pushes it further in that direction. See, generally, Waldrop, M. M. (1992). *Complexity, the Emerging Science at the Edge of Order and Chaos*. New York, NY: Simon & Schuster, 144–48, 250–55, 284–313, 325–27. Since the author drafted this article and footnote in 2006, another book has been published discussing some of these points: Beinhocker, E. D. (2007). *Origin of Wealth-Evolution, Complexity and the Radical Remaking of Economics*. Cambridge, MA: Harvard Business School; see also, Tassey, G. (2016) "The Technology Element Model, Path-Dependent Growth and Innovation Policy", *Economics of Innovation and New Technology* 26/6: 594–612, http://dx.doi.org/10.1080/10438599.2015.1100845; Bonvillian, W. B. and Singer, P. (2018). *Advanced Manufacturing: The New American Innovation Policies*. Cambridge, MA: MIT Press (chapter 4), https://doi.org/10.7551/mitpress/9780262037037.001.0001

It is not enough to have the ingredients of R&D and talent; they have to collaborate in an effective way for a highly productive innovation system. We'll call this third factor innovation organization. Linking two factors together, AB, is shorthand in math for multiplying them; arguably, there is a multiplier factor here, too—the way R&D and talent join and are organized can be a multiplier for each. If innovation organization is a kind of multiplier for the two key direct innovation factors, then the way defense and civilian innovation systems organize R&D and talent, and the massive areas where the two systems overlap, will be profoundly determinative of innovation advance for the two systems, and therefore of economic and military leadership.

Governmental science and technology organization in the U.S. largely dates from World War II and the immediate post-war. As suggested earlier, technology evolution in this country comes from a kind of "PushMi-Pullyu" relationship between civilian economic and defense sectors, and World War II was a transformative period where the pressure for military technology advance later led to a dramatic economy-wide advance.

Vannevar Bush led this charge,[12] acting as President Franklin D. Roosevelt's personal science executive during the war. He was allied to a remarkable group of fellow science organizers, including Alfred Loomis, an investment banker and scientist, physicist Ernest Lawrence of Berkeley, and two university presidents, James Conant of Harvard and Karl Compton of MIT. Successively, Bush created and took charge of the two leading organizing entities for U.S. science and technology, the National Defense Research Council (NDRC) and then the Office of Science Research and Development (OSRD). These became the coordinating entities for U.S. wartime R&D, creating crash research projects in critical areas, such as the Rad Lab at MIT and Los Alamos, and they, in turn, insured interaction and coordination with a rich mix of research components.

12 Zachary, G. P. (1999). *Endless Frontier, Vannevar Bush, Engineer of the American Century*. Cambridge, MA: The MIT Press. See also, Conant, J. (2002). *Tuxedo Park: A Wall Street Tycoon and the Secret Palace of Science that Changed the Course of World War II*. New York, NY: Simon & Shuster (a biography of Alfred Loomis, founder of MIT's Rad Lab). For a discussion of U.S. pre-WWII science organization see, Hart, D. (1998). *Forged Consensus*. Princeton, NJ: Princeton University Press.

Influenced by the frustrations of his WW1 military research experience, where technology breakthrough could not transition past bureaucratic barriers into defense products, Bush kept civilian science control of critical elements of defense research, insisting that his science teams stay out of uniform and separate from military bureaucratic hierarchies, which he found unsuited to the close-knit interaction needed for technology progress.

To summarize, Bush brought all defense research efforts under one loose coordinating tent, NDRC then OSRD, and set up flat, non-bureaucratic, interdisciplinary project teams oriented to major technology challenges, like radar and atomic weapons, as implementing task forces. He created "connected" science, where technology breakthroughs at the fundamental science stage were closely connected to the follow-on applied stages of development, prototyping and production, operating under what we will call a technological "challenge" model. Because Bush (and his ally Loomis) could go directly to the top for backing from Roosevelt, through Secretary of War Henry Stimson and Presidential Aide Harry Hopkins, Bush made his organizational model stick during the war, despite relentless military pressure, from the Navy in particular, to capture it.

Then, immediately after the war, he systematically dismantled his remarkable connected science creation.

Envisioning a period of world peace, convinced that the wartime levels of government science investment would be slashed, and probably wary of a permanent alliance between the military and science, Bush decided to try and salvage some residual level of federal science investment. He wrote the most influential polemic in U.S. science history, *Science: The Endless Frontier*, for Roosevelt, arguing that the federal government should fund basic research, which would deliver ongoing progress in economic well-being, national security and health to the country.[13] In other words, he proposed ending his model of connected science, and dropping his challenge model, in favor of making the federal role one of funding one stage of technology advance: exploratory basic research. His approach would become known as the "pipeline" model for science investment. The federal government would dump basic science into one end of

13 Bush, V. (1945). *Science: The Endless Frontier*. Washington, DC: Government Printing Office, 1–11, https://www.nsf.gov/od/lpa/nsf50/vbush1945.htm.

an innovation pipeline, and somehow early and late state technology development and prototyping would occur inside the pipeline, with new technology products emerging, genie-like, at the end. Because he assembled a connected science model during World War II, Bush no doubt realized the deep connection problems inherent in this pipeline model, but likely felt that salvaging federal basic research investment was the best he could achieve in a period of anticipated peace.

He did argue that this basic research approach should be organized and coordinated under "one tent" to direct all the nation's research portfolios, proposing what would become the National Science Foundation (NSF). Because he wanted this entity controlled by a scientific elite separated from the nation's political leadership, Bush got into a battle with Roosevelt's successor, Harry Truman. In his typical, take-charge way, Truman insisted that the scientific buck would stop on his desk, not on some Brahmin scientist's desk, and that NSF appointments would be controlled by the President. Bush disagreed.

Truman therefore vetoed Bush's NSF legislation, stalling its creation for another five years.[14] Meanwhile, science did not stand still. New agencies proliferated, and the outbreak of the Korean War led to a renewal of defense science efforts. By the time NSF was established and funded, its potential coordinating role had been bypassed. It also became a much smaller agency than Bush anticipated, only one among many. Despite Bush's support for one tent where scientific disciplines and agencies could coordinate their work, as they did in World War II, the U.S. thus adopted a highly decentralized model for its science endeavor.[15]

Bush's concept of federal funding focused on basic science did prevail, however, with most of the new science agencies adopting

---

14 Blanpied, W. A. (1998). "Inventing U.S. Science Policy", *Physics Today* 51/2: 34–40, https://doi.org/10.1063/1.882140 (an article examining the post-WWII evolution of U.S. science organization and NSF); Mazuzan, G. (1988). *The National Science Foundation: A Brief History (1950–85).* Arlington, VA: The National Science Foundation, 1–25, https://www.nsf.gov/pubs/stis1994/nsf8816/nsf8816.txt (a history of NSF in the context of post-WWII science).

15 It must be emphasized that there are major advantages to decentralized science. It creates a variety of pathways to science advance and a series of safety nets to ensure multiple routes can be explored. Since science success is largely unpredictable, the "science czar" approach risks major failures that a broad front of advance does not. Nonetheless, the U.S. largely lacks the ability to coordinate its science efforts across agencies particularly where advances that cut across disciplines require coordination and learning from networks.

this model for the federal science role. These twin developments left U.S. science fragmented at the institutional level in two ways: overall science organization would be fragmented among numerous science agencies, and federal investment would be focused on only one stage of technological development: exploratory basic research.[16] Remarkably, Bush left a legacy of two conflicting models for scientific organizational advance: the connected, challenge model of his World War II institutions, which he dismantled after the war,[17] and the fundamental-science focused, disconnected, multi-headed model of post-war U.S. science institutional organization.

Summary of the Innovation Analytical Framework

To summarize the discussion thus far, innovation is not only about R&D investment levels, it's about content and efficiency.[18] U.S. post-war policy institutionally severed R from D, which had been connected in the wartime model, and posited a pipeline theory of innovation where the federal government dumped research funding into one end of the pipeline, then mysterious things occurred within the innovation pipeline, then remarkable products emerged at the other end. Neoclassical economics, through the work of Robert Solow, came to realize the central role of innovation in economic growth but was unable to apply existing economic models to the mystery inside the pipeline, and therefore treated innovation as "exogenous" to the economy. That response was ultimately unacceptable—it is as though

16 See the discussion of these developments in Stokes, D. E. (1997). *Pasteur's Quadrant, Basic Science and Technological Innovation*. Washington, DC: Brookings Institution Press.

17 The term "dismantled" is used to indicate that the structure for science management in World War II was ended, and many wartime science entities were shut down, including MIT's Rad Lab. Obviously, other existing science entities continued in operation, such as NACA, which Bush chaired before the war, and was an early example of a connected, challenge model approach. See Roland, R. (1985). Model Research: The National Advisory Committee for Aeronautics, 1915–1958. Washington DC: Government Printing Office, 225–58 (chapter 10), https://history.nasa.gov/SP-4103/. However, even within DOD, the Office of Naval Research was largely stood up after the war around a fundamental science model. Sapolsky, H. M. (1990). Science and the Navy—The History of the Office of Naval Research. Princeton, NJ: Princeton University Press, 9–81 (chapters 2–4).

18 Tassey, G. (2007). *The Innovation Imperative*. Cheltenham, UK: Edward Elgar (chapters 3, 7, 8).

economics, after finally discovering the innovation monster in the economic growth room, then declined to look at it. A group of growth economists, initially led by Paul Romer, gradually began to whittle away at the monster, treating it as "endogenous", slowly delineating its economic attributes. However, this delineation process still has barely begun.[19] Economic institutions still collect extensive data on the two factors classical economics tied to economic growth—capital supply and labor supply, and data on R&D investment totals. We have little data on the monster, the content and efficiency of the innovation system.[20] Few are searching for and analyzing the new factors and metrics for innovation evaluation. Interestingly, two decades after Solow won the Nobel Prize for identifying the innovation monster, the U.S. Department of Commerce has announced the need to begin an intensive data collection process around innovation.[21] The National Science Foundation, which has long collected data on innovation investment levels and science education,[22] has begun an effort to look at data and analysis around innovation with a program entitled the Science of Science and Innovation Policy.

But what is the framework for the innovation metrics and analysis? Although we track R&D investment, what about the composition and efficiency factors? This chapter attempts to identify some of the elements lurking inside the innovation pipeline. Following Solow and Romer, it argues, as noted, that R&D and talent (shorthand terms for

19 For a critical view of the progress of endogenous growth theory in economics, see Solow, R. M. (2000). "Toward a Macroeconomics of the Medium Run", *Journal of Economic Perspectives* 14/1: 151–58.

20 Despite the emergence over two decades ago of growth economics and its doctrine that growth is predominantly innovation based, the two U.S. political parties are still largely organized around the old factors posited by classical economics as responsible for growth, capital supply and labor supply.

21 U.S. Department of Commerce. (2008). *Innovation Measurement, Tracking the State of Innovation in the American Economy*. Report to the Secretary of Commerce. Washington, DC: U.S. Department of Commerce, http://users.nber.org/~sewp/SEWPdigestFeb08/InnovationMeasurement2001_08.pdf; Mandel, M. (2008). "A Better Way to Track the Economy, A Groundbreaking Commerce Dept. Report Could Lead to New Yardsticks for Measuring Growth", *Business Week*, 28 January, p. 29.

22 National Science Board. (2006). *Science and Engineering Indicators*. Arlington, VA: The National Science Foundation, https://wayback.archive-it.org/5902/20160210153725/http://www.nsf.gov/statistics/seind06/ At the time this DARPA book was published, the latest version of *Science and Engineering Indicators* was from 2018: https://www.nsf.gov/statistics/2018/nsb20181/.

their extended ideas) can be considered *two direct innovation factors*, indispensable to innovation, and are surrounded by an ecosystem of indirect factors, less critical but nonetheless significant. This chapter further posits that there is a *third direct innovation factor*, innovation organization, the space where the talent and R&D converge. An essential aspect of innovation organization requires evaluation at the institutional level. Summarized above is the brilliant success the U.S. experienced at the institutional level during World War II with a connected science model built around technological challenges, formed under one organizational tent.

The U.S., following the war, shifted to a highly decentralized model, scattering government-funded research among a series of mission agencies. It was predominantly a basic-science focused model, not connected science, and left what later became known as a "valley of death" between research and development stages. The handoff from publicly-funded research and to private sector development therefore lacked institutional bridging mechanisms. As we will see, the major exception to that U.S. institutional rule was DARPA.[23]

We turn now from a review of innovation at the institutional level to a second analytical perspective on innovation organization, innovation at the personal, face-to-face level. Following this review, we will examine how these twin perspectives on innovation organization have operated within an arguably critical U.S. innovation organization, DARPA, evaluating how it has worked at both levels, institutional and personal.

23 This is not to assert that the fundamental science mission agencies dating from the 1940's have remained frozen in time. While the basic science mission remains paramount at agencies such as NSF, NIH and the DOE Office of Science, at the National Science Foundation, for example, there is funding not only for small individual investigator basic research but larger areas of interdisciplinary advance, such as nanotechnology, which can incorporate grand challenges. For example, NSF's issue workshops and similar organizing mechanisms bring in ideas for coordinated science-engineering advance for initial buy-in and research program design by fundamental and applied communities. As another example, NSF's engineering directorate supports engineering centers tying science advance to fundamental engineering advance. Somewhat similar efforts around interdisciplinary centers have evolved at NIH and DOE. The point remains that these functions supplement established fundamental science efforts.

Innovation Systems at the Personal Level: Great Groups

Innovation organization should be analyzed at the institutional level, as discussed above. However, it also requires understanding at the ground level, from the personal, face-to-face point of view. Innovation is different from scientific discovery and invention, which can involve solo operators. Instead, innovation requires taking both scientific discovery and invention and piling applications on a breakthrough invention or group of inventions to create disruptive productivity gains that transform significant segments of an economy and/or defense system. So, innovation is a third phase built on phases of discovery and invention. Innovation requires not only a process of creating connected science at the *institutional level,* it also must operate at the *personal level.* People are innovators, not simply the overall institutions where talent and R&D come together. Warren Bennis and Patricia Biederman have argued that innovation, because it is much more complex than the earlier stages of discovery and invention, requires "great groups", not simply individuals.[24] Robert W. Rycroft and Don E. Kash make a similar argument but use a different term: innovation requires collaborative networks[25] which can be less face-to-face and more virtual. As we look at innovation organization at the personal level, we will explore the rule sets for three sample "great groups" of innovators.

Edison's "Invention Factory" at Menlo Park, New Jersey

Thomas Edison formed the prototype for innovator great groups.[26] Edison placed his famous Menlo Park laboratory in a simple 100-foot long wooden frame building, a lab, on his New Jersey farm. In it he placed a team of a dozen or so artisans, mixing a wide range of skills with a few trained scientists. They worked intensely, sometimes 24/7, and took midnight breaks together, eating pies, reciting poems and

24 Bennis, W. and Biederman, P. W. (1997). *Organizing Genius: The Secrets of Creative Collaboration*. New York, NY: Basic Books.

25 Rycroft, R. W., and Kash, D. E. (1999). "Innovation Policy for Complex Technologies", *Issues in Science and Technology* 16/1, https://issues.org/byline/robert-w-rycroft/

26 See discussion in Evans, H. (2005). *They Made America*. Sloan Foundation Project. New York, NY: Little, Brown and Company, 152–71.

singing songs. They mixed a range of disciplines and organized their intense effort around the challenge of electric light. They were a great group, highly collaborative. Great groups also require collaboration leaders, and Edison was a remarkable team leader. They worked on the idea of filling the gap between electric poles with a filament placed in a vacuum tube. But that was only the breakthrough invention, not the innovation. To make their light usable, Edison and his team then had to invent much of the infrastructure for electricity—from generators to wiring to fire safety to the structure of a supporting electric utility industry. Edison and his team become inventors and innovators, visionaries and (as initiators of a network of companies with Wall Street backing) vision enablers.

Interestingly, as part of this process, Edison had to derive elements of electron theory to explain his results—his "Edison Effect" helped lead to atomic physics advances. There is a major lesson in this: science is not simply a linear pipeline going from basic to applied. Rather, it goes both ways: basic to applied and applied to basic. Menlo Park teaches us parts of the rule set for great groups. It is organized around a challenge model, with the group trying to solve a specific challenge or goal; it applies an interdisciplinary mix of both practical and basic science to get there; and it uses a connected science model, tying invention to innovation and incorporating all stages of innovation advance. While the group is under Edison's clear leadership—and that leadership factor is vital—it is nonetheless a non-hierarchical, relatively flat, two-level, highly collaborative effort. The team mixes experimentalists and theorists, artisans and trained scientists and engineers, for a blend of experimental and theoretical capability and disciplines.

Alfred Loomis and the Rad Lab at MIT, 1940–1945

Alfred Loomis loved science but family needs compelled him to become lawyer; he combined his science and legal skills to become a leading Wall Street financier for the emerging electric utility industry in the 1920's.[27] Anticipating the market crash, he sold out in 1928 with his great fortune intact. He used it to pursue science, setting up his own private lab at his Tuxedo Park, New York estate in the 1930's and

27 Details from Loomis' biography, Conant. (2002). *Tuxedo Park.*

assembling there a who's who of pre-war physics. Loomis' own field of study there was microwave physics. As World War II loomed, Vannevar Bush, respecting Loomis' industrial organizing skills, asked him to join Roosevelt's NDRC to mobilize science for the war.

Because the American military was initially uninterested, the British handed over to Loomis a suitcase with their secrets to microwave radar in his penthouse in the Shoreham Hotel in Washington in 1940. As the Battle of Britain raged, Loomis' microwave expertise enabled him to grasp immediately that this was a war winning technology for air warfare. He promptly persuaded his cousin and mentor, Secretary of War Henry Stimson, that this technology must be developed and exploited without delay. With Bush's and Roosevelt's immediate approval, Loomis within two weeks established the Radiation Laboratory (Rad Lab) at MIT. Because he knew them from his Tuxedo Park lab, Loomis and his ally and friend Ernest Lawrence of Berkeley called in the whole talent base of U.S. physics to join the Rad Lab, and nearly all came. Because the government was not used to establishing major labs literally overnight, Loomis personally funded the startup while government approvals and procurement caught up.

The Rad Lab was non-hierarchical and flat, with only two levels, project managers and project teams, each devoted to a particular technology path. It was characterized by intense work, often around the clock, and by high spirits and morale. Loomis and Bush purposely kept it out of the military. The Rad Lab used a talent base with a mix of science disciplines and technology skills. It was highly collaborative, it was organized around the challenge model, and it used connected science, moving from fundamental breakthrough to development, prototyping and initial production. Interestingly, the Rad Lab organizational model was systematically adopted at Los Alamos, and ten leading Rad Lab scientists shifted to Los Alamos to implement it.[28] The Rad lab developed great advances in microwave radar and the proximity fuse, technologies vital to success for the allies. Eight Nobel prizewinners came out of the Rad Lab and it ended up laying the foundations for important parts of modern electronics. It also embodied another feature key to successful

28 See discussion of Los Alamos in Sherwin, M., and Bird, K. (2005). *American Prometheus, The Triumph and Tragedy of J. Robert Oppenheimer*. New York, NY: Alfred A. Knopf; and Conant, J. (2005). *109 East Palace*. New York, NY: Simon & Shuster.

great groups—through Loomis and Bush, the Rad Lab had direct access to the top decision-makers able to mandate the execution and adaptation of its findings, Stimson and Roosevelt.

The Transistor Team at Bell Labs (1947)

Bell Labs' Murray Hill facility was consciously set in the New Jersey countryside after Edison's Menlo Park model and also drew from the great military labs of World War II, the Rad Lab and Los Alamos. AT&T's R&D Vice President, Mervin Kelly, and his lead researcher, William Shockley, wanted a solid-state physics team of fifty scientists and technicians from various fields with capability for fundamental research leading to practical applications. Their task was to develop a solid-state physics-based replacement for vacuum tubes so that AT&T's switching capability could continue to advance telephone speed and capacity. John Bardeen and Walter Brattain, two of the leading solid-state physics researchers who joined this team, developed a profoundly close collaboration, where the scientific and personal skills of one matched the other's—one a theorist, the other an experimentalist, one outgoing, the other reflective. They were social friends and held a strong mutual respect. Backed-up by Bell Labs' deep industrial technical support system, with the latest equipment and very strong technical staff, the two entered into a "magic month" from mid-November to 16 December 1947, and developed the first transistor.

As Bardeen's biographers put it, "The solid-state group divided up the tasks: Brattain studied surface properties such as contact potential; Pearson looked at bulk properties such as the mobility of holes and electrons; and Gibney contributed his knowledge of the physical chemistry of surfaces. Bardeen and Shockley followed the work of all members, offering suggestions and conceptualizing the work".[29] Brattain later commented, "It was probably one of the greatest research teams ever pulled together on a problem… I cannot overemphasize the rapport of this group. We would meet together to discuss important

29 Huddleson, L., and Daitch, V. (2002). *True Genius—The Life and Science of John Bardeen*. Washington, DC: Joseph Henry Press of the National Academies of Sciences, 127–28.

steps almost on the spur of the moment of an afternoon. We would discuss things freely. I think many of us had ideas in these discussion groups, one person's remarks suggesting an idea to another. We went to the heart of many things during the existence of this group, and always when we got to the place where something needed to be done, experimental or theoretical, there was never any question as to who was the appropriate man in the group to do it".[30]

Unfortunately, Shockley's reaction wrecked further working collaboration in the group. He attempted to garner credit for Bardeen's and Brattain's work, then worked secretly at his home designing a further break-through improvement, where a semiconductor "sandwich" replaced the transistor's electrical contact point, without telling the rest of the group. Before distrust descended, however, the group followed many of the rules of the other groups cited above—it was highly talented, relatively non-hierarchical, organizationally flat with essentially two levels, highly collaborative, and brought to bear a range of expertise and disciplines, including theorists and experimentalists, with each participant working in his strongest skill area. It was organized on a challenge model and the connection to AT&T's VP Mervin Kelly assured a tie to a decisionmaker who could enable development of breakthroughs. The group traded ideas on a continuous basis, meeting frequently with each providing thoughts to assist the others' progress, and Bardeen and Shockley played a leadership role by continually moving conceptual ideas among the group.

Many of the organizational features of these three "great groups" are common to others, including the development of atomic weapons at Los Alamos, the integrated circuit and microchip at Fairchild Semiconductor and Intel, the aeronautics and stealth advances at Lockheed's Skunk Works, the personal computer at Xerox PARC and Apple, biotech at Genentech and J. Craig Venter's genomics projects.[31] These projects are not unique.

---

30 *Ibid.*

31 Sherwin and Bird. (2005). *American Prometheus*, 205–28, 255–59, 268–85, 293–97; Conant. (2005). *109 East Palace*, 106, 108, 110, 255; Berlin, L. (2005). *The Man Behind the Microchip, Robert Noyce and the Invention of Silicon Valley*. Oxford: Oxford University Press (chapters 3–8); Rich, B, and Janos, L. (1994). *Skunk Works: A Personal Memoir of My Years of Lockheed*. Boston: Little, Brown & Company; Evans. (2005). *They Made America*, 420–31 (on Boyer and Swanson founding Genetech and starting

A venture capitalist has commented that he looks for these same kinds of characteristics every time he funds a startup. To summarize, a common rule set seems to characterize successful innovation at the personal and face-to-face level. The rules include ensuring: a highly-collaborative team or group of great talent; a non-hierarchical, flat and democratic structure where all can contribute; a cross-disciplinary talent mix, including experimental and theoretical skills sets networked to the best thinking in relevant areas; organization around a challenge model; using a connected science model able to move breakthroughs across fundamental, applied, development and prototype stages; cooperative, collaborative leaders able to promote intense, high morale; and direct access to top decisionmakers able to implement the group's findings.[32]

DARPA as a Unique Model—Combining Institutional Connectedness and Great Groups

We have discussed the concept of innovation organization as a third direct innovation factor, and noted that it operates in macro and micro ways, at both the institutional level and the personal level. Our focus now shifts to the Defense Department's Defense Advanced Research Projects Agency. Created in 1958 by Eisenhower as a unifying force for defense R&D in light of the stove-piped military services' space programs that had helped lead to America's Sputnik failure, DARPA became a unique entity. In many ways, DARPA directly inherited the connected science, challenge and great group organization models of the Rad Lab and Los Alamos stood up by Bush, Loomis and Oppenheimer. However, unlike the personal-level models discussed above, DARPA has operated at *both* the institutional and personal levels. DARPA became a bridge organization connecting these two institutional and personal organizational elements, unlike any other R&D entity stood up in government.

biotech); Bennis and Biederman. (1997). *Organizing Genius*, 63–86 (on Xerox PARC and Apple); Morrow, D. S. (2003). "Dr Craig Venter: Oral History", *Computerworld Honors Program*, 3–53, 56–58; Venter, J. C. (2007). *A Life Decoded: My Genome, My Life*. New York, NY: Viking Press (chapter 12).

32 For discussion of additional great groups and variations in this suggested rule set, see Bennis and Biederman. (1997). *Organizing Genius*.

J. C. R. Licklider and the Beginnings of the DARPA Model

The DARPA model is perhaps best illustrated by one of its most successful practitioners, J. C. R. Licklider, who, as an office director at DARPA working with and founding a series of great technology teams, laid the foundations for two of the twentieth century's technology revolutions, personal computing and the Internet.[33] In 1960, Licklider, trained in psychology with a background in physics and mathematics, wrote about what he called the "Man-Machine Interface" and "Human-Computer Symbiosis": "The hope is that in not too many years, human brains and computing machines will be coupled together very tightly, and that the resulting partnership will think as no human brain has ever thought".[34] By 1960, Licklider envisioned timesharing as a path to real time personal computing (as opposed to the then-dominant main-frame computing), digital libraries, the Internet (the "Intergalactic Computer Network"), what we now call the World Wide Web, and most of the features—like computer graphing, simulations and modeling—that we are still evolving to implement those revolutions. Licklider was hired by DARPA[35] to work on what was being called the "command and control" problem, and then that problem took off in importance. This was because John F. Kennedy and Robert McNamara had become deeply frustrated with a profound command and control problem, namely, their inability to obtain and analyze real time data and interact with on-scene military commanders during the Cuban Missile Crisis. DARPA gave Licklider the major resources to tackle this problem. It was the rare case of the visionary being placed in the

33 Discussion in this section drawn from Licklider's biography by Waldrop, M. M (2001). *The Dream Machine: J. C. R. Licklider and the Revolution that Made Computing Personal*. New York, NY: Viking Press. For discussions of DARPA's and DOD's central role in fostering the many phases of the IT revolution, see, Ruttan. (2006). *Is War Necessary*, 91–129; Fong, G. R. (2001). "ARPA Does Windows; the Defense Underpinning of the PC Revolution", *Business and Politics* 3/3: 213–37, https://doi.org/10.2202/1469-3569.1025 (Chapter 6 in this volume); National Research Council, Science and Telecommunications Board. (1999). *Funding a Revolution, Government Support for Computing Research*. Washington, DC: National Academy Press, 85–187, https://doi.org/10.17226/6323

34 Licklider, J. C. R. (1960). "Man-Computer Symbiosis", *IRE Transactions on Human Factors in Electronics* 1: 4–11, https://doi.org/10.1109/thfe2.1960.4503259

35 DARPA Director Jack Ruina later concluded that hiring Licklider was his most significant act at DARPA. In seeking an office director, Ruina realized he had found a visionary. See Waldrop. (2001). *The Dream Machine*.

position of vision-enabler. Strongly backed by noted early DARPA Directors Jack Ruina and Charles Herzfeld, Licklider found, selected, funded, organized and stood up a remarkable support network of early information technology researchers at universities and firms that over time built personal computing and the Internet. He served at two different periods in DARPA.

At the institutional organization level, DARPA and Licklider became a collaborative force among the Defense Department's research agencies controlled by the services, using DARPA IT investments to leverage participation by the agencies to solve common problems under connected science and challenge models. DARPA and Licklider also kept their own research bureaucracy to a bare-bones minimum, using the service R&D agencies to carry out project management and administrative tasks, so that DARPA's efforts created co-ownership with the service R&D stovepipes. Institutionally, although it certainly did not always succeed, DARPA attempted to become a research supporter and collaborator, not a rival competitor to the DOD service research establishment.[36]

At the personal level of innovation organization, Licklider created a remarkable base of information technology talent both within DARPA and in a collaborative network of great research groups around the country. This team of apostles, including Doug Engelbart, Ivan Sutherland, Robert Taylor, Larry Roberts, Vint Cerf, Robert Kahn, and their many comrades, are a who's who of personal computing and internet history. Because of ongoing progress, DARPA was willing to be patient and able to look at the long term in these IT talent and R&D investments in a way that corporations and venture capital firms are not structured to undertake.[37] Licklider's DARPA model was also not

36 The military service R&D organizations initially saw DARPA as a usurper and competitor for scarce research funds. DARPA's efforts over the decades to link with the service R&D organizations and become their collaborator and banker for advanced projects they might not otherwise obtain approval for has helped defuse service hostility, and frequently the collaboration has been highly mutual and beneficial. But resentment remains of DARPA as a favored child, even after a half century. Licklider's efforts mark an early success at cross-stovepipe collaboration, although such success is not uniform.

37 Licklider, as DARPA's IPTO head, received strong backing from DARPA Directors Jack Ruina and Charles Herzfeld, who bet on his vision, which enabled Licklider to build a cadre of successors—Ivan Sutherland, Bob Taylor and Larry Roberts—who shared and enhanced his vision for a coherent program with ongoing technical process steps that led to the Internet and personal computing and a network of

a flash in the pan—internally it was able to institutionalize innovation so that successive generations of talent sustained and kept renewing the technology revolution over the long term. At the personal level of innovation, the great groups Licklider started, in turn, shared key features of the Menlo Park, Rad Lab and other groups previously discussed. Licklider's Information Processing Techniques group was the first and greatest success of the DARPA model, but this success was not unique; DARPA was able to achieve similar accomplishments in a series of other technology areas.[38]

There is one further key point to consider: DARPA has been willing to spawn technology advances not only in the defense sector but also in the non-defense economy, recognizing that an economy-wide scale as opposed to a defense sector-only scale may be needed to speed the advance. DARPA has made specific choices to encourage and support technology advances with non-defense organizations, both academic and commercial, rather than defense-only organizations, as its best means of gestating new concepts into implementation.[39] This enables the Department of Defense (DOD) at a later stage to take advantage of this technology evolution speed up, with corresponding shared and therefore reduced development and acquisition costs. This was exactly

---

related advances. There was no special management doctrine at DARPA that enabled this successive effort but it was allowed by DARPA leaders to proceed full throttle for a decade, until scrutinized somewhat by DARPA Director George Heilmeier. Fluent with practical electronics, he imbedded the "Heilmeier Catechism" which insisted on more application relevance, to Licklider's frustration during his second DARPA tour. See Waldrop. (2001). *The Dream Machine.*

38 Van Atta, R., Deitchman, S., and Reed, S., (1990–1991). *DARPA Technical Accomplishments.* 3 Volumes. Alexandria, VA: Institute for Defense Analyses. See, also, Van Atta, R. (2008). "Fifty Years of Innovation and Discovery", in *DARPA, 50 Years of Bridging the Gap,* ed. C. Oldham, A. E. Lopez, R. Carpenter, I. Kalhikina, and M. J. Tully. Arlington, VA: DARPA. 20–29, https://issuu.com/faircountmedia/docs/darpa50 (Chapter 2 in this volume). Dr. Van Atta has been generous to the author with his insights on DARPA, which are reflected at a number of points in this chapter.

39 Licklider and his colleagues largely relied on universities for idea—creation and the subsequent spin-out of these ideas into new commercial firms (such as Digital or Sun) for their application. While existing smaller commercial firms, such as BB&N, which stood up the Internet for DARPA, also played a role, the larger commercial firms, defense contractors and defense R&D organizations were usually not the source of new concepts or their implementation. DARPA thus played a vital role in creating the highly productive pathway in the U.S.'s late twentieth-century IT economy of academic research, start-up companies, venture funding, commercialization, and the institutions that grew up to line this pathway.

the case with the IT revolution that Licklider and DARPA made crucial contributions to. Although IT has been in a thirty-year development process which is still ongoing, DARPA's support for and reliance on a primarily civilian sector development process enabled DOD to obtain much more quickly and cheaply the tools it needed to solve its initial command and control problem.

Actually, DOD got many more benefits than just these tools for command and control. When Andy Marshall, DOD's legendary in-house defense theorist and head of its Office of Net Assessment, argued in the late 1980's that that U.S. forces were creating a "Revolution in Military Affairs",[40] this defense transformation was built around many of the IT breakthroughs DARPA initially sponsored.[41] Admirals Bill Owens and Art Cebrowski, and others, in turn, translated this IT revolution into a working concept of "network centric warfare"[42] which further

40 Marshall, A. W. (1993). "Some Thoughts on Military Revolutions—Second Version", DOD Office of Net Assessment, Memorandum for the Record, 23 August; Lehman, N. (2001). "Dreaming about War", *The New Yorker*, 16 July, http://www.comw.org/qdr/0107lemann.html

41 William Perry and Harold Brown, Defense Department leaders during the Carter Administration, for example, developed what Perry later called an "offsets" theory of defense technology. During the Cold War, the Soviet Union held a roughly three to one advantage in numbers of troops, tanks, and aircraft. Perry has argued that the U.S. at first accepted that disparity because it held an advantage in nuclear weapons. When the Soviets achieved rough parity in nuclear weapons and the missiles to deliver them, U.S. deterrence theory was at risk, so Brown and Perry decided to achieve parity in conventional battle through systematic technological advance. They began a process of translating advances in computing, information technology, and sensors, which had been initiated and long-supported by defense research investments, including DARPA's in particular, into precision weapons at the service level. First exhibited in the Gulf War, these became a massive "force multiplier" for U.S. conventional forces. See, generally Van Atta, R., Lippitz, M., et al. (2003). *Transformation and Transition, DARPA's Role in Fostering a Revolution in Military Affairs. Volume 1.* Alexandria, VA: Institute for Defense Analyses, https://doi.org/10.21236/ada422835, https://fas.org/irp/agency/dod/idarma.pdf, which discusses fifteen years of DARPA research in areas such as stealth and precision strike that in turn enabled the implementation in the 1990's of the offsets theory of Brown and Perry.

42 Owens, W., with Offley, E. (2000). *Lifting the Fog of War*. New York, NY: Farrar, Straus & Giroux (chapter 3); Alberts, D., Garska, J., and Stein, F. (1999). *Network Centric Warfare: Developing and Leveraging Information Technology*. Washington, DC: CCRP Publication Series, Department of Defense, http://www.dodccrp.org/files/Alberts_NCW.pdf; Cebrowski, A., and Garska, J. (1998). "Network Centric Warfare: Its Origin and Future", *US Naval Institute Proceedings*, January. See, generally, Hundley, R. O. (1999). *Past Revolutions, Future Transformations: What Can the History of Revolutions in Military Affairs Tell Us About Transforming the U.S. Military*. Santa Monica, CA: Rand.

enabled the U.S. in the past decade to achieve unparalleled dominance in conventional warfare. And the foundation of this IT revolution, enabling this defense transformation, was a great innovation wave that swept into the U.S. economy in the 1990's, creating strong productivity gains and new business models that led to new societal wealth creation[43] which, in turn, provided the funding base for the defense transformation. To summarize, the DARPA model can support traditional technology development within the defense sector where that technology is primarily or overwhelmingly defense-relevant (like stealth). Alternatively, it can support joint defense-civilian sector technology development where the technology is relevant to both. This enables DOD potentially to take major advantage of academia's openness to new ideas, the willingness of entrepreneurs to commercialize these innovations, and the corresponding scale of an economy-wide advance.

Elements of the DARPA Model

At the Institutional level, DARPA undertakes connected science, rather than simply fundamental research. Its model focuses on revolutionary technology development, not simply incremental advance,[44] moving a technology from fundamental science connected through the development up to prototyping stages, then encouraging and promoting its concepts with partners who move it into service procurement and/or the civilian sector for initial production, enabling full innovation not simply invention.

43 See for example, Jorgenson, D. (2001). "U.S. Economic Growth in the Information Age", *Issues in Science and Technology* 18/1: 42–50, http://www.issues.org/18.1/jorgenson.html (on the role of IT drivers in growth in the 1990s).

44 Looked at in another way, DARPA historically has had two significant roles, breakthrough military applications and systems, such as stealth or precision strike, and broad generic emerging technologies, such as information processing, microsystems or advanced materials. Both roles interrelate and both have transformational effects. See Van Atta, R. (2005). *Energy and Climate Change Research and the DARPA Model*. Presentation at the Washington Roundtable on Science and Public Policy, November. DARPA has also developed concept prototypes and demonstrations to meet established military needs which have not yet been defined as military requirements, aside from its breakthrough technology role. Van Atta, R. (2008). "Fifty Years of Innovation and Discovery", in *DARPA, 50 Years of Bridging the Gap*, ed. C. Oldham, A. E. Lopez, R. Carpenter, I. Kalhikina, and M. J. Tully. Arlington, VA: DARPA. 20–29, https://issuu.com/faircountmedia/docs/darpa50 (Chapter 2 in this volume).

There are other ways DARPA assures connectedness, as suggested above. DARPA developed the ability to make technology development connections across the DOD R&D stove-pipes by using its funding to leverage contributions from other DOD military service technology development organizations, which in turn promotes service adaptation and procurement of its prototypes. DARPA also uses the other DOD R&D agencies as its administrative agents which, on those days when these stars get aligned, likewise promotes cross-institution collaboration and follow-on procurement.

Other DARPA characteristics enhance its ability to operate at both the Institutional and personal innovation organization levels. The following list, which we will call the twelve commandments, is largely drawn from DARPA's own descriptions of its organizing elements:[45]

1) *Small and flexible:* DARPA consists of only 100–150 professionals; one unknown commentator described DARPA as "100 geniuses connected by a travel agent".
2) *Flat organization*: DARPA avoids military hierarchy, essentially operating at only two levels to ensure participation.
3) *Autonomy and freedom from bureaucratic impediments*: DARPA operates outside the civil-service hiring process and standard government contracting rules, which gives it unusual access to talent, plus speed and flexibility in organizing R&D efforts. Stated technically, DARPA has "IPA" hiring authoring authority, which gives it the ability to take personnel employed by industry or universities, and it invented "other transactions authority" in contracting which gives it great flexibility and speed in contracting outside the normally lengthy federal procurement process.
4) *Eclectic, world-class technical staff*: DARPA seeks great talent, drawn from industry, universities, and government laboratories and R&D centers, mixing disciplines and theoretical and experimental strengths. This talent has been hybridized through joint corporate-academic collaborations.

45 DARPA. (2008). *DARPA—Bridging the Gap, Powered by Ideas*. Arlington, VA: Defense Advanced Research Projects Agency, http://www.dtic.mil/cgi-bin/GetTRDoc?Location=U2&doc=GetTRDoc.pdf&AD=ADA433949; DARPA. (2003). *DARPA Over the Years*. Arlington, VA: Defense Advanced Research Projects Agency.

5) *Teams and networks*: At its very best, DARPA creates and sustains great teams of researchers that are networked to collaborate and share in the team's advances, so that DARPA operates at the personal, face-to-face level of innovation. It isn't simply about funding research; its program managers are dynamic playwrights and directors.

6) *Hiring continuity and change*: DARPA's technical staff are hired or assigned for three- to five-years. Like any strong organization, DARPA mixes experience and change. It retains a base of experienced experts that know their way around DOD, but rotates most of its staff from the outside to ensure fresh thinking and perspectives.

7) *Project-based assignments, organized around a challenge model*: DARPA organizes a significant part of its portfolio around specific technology challenges. It works "right-to-left" in the R&D pipeline, foreseeing new innovation-based capabilities and then working back to the fundamental break-throughs that take them there. Although its projects typically last three to five years, major technological challenges may be addressed over much longer time periods, ensuring patient long-term investment on a series of focused steps and keeping teams together for ongoing collaboration.

8) *Outsourced support personnel*: DARPA uses technical, contracting and administrative services from other agencies on a temporary basis. This provides DARPA the flexibility to get into and out of a technology field area without the burden of sustaining staff, while building cooperative alliances with the line agencies it works with.

9) *Outstanding program managers*: In DARPA's words, "The best DARPA program managers have always been freewheeling zealots in pursuit of their goals". The DARPA director's most important job historically has been to recruit highly talented program managers and then empower their creativity to put together great teams around great advances. In particularly fruitful areas, DARPA has created a succession of project

leaders that share and build a common vision for progress over time, as in the case of Licklider and his successors.

10) *Acceptance of failure*: At its best, DARPA pursues a high-risk model for breakthrough opportunities and is very tolerant of failure if the payoff from potential success is great enough.

11) *Orientation to revolutionary breakthroughs in a connected approach*: DARPA historically has focused not on incremental but radical innovation. It emphasizes high-risk investment, moves from fundamental technological advances to development, and then encourages the prototyping and production stages in the armed services or the commercial sector. From an institutional innovation perspective, DARPA is a connected model, crossing the barriers between innovation stages.

12) *Mix of connected collaborators*: DARPA typically builds strong teams and networks of collaborators, bringing in a range of technical expertise and applicable disciplines and involving university researchers and technology firms that are often new and small and not significant defense contractors (which generally do not focus on radical innovation).[46] The aim of DARPA's "hybrid" approach, unique among American R&D agencies, is to ensure strong collaborative "mindshare" on the challenge and the capability to connect fundamentals with applications.

These DARPA "twelve commandments" provide important R&D organizing lessons for any innovation entity, whether in the private or public sectors.

DARPA Today—The Future of the Model

Economic innovation sectors are best described as ecosystems. Marco Iansati and Roy Levien have argued that within these systems frequently

46 There are, of course, exceptions to this, particularly in projects involving systems engineering. Stealth, stand-off precision weapons, and night vision were projects contracted to major defense contractors. Lockheed's *Skunk Works* has long worked with DARPA as well as the Air Force, and represents a radical innovation model operated within a more standard defense firm.

are keystone firms that, like critical species, take on the task of sustaining the whole ecosystem by connecting participants and promoting the progress of the whole system.[47] Iansati and Levien have also argued that these innovation systems start to decline or shift elsewhere when the keystone firms cease being thought of as leaders and instead shift to what they call "landlord" status. In this state, the "landlord" firm shifts to simply extracting value from the existing system rather than continuously attempting to renew and build the system. There have been concerns voiced in recent years and considered below, that DARPA could be moving away from its keystone role, particularly in IT.

Questions about the DARPA Role

DARPA since September 2001 has been increasingly focused on wars in Iraq and Afghanistan, asymmetric conflicts against terrorism requiring different approaches from the symmetric nation state conflict technologies it evolved in the past. While DARPA had been concerned with asymmetric conflicts at least since the demise of the Soviet Union, many noted that the two wars created a significant shift in emphasis at DARPA toward shorter-term military issues and away from some longer-term technology support areas. Concerns about a change in DARPA's role in IT areas, where it has played a keystone role, came up in a series of forums: in a 2005 House Science Committee hearing reviewing DARPA's continuing role in its computer science mission, in a discussion in a Defense Science Board report over its shifting role in microprocessors, in concerns over DARPA's role from PITAC (the President's Information Technology Advisory Council, which was subsequently disbanded by the White House) in IT and cybersecurity, and in papers from a number of IT sector R&D leaders.[48] DARPA has

---

47 Iansati, M., and Levien, R. (2004). *The Keystone Advantage: What the New Dynamics of Business Ecosystems Mean for Strategy, Innovation, and Sustainability.* Boston, MA: Harvard Business School Press.

48 U.S. Congress. (2005). *House Science Committee Hearing on the Future of Computer Science Research in the U.S., 12 May 2005* (Testimony by Wm. A. Wulf, Pres., National Academy of Engineering, Prof. Thomas F. Leighton, Chief Scientist Akamai Tech. Inc., Joint Statement of the Computing Research Community, and Letters in Response to Committee Questions from W. Wulf and T. Leighton), July, http://commdocs.house.gov/committees/science/hsy20999.000/hsy20999_0.htm; Lazowska, E. D., and Patterson, D. (2005). "An Endless Frontier Postponed", *Science* 308: 757, https://doi.org/10.1126/science.1113963; Markoff, M. (2005). "Clouds Over

long been famed as the most successful U.S. R&D agency, so these concerns appear worth weighing.

Let's review some of the questions raised about DARPA's future role. Most involve arguments that DARPA has been shifting out of the IT field it played an historic role in creating, even though this technology revolution is still in its youth—after all, we are still not even close to artificial intelligence. DOD's Defense Science Board (DSB) of leading defense technologists issued a report that recognized the critical gains DOD achieved from DARPA's historic role supporting university and industry-led R&D in microprocessor advances. But it concluded that DOD and DARPA were "no longer seriously involved in…research to enable the embedded processing proficiency on which its strategic advantage depends".[49] Since DOD's strategic superiority in symmetric and potentially asymmetric warfare has become in significant part its network-centric capability, and secure semiconductor microprocessors are the base technology for this capability, DSB found that DOD faces a serious strategic problem as the newest generation of semiconductor production facilities is increasingly shifting to China and other Asian nations. In fact, the U.S. share of the world's leading-edge semiconductor manufacturing capacity dropped from 36 percent to 11 percent in the past seven years.[50] This problem may be compounded if semiconductor design and research, which historically have had to be collocated with production facilities, shift abroad as well. DARPA's departure from its systematic support of U.S. technology leadership in this field appears to present a serious defense issue if other parts of the Department do not absorb some of this function. DARPA's view in recent years has been that semiconductor advance should be led by industry, increasingly dominated in the U.S. by mature, large-scale firms that DARPA's

---

'Blue Sky' Research Agency", *New York Times*, 4 May, p. 12. President's Information Technology Advisory Committee (2005). *Cybersecurity: A Crisis of Prioritization*, Report to the President, February; Defense Science Board. (2005). *High Performance Microchip Supply*. Washington, DC: National Academies Press, 87–88, https://www.hsdl.org/?view&did=454591. Compare DARPA's responses, in U.S. Congress (2005). House Science Committee. (DARPA Testimony with Appendices A-D). NB: the issues raised about DARPA in this section of the chapter concern policies in the George W. Bush Administration; subsequent DARPA leaders attempted to move DARPA back to more of its historic program manager-led model.

49 Defense Science Board. (2005). *High Performance Microchip Supply*.

50 Augustine, N. (2007). *Is America Falling Off the Flat Earth?* Washington, DC: The National Academies Press, 17.

leaders feel should manage their own problems. But if industry increasingly is being forced to shift abroad because of cost pressure from massive industrial subsidies available there,[51] DOD has a long-term problem with what still appears to be a foundation technology. It is serious enough that a 2005 Defense authorization bill directed DOD to implement DSB's proposals to try to control the problem and retain U.S. technology leadership in this area.[52] A DARPA chip strategy, some would argue, should be to try to secure leadership in a post-silicon, post-Moore's Law world in bio-nano-quantum-molecular computing; DARPA would respond that it is working in a number of those fields. Others would dispute whether it is doing enough to nurture leadership in these emerging areas.

Status of the Hybrid Model

More broadly, DSB notes that one of DARPA's critical roles was to fund through its applied research portfolio (known in DOD as "6.2") "hybridized" university and industry efforts through a process that envisioned revolutionary new capabilities, identified barriers to their realization, focused the best minds in the field on new approaches to overcome those barriers, and fostered rapid commercialization and DOD adoption. The hybrid approach bridged the gaps between academic research and industry development, keeping each side knowledgeable about DOD's needs, with each acting a practical prod to spur on the other. DSB expressed concern that this fundamental DARPA approach was breaking down as it cut back its 6.2 university computer science investments, and shifted more of its portfolio to classified "black" research, under pressure from the ongoing war, which cannot include

---

51 Howell, T. (2003). "Competing Programs: Government Support for Microelectronics", in *Securing the Future—Regional and National Programs to Support the Semiconductor Industry*, ed. C. W. Wessner. Washington, DC: The National Academies Press, https://doi.org/10.17226/10677; Howell, T., et al. (2003), https://www.nap.edu/catalog/10677/securing-the-future-regional-and-national-programs-to-support-the. *China's Emerging Semiconductor Industry*. San Jose, CA: Semiconductor Industry Association.

52 Defense Auth. Act for 2005, H.R. 1815 (Sen. Amend. 1361). DOD has established a "trusted foundry" program, initiated in cooperation with IBM, to try to protect its own access to a stable supply of secure semiconductor chips, a particular concern of intelligence agencies, but this does not assure it long term access to technology leadership in what many continue to argue remains a critical technology.

most universities and non-defense tech firms, and, so DSB suggested, reduces DARPA's intellectual mindshare on critical technology issues.[53]

Grid Security

PITAC's report on cybersecurity[54] noted that DARPA plans to terminate funding for its High Confidence Software and Systems development area, aiming to curtail cybersecurity funding except for classified work. Historically, one of Eisenhower's key aims in establishing DARPA was to make sure the U.S. was never again subject to a major technological surprise like Sputnik, and it is widely acknowledged that defense and critical private sector IT systems remain vulnerable to cybersecurity attack. Defense theorists, noting the major economic consequences of the 9/11 attack on financial markets and the insurance sector have argued that asymmetric cyber-attacks on fundamental financial infrastructure by largely unidentifiable state or non-state actors could be devastating to the developed world, potentially striking a powerful blow to the world economy. PITAC has noted that because IT is dominated by the private sector, and even DOD's proposed secure high-speed Global Information Grid must interact with the Internet, shared solutions between defense and private sectors must be developed. Thus, classified research in many cases cannot be effectively implemented. PITAC identified ten defense-critical IT research areas, from authentication technologies to holistic security systems, it believes require future DARPA investment.

Altering the Ecosystem

Dr. Thomas Leighton, Chief Scientist of Akamai Corp., in response to questions from the House Science Committee, argued that DARPA's most important contribution to IT has been "its unique approach

53 Total DARPA university funding as a percentage of DARPA science and technology funding fell from 23.7 percent in FY2000 to 14.6 percent in FY2004 according to 2005 DARPA data, supplied with hearing testimony, (see Footnote 48). A series of major university computer science research department underwent DARPA funding cutbacks of 50 percent and more in the past six years; some observers have argued that new generations of graduate students are no longer trained in DARPA Hard problems and tied to the agency, so that DARPA has reduced connections to its future talent base.

54 PITAC. (2005). *Cybersecurity.*

(and commitment) to developing communities of researchers in both industry and academia" focused on "'pushing the envelope' of computer science".[55] Although DARPA continues to look at some IT problems, "its growing failure to support the university elements of that community is altering the innovation ecosystem" that it created "in an increasing negative way, with no other agency ready or able to pick up that role". Some university computer science departments and labs report that although the DARPA cutbacks in funding have been at least partially made up by industry support, this is often short-term and not breakthrough-oriented, and often is from Asian firms that control the IP for technology developed and for obvious competitive reasons preclude it going into U.S. spinoffs. It should be noted that an increase in NSF computer science funding has offset some of the effects of the decline in DARPA university funding. DARPA's leadership has argued, as justification for the cutback, that it was not seeing enough new ideas from this sector.

Dr. William Wulf, a computer scientist and, until recently, President of the National Academy of Engineering, told the House Science Committee that, "There is now no DOD organization like the 'old DARPA'...that fills the role of discovery of breakthrough technologies".[56] Although he acknowledged that DARPA was looking at cognitive computing, he argued that there were problems in the subjects DARPA was selecting for IT research because it was not confronting key security areas. For example, "our basic model of computer security (perimeter security) is fatally flawed" and will not be solved by the "short term, risk-adverse approach being currently taken by DARPA". He argued that our "ability to produce reliable, effective software" is tottering on "the brink of disaster" but DARPA has not focused on solutions, and also is not reviewing the fact that our basic model for computing is not yet close to human brain capability, and requires a new model "of parallel computing" with "architectures and algorithms of immense power". He also argued that the "use of computers in education has progressed little from the 'automated drill' model of the Plato system of the 1960's."

55 Response of Dr. Tom Leighton to Questions from the House Science Committee Hearing (U.S. Congress. (2005)).

56 Dr. William A. Wulf, Response to Questions from the House Science Committee Hearing (U.S. Congress. (2005)).

This is the case even though "we know much more about how people learn physiologically and psychologically," including how "emotion interacts with learning." Wulf argued that we could put this newer knowledge to good use in quickly training troops in urban combat and counterinsurgency, and DARPA should also be more involved in this area. DARPA spokesmen have noted in response to these arguments that DARPA has funded, as has the Army, soldier training simulation systems at USC's center for this work, and that it was the primary initial funder of grid computing. Perhaps one part of the answer is that DARPA may lack a Licklider with the vision to see and evolve a new IT territory. Critics respond that because of a top-down management style in recent years at DARPA, office directors and program managers lack the authority to initiate in this way.

It is generally understood that DARPA has had to be increasingly focused on solving a problem it ran into at the end of the Cold War with its resulting cuts in defense procurement starting in 1986: the breakdown of technology transition from DARPA into services. DARPA, even during the Cold War, had a transition problem with the services as it focused on disruptive, change-state, radical innovation. It solved some of these problems in the past by transitioning technology, such as IT, into the civilian economy. In other areas, it had to rely on the clout of the Secretary of Defense and, when available, a strong Director of Defense Research & Engineering (DDR&E). DARPA typically did not enjoy a consensus with the military unless it was hammered out by the Office of the Secretary of Defense and the service secretaries. Nonetheless, following the Cold War, technology transition declined. Unsuccessful in building a new consensus with the military services for transferring the results of revolutionary technology investment into service procurement, DARPA technology strategy has been moving from its history of radical innovation to more incremental innovation, shifting a larger part of its investment into later stage development efforts that the services are more ready to invest in. Defense budget analysts report that shorter term incremental work, space launch, and satellite "repair" are requiring growing parts of the DARPA budget. A new DARPA review process, mandated by improving transition to the services, of frequent "up or out" decisions with limited development time is placing more of its R&D on a shorter-term course. Congress may be playing a role in this, as well, focusing more on DARPA's record rather than its overall impact. The current emphasis on a pre-agreed transition plan may further limit disruptive work. Some believe that resulting more frequent policy reversals and turns may limit DARPA's ability to mount enough creative, longer-term investment programs so important to past development. Although the heart of DARPA's creativity in the past was in highly talented and empowered project managers, some believe that the role of project managers has been significantly limited by this short-term review approach. Although DARPA has always been able to pick among the brightest technologists in the nation, its larger focus on classified programs[57] may limit its access to some of the university researchers it has relied on in the past, creating difficulty over time in attracting talent.

DARPA in the past has operated in both the civilian and defense economies, understanding they are the same economy. As noted, it has built "great groups" and spun off civilian-relevant technology, such as in computing, to the civilian sector where it evolved further, enabling DOD to buy it back at radically lower costs and to take advantage of civilian development advances. Alternatively, it has spun off to the defense sector defense-only technologies like stealth and unmanned aerial vehicles (UAV's). DARPA's need to focus on the current asymmetric conflict and corresponding classified work, as well as shorter term technology transition, may make it less able to spin off technology to the civilian economy, despite DOD's growing capital plant cost crisis and its need to take better advantage of advances in that sector.[58] Given DARPA's historic role in successfully straddling both sectors, DARPA needs to protect its ability to play in both worlds.

Much of the above debate is driven by IT sector concerns. But there is a larger debate emerging over DARPA's role in IT, because DARPA, starting with Licklider, played a profound role at the center of most aspects of the IT revolution.

---

57 DARPA has always had, of course, a large classified program base separate from its academic research. The assertion here is that the balance has changed with more of a tilt toward classified work.

58 Research investment also affects defense capability. With defense R&D, nations generally "get what they pay for", with weapon system capability and quality directly corresponding to intensity of research investment. Middleton, A., and Bown, S., with Hartley, K. and Reid, J. (2006). "The Effect of Defense R&D on Military Equipment Quality", *Defense and Peace Economics* 17/2: 117–39.

There is a question whether its current focus on shorter term and classified programs due to the war inevitably will signal a broader retreat from the IT sector,[59] and whether the state of the sector can justify such a retreat?[60] The first question that must be asked is where are we in the IT revolution? In the past, innovation waves fully matured in forty or fifty years and society moved on to the next innovation stage. Accordingly, some argue that the IT revolution is maturing and that we need to move on to the next big things.[61] Where do we measure the IT wave from? If we measure it from the first post-World War II mainframe, ENIAC, the half-century mark for the revolution ran out in 1995. 1995, however, was the period when we were bringing on personal computing and internet access at levels that reached a major portion of our society. If we measure the IT innovation wave from around 1995, when real time and networked computing took off with the public, then we are still a decade into an IT revolution wave. Perhaps DARPA should be moving on to another innovation wave?

On the other hand, the IT revolution may be different from steam engines or electricity. The four- or five-decade model for past innovation waves may not be fully relevant to the IT revolution. When we work with the information domain, we have to keep in mind that we are working with a fundamental force that Norbert Wiener suggested in 1948 was a coequal to mass and energy.[62] We have already been through a succession of unfolding and sometimes parallel IT waves, from business (and military) computational capability, to data retrieval, processing and display, to advanced digital communications, to data mining and using mass data as a predictive tool, and we are beginning to make progress on symbolic manipulation and computer theorem proving and are thinking about quantum computing. The grail quest of computing is true artificial intelligence. This is not a technology pursuit similar to

59 Vernon Ruttan has raised the concern that with the post-Cold War decline in defense innovation, the U.S. innovation system may not now be strong enough to launch new breakthrough technologies in either the public or the private sector. Ruttan, V. W. (2006). "Will Government Programs Spur the Next Breakthrough?", *Issues in Science and Technology* 22/2: 55–61.

60 *Ibid.*

61 Atkinson, R. (2006). "Is the Next Economy Taking Shape?", *Issues in Science and Technology* 22/2: 62–67 at 62, https://issues.org/atkinson-3/

62 Wiener, N. (1948). *Cybernetics or Control and Communication in the Animal and the Machine.* Cambridge, MA: The MIT Press.

past efforts because it is ultimately a quest to take on a god-like power.[63] We have a long, long way to go in achieving this stage. Progress on the Turing Test—can a computer's thinking be mistaken for a human's—has been limited.[64] Although computers now play chess at the highest level and drive SUVs through DARPA's desert and urban obstacle courses, computing isn't even close yet to the intuitive powers of the human brain. Although an artificial intelligence quest may ultimately be futile or only partially achievable, even if we have to settle for Licklider's "Man-Computer Symbiosis" we have a long way to go before this more limited vision is close to being played out. In other words, there may be decades of radical, breakthrough innovation to go in IT, not simply incremental advances. If this is right then DARPA, given its historic breakthrough technology mission and responsibility to avoid Sputnik-like technological surprises, continues to have a future in IT.

Even setting aside the ultimate artificial intelligence challenge, Victor Zue has argued that the next generation of computing challenges are more profound than ever.[65] While yesterday's problem was computation of static functions in a static environment within well-understood specification, today, adaptive systems are needed that operate in environments that are dynamic and uncertain. While computation was the main past goal, communication, sensing and control are also now critical. While computing used to focus on the single operating agent, it must now focus on multiple agents that may be cooperative, neutral or adversarial. While batch processing of text and homogeneous data used to be the task, stream processing of massive heterogeneous data now is. While stand-alone applications once prevailed, deep interaction with humans is now key.

While there was a binary notion of correctness in computing, now there is a trade-off between multiple criteria. In today's computing world these opportunities arise in a far more complex environment of cheap communication, ubiquitous communication, overwhelming data, and limited human resources. Major IT tasks for the military become, for example, much deeper human computer interface, social and cultural

63 Foerst, A. (2005). *God in the Machine*. New York, NY: Penguin Books.

64 Halpern, M. (2006). "The Trouble with the Turing Test", *The New Atlantis* 11: 42–63.

65 Zue, V. (2008). *Introduction to CSAIL*. MIT. 15 April, 6, 14. (Details about Professor Zue and MIT's Computer Science and Artificial Intelligence Laboratory (CSAIL) are available at: https://www.csail.mit.edu/person/victor-zue).

modeling; far more robust and secure computation; smart, self-directed autonomous surveillance; and robots ready for human interaction.

DARPA strongly maintains it is funding IT, even though an increasing amount of its work must be classified. It is also funding what it believes is a critical breakthrough area in computing, cognitive computing, and supports biocomputing and robotics. The ongoing wars in Iraq and Afghanistan appropriately force DARPA toward shorter term solutions for the military; it went through a similar evolution during the Vietnam War. DARPA has had, as noted, a profound problem with technology transition with the military services and, to solve it, must focus on better meeting service needs. Still, the question must be asked whether there is a danger that DARPA may be, over time, retreating into Iansati's and Levien's "landlordism"—not continuously renewing but living off incremental improvements on past advances. For example, it is felt by some observers that DARPA lacks a tactical technology vision as that program has become increasingly smaller-scale, less coherent and non-tactical. DARPA should also evaluate the emerging new dimensions of whether it has a coherent IT vision for approaching some of the challenges Zue and others suggest. Given DARPA's unique historical role in U.S. technology advance, this is a significant issue. Because even great technology advances take a decade or two to produce, the pipeline of advance is hard to see, but problems we may have now in filling that pipeline will have a profound effect on our future a decade or more out.[66]

DARPA is not the only aspect of DOD technology leadership facing difficulties. DOD depends on a strong fundamental physical science research to support its breakthrough potential, but these programs and funding levels are in decline.[67] Boomer generation scientists have been the mainstay of DOD science talent in its labs and research centers, but are now retiring in droves, and are not being adequately replaced. DOD faces a very serious science talent supply problem and needs hiring and retention flexibility beyond civil service limits, but a rigid position in the past by DOD personnel staff that there must be only one personnel

66 Van Atta et al. (1990–1991). *DARPA Technical Accomplishments*. 3 Volumes.

67 Lewis, J. A. (2006). *Waiting for Sputnik*. Washington, DC: Center for Strategic and International Studies, https://www.csis.org/analysis/waiting-sputnik; See, also, Young, J. (2007). "Info Memo for Secretary of Defense Robert M. Gates", DOD Science and Technology Program, 24 August. (on the need and corresponding proposal for increased DOD S&T funding, listing potential high pay-off research areas).

system for all at DOD has thwarted Congressional reform efforts to create more flexibility for scientists. The pressure of the tempo of ongoing military operations is, in turn, putting pressure on funding for science in the military services. The pattern of technology leadership in DOD may not be as strong as in the past. DDR&E leaders of the caliber of John Foster, Malcolm Currie and William Perry have been infrequent, and the overall depth of technical competence in the Office of the Secretary of Defense to backup DARPA and push for technology implementation has declined. Overall, the picture for DOD science is not getting prettier, and this is against a backdrop of serious problems in U.S. physical science in general, as explored in recent major reports by the National Academies.[68]

Yet, our security challenges are growing. The emergence of the terrorist model, of non-state actors relatively immune to state-to-state pressure, represents a profound asymmetric challenge to a Western military model that has been world-dominant since the fifteenth century. In parallel is the emergence of other peer competitors, working on both symmetric and asymmetric approaches, pursuing a technology innovation model for economic development which, as discussed, has significant military implications.

This raises a fundamental concern: can U.S. technological superiority be the continuing basis of U.S. security in an increasingly globalized technological and economic world? Since U.S. economic and military success, as argued at the outset, has relied on profound integration between defense and civilian elements of its innovation system for technological superiority both military and economic, consequences on one side of this equation, such as long-term DARPA capability, have major effects on the other side.

Summary

Arguably innovation organization—the way in which the direct innovation factors of R&D and talent come together, how R&D and

68 National Academy of Sciences. (2007). *Rising above the Gathering Storm: Energizing and Employing America for a Brighter Economic Future*. Washington, DC: The National Academies Press, https://doi.org/10.17226/12537, https://www.nap.edu/catalog/11463/rising-above-the-gathering-storm-energizing-and-employing-america-for#toc; Augustine. (2007). *Is America Falling Off the Flat Earth*.

talent are joined in an innovation system—is a third direct innovation factor.

DARPA emerged as a unique model—operating at both the institutional and personal level of science organization. Building on the Rad Lab example, it built a deeply collaborative, flat, close-knit, talented, participatory, flexible system, oriented to breakthrough radical innovation. It has used a challenge model for R&D, focusing on trying to meet a particular technical challenge, then moving from fundamental research to applied research. Then it would link this research with the follow-on stages of development, prototyping, and access to initial production. In other words, it followed an innovation path, not simply a discovery or invention path. We call this approach the connected science model.

Like all human institutions, these organizational models are transitory. The DARPA model has been one of the longest lasting, unique in the federal government, and seemed to be the most capable of ongoing renewal.

But that DARPA model now may be shifting under pressure of ongoing operations, particularly regarding DARPA's role in the IT sector, with potential long-term effects on U.S. defense as well as civilian sector technology superiority. This shift occurs against a backdrop of overall problems in U.S. physical science strength. DARPA has long served a keystone function in the U.S. innovation system and it is in the nation's national security and economic interest that it continues to avoid "landlord" status.

References

Alberts, D., Garska, J., and Stein, F. (1999). *Network Centric Warfare: Developing and Leveraging Information Technology*. Washington, DC: CCRP Publication Series, Department of Defense, http://www.dodccrp.org/files/Alberts_NCW.pdf

Atkinson, R. (2006). "Is the Next Economy Taking Shape?", *Issues in Science and Technology* 22/2: 62–67, https://issues.org/atkinson-3/

Atkinson, R. D. (2004). *The Past and Future of America's Economy—Long Waves of Innovation that Power Cycles of Growth*. Northampton, MA: Edward Elgar.

Augustine, N. (2007). *Is America Falling Off the Flat Earth?* Washington, DC: The National Academies Press.

Beinhocker, E. D. (2007). *Origin of Wealth—Evolution, Complexity and the Radical Remaking of Economics*. Cambridge, MA: Harvard Business School.

Bennis, W., and Biederman, P. W. (1997). *Organizing Genius: The Secrets of Creative Collaboration*. New York, NY: Basic Books.

Berlin, L. (2005). *The Man Behind the Microchip, Robert Noyce and the Invention of Silicon Valley*. Oxford: Oxford University Press.

Blanpied, W. A. (1998). "Inventing U.S. Science Policy", *Physics Today* 51/2: 34–40 https://doi.org/10.1063/1.882140

Bonvillian, W. B., and Singer, P. (2018). *Advanced Manufacturing, The New American Innovation Policies*. Cambridge, MA: MIT Press, https://doi.org/10.7551/mitpress/9780262037037.001.0001

Bonvillian, W. B. (2006). "Power Play, The DARPA Model and U.S. Energy Policy", *The American Interest* 2/2, November/December, 39–48, https://www.the-american-interest.com/2006/11/01/power-play/

Bush, V. (1945). *Science: The Endless Frontier*. Washington, DC: Government Printing Office, https://www.nsf.gov/od/lpa/nsf50/vbush1945.htm

Chambers, J., ed. (1999). *The Oxford Companion to American Military History*. Oxford: Oxford University Press.

Cebrowski, A., and Garska, J. (1998). "Network Centric Warfare: Its Origin and Future", *US Naval Institute Proceedings*, January.

Conant, J. (2005). *109 East Palace*. New York, NY: Simon & Shuster.

Conant, J. (2002). *Tuxedo Park: A Wall Street Tycoon and the Secret Palace of Science that Changed the Course of World War II*. New York, NY: Simon & Shuster.

DARPA. (2005). *DARPA—Bridging the Gap, Powered by Ideas*. Arlington, VA: Defense Advanced Research Projects Agency, http://www.dtic.mil/cgi-bin/GetTRDoc?Location=U2&doc=GetTRDoc.pdf&AD=ADA433949

DARPA. (2003). *DARPA Over the Years*. Arlington, VA: Defense Advanced Research Projects Agency.

Defense Science Board. (2005). *High Performance Microchip Supply*. Washington, DC: National Academies Press.

Evans, H. (2005). *They Made America*. Sloan Foundation Project. New York, NY: Little, Brown and Company.

Goodwin, J. C., et al. (1999). *DARPA, Technology Transition*. Arlington, VA: Defense Advanced Research Projects Agency.

Foerst, A. (2005). *God in the Machine*. New York, NY: Penguin Books.

Fong, G. R. (2001). "ARPA Does Windows; the Defense Underpinning of the PC Revolution", *Business and Politics* 3/3: 213–37, https://doi.org/10.2202/1469-3569.1025 (Chapter 6 in this volume).

Halpern, M. (2006). "The Trouble with the Turing Test", *The New Atlantis* 11: 42–63.

Hart, D. (1998). *Forged Consensus*. Princeton, NJ: Princeton University Press.

Howell, T., et al. (2003). *China's Emerging Semiconductor Industry*. San Jose, CA: Semiconductor Industry Association.

Howell, T. (2003). "Competing Programs: Government Support for Microelectronics", in C. W. Wessner, ed., *Securing the Future—Regional and National Programs to Support the Semiconductor Industry*. Washington, DC: The National Academies Press, https://doi.org/10.17226/10677

Huddleson, L., and Daitch, V. (2002). *True Genius—The Life and Science of John Bardeen*. Washington, DC: Joseph Henry Press of the National Academies of Sciences.

Hundley, R. O. (1999). *Past Revolutions, Future Transformations: What Can the History of Revolutions in Military Affairs Tell Us About Transforming the U.S. Military*. Santa Monica, CA: RAND.

Iansati, M., and Levien, R. (2004). *The Keystone Advantage: What the New Dynamics of Business Ecosystems Mean for Strategy, Innovation, and Sustainability*. Boston, MA: Harvard Business School Press.

Jorgenson, D. (2001). "U.S. Economic Growth in the Information Age", *Issues in Science and Technology* 18/1: 42–50, http://www.issues.org/18.1/jorgenson.html

Lazowska, E. D., and Patterson, D. (2005). "An Endless Frontier Postponed", *Science* 308:757, https://doi.org/10.1126/science.1113963

Lehman, N. (2001). "Dreaming about War", *The New Yorker*, 16 July, http://www.comw.org/qdr/0107lemann.html

Lewis, J. A. (2006). *Waiting for Sputnik*. Washington, DC: Center for Strategic and International Studies, https://www.csis.org/analysis/waiting-sputnik

Licklider, J. C. R. (1960). "Man-Computer Symbiosis", *IRE Transactions on Human Factors in Electronics* 1: 4–11, https://doi.org/10.1109/thfe2.1960.4503259

Mandel, M. (2008). "A Better Way to Track the Economy, A Groundbreaking Commerce Dept. Report Could Lead to New Yardsticks for Measuring Growth", *Business Week*, 28 January, p. 29.

Markoff, J. (2005). "Clouds Over 'Blue Sky' Research Agency", *New York Times*, 4 May, p. 12.

Marshall, A. W. (1993). "Some Thoughts on Military Revolutions—Second Version", DOD Office of Net Assessment, Memorandum for the Record, 23 August.

Mazuzan, G. (1988). *The National Science Foundation: A Brief History (1950–85)*. Arlington, VA: The National Science Foundation, https://www.nsf.gov/pubs/stis1994/nsf8816/nsf8816.txt

Middleton, A., and Bowns, S., with Hartley, K. and Reid, J. (2006). "The Effect of Defense R&D on Military Equipment Quality", *Defense and Peace Economics* 17/2: 117–39.

Morrow, D. S. (2003). "Dr Craig Venter: Oral History", *Computerworld Honors Program*.

National Academy of Sciences. (2007). *Rising above the Gathering Storm: Energizing and Employing America for a Brighter Economic Future*. Washington, DC: The National Academies Press, https://doi.org/10.17226/12537, https://www.nap.edu/catalog/11463/rising-above-the-gathering-storm-energizing-and-employing-america-for#toc

National Science Board. (2006). *Science and Engineering Indicators*. Arlington, VA: The National Science Foundation, https://wayback.archive-it.org/5902/20160210153725/http://www.nsf.gov/statistics/seind06/

National Research Council, Science and Telecommunications Board. (1999). *Funding a Revolution, Government Support for Computing Research*. Washington, DC: National Academy Press, https://doi.org/10.17226/6323

Owens, W., with Offley, E. (2000). *Lifting the Fog of War*. New York, NY: Farrar, Straus & Giroux.

Perez, C. (2002). *Technological Revolutions and Financial Capital: The Dynamics of Bubbles and Golden Ages*. Northampton, MA: Edward Elgar.

President's Information Technology Advisory Committee (2005). *Cybersecurity: A Crisis of Prioritization*, Report to the President, February.

Rich, B, and Janos, L. (1994). *Skunk Works: A Personal Memoir of My Years of Lockheed*. Boston: Little, Brown & Company.

Roland, R. (1985). *Model Research: The National Advisory Committee for Aeronautics, 1915–58*. Washington, DC: Government Printing Office, http://history.nasa.gov/SP-4103/

Romer, P. (1990). "Endogenous Technological Change", *Journal of Political Economy* 98: 72–102.

Ruttan, V. W. (2006). *Is War Necessary for Economic Growth? Military Procurement and Technology Development*. New York, NY: Oxford University Press.

Ruttan, V. W. (2006). "Will Government Programs Spur the Next Breakthrough?", *Issues in Science and Technology* 22/2: 55–61.

Rycroft, R. W., and Kash, D. E. (1999). "Innovation Policy for Complex Technologies", *Issues in Science and Technology* 16/1, https://issues.org/rycroft/

Sapolsky, H. M. (1990). *Science and the Navy—The History of the Office of Naval Research*. Princeton, NJ: Princeton University Press.

Sherwin, M., and Bird, K. (2005). *American Prometheus, The Triumph and Tragedy of J. Robert Oppenheimer*. New York, NY: Alfred A. Knopf.

Solow, R. M. (2000). *Growth Theory, An Exposition.* 2nd ed. Oxford: Oxford University Press, http://nobelprize.org/nobel_prizes/economics/laureates/1987/solow-lecture.html

Solow, R. M. (2000). "Toward a Macroeconomics of the Medium Run", *Journal of Economic Perspectives* 14/1: 151–58.

Stokes, D. E. (1997). *Pasteur's Quadrant, Basic Science and Technological Innovation.* Washington, DC: Brookings Institution Press.

Tassey, G. (2016) "The Technology Element Model, Path-Dependent Growth and Innovation Policy", *Economics of Innovation and New Technology* 26/6: 594–612, http://dx.doi.org/10.1080/10438599.2015.1100845

Tassey, G. (2007). *The Innovation Imperative.* Cheltenham, UK: Edward Elgar.

U.S. Congress. (2005). *House Science Committee Hearing on the Future of Computer Science Research in the U.S., 12 May 2005* (Testimony by Wm. A. Wulf, Pres., National Academy of Engineering, Prof. Thomas F. Leighton, Chief Scientist Akamai Tech. Inc., Joint Statement of the Computing Research Community, and Letters in Response to Committee Questions from W. Wulf and T. Leighton), July, http://commdocs.house.gov/committees/science/hsy20999.000/hsy20999_0.htm

U.S. Department of Commerce. (2008). *Innovation Measurement, Tracking the State of Innovation in the American Economy.* Report to the Secretary of Commerce. Washington, DC: U.S. Department of Commerce, http://users.nber.org/~sewp/SEWPdigestFeb08/InnovationMeasurement2001_08.pdf

Van Atta, R. (2008). "Fifty Years of Innovation and Discovery", in *DARPA, 50 Years of Bridging the Gap,* ed. C. Oldham, A. E. Lopez, R. Carpenter, I. Kalhikina, and M. J. Tully. Arlington, VA: DARPA. 20–29, https://issuu.com/faircountmedia/docs/darpa50 (Chapter 2 in this volume).

Van Atta, R. (2005). *Energy and Climate Change Research and the DARPA Model.* Presentation at the Washington Roundtable on Science and Public Policy, Washington, DC, November.

Van Atta, R., Lippitz, M., et al. (2003). *Transformation and Transition, DARPA's Role in Fostering a Revolution in Military Affairs. Volume 1.* Alexandria, VA: Institute for Defense Analyses, https://doi.org/10.21236/ada422835, https://fas.org/irp/agency/dod/idarma.pdf

Van Atta, R., et al. (1991). *DARPA Technological Accomplishments, An Historical Review of DARPA Projects.* Alexandria, VA: Institute for Defense Analyses.

Venter, J. C. (2007). *A Life Decoded: My Genome, My Life.* New York, NY: Viking Press.

Waldrop, M. M (2001). *The Dream Machine: J. C. R. Licklider and the Revolution that Made Computing Personal.* New York, NY: Viking Press.

Waldrop, M. M. (1992). *Complexity, the Emerging Science at the Edge of Order and Chaos.* New York, NY: Simon & Schuster.

Warsh, D. (2006). *Knowledge and the Wealth of Nations.* New York, NY: W. W. Norton.

Wiener, N. (1948). *Cybernetics or Control and Communication in the Animal and the Machine.* Cambridge, MA: The MIT Press.

Young, J. (2007). "Info Memo for Secretary of Defense Robert M. Gates", DOD Science and Technology Program, 24 August.

Zachary, G. P. (1999). *Endless Frontier, Vannevar Bush, Engineer of the American Century. Cambridge,* MA: The MIT Press.

Zue, V. (2008). *Introduction to CSAIL.* MIT. 15 April. (Details about Professor Zue and MIT's Computer Science and Artificial Intelligence Laboratory (CSAIL) are available at: https://www.csail.mit.edu/person/victor-zue)

5. The Value of Vision in Radical Technological Innovation[1]

Tamara L. Carleton

The Value of Vision in Radical Technological Innovation

This study provides empirical evidence of the role of vision in fostering technological invention, adding to the existing literature about radical innovation.[2] DARPA provides a long history of examples of technical program visions and how these visions are formed and communicated time after time. In this section, the four main findings of the study are discussed in detail and in context of the literature.

First, this study shows a relationship between the formation of a technological vision and the sustained creation of radical innovation, providing new knowledge about the role of vision in radical innovation. Since its inception in 1958, new programs at DARPA have required a vision to be started, which then guides subsequent work and development. Several dimensions arise regarding the role of vision, which entail functioning primarily at the program level, characterized as "DARPA

1 This chapter is an excerpt from Tamara L. Carleton's PhD thesis: Carleton, T. L. (2010). "The Value of Vision in Radical Technological Innovation", PhD thesis, Stanford University, Palo Alto.

2 E.g., Roberts, E. B., ed. (1987). *Generating Technological Innovation*. New York, NY: Oxford University Press; Tornatzky, L. G., Fleischer, M., and Chakrabarti, A. K. (1990). *The Processes of Technological Innovation*. Lexington, MA: Lexington Books; O'Connor, G. C., Leifer, R., Paulson, A. S., and Peters, L. S. (2008). *Grabbing Lightning: Building a Capability for Breakthrough Innovation*. San Francisco, CA: Jossey-Bass.

 https://doi.org/10.11647/OBP.0184.05

Hard", and relying on the program manager as a vision champion. Second, this study describes the use of expert workshops and proof-of-concepts, used steadily by DARPA to shape partial visions into complete visions, which demonstrates critical efforts occurring prevision. Third, this study describes the importance of socialization in order to prepare and instruct program managers in their envisioning skills. Immersed in the culture at DARPA, new program managers learn from each other and their network connections. Fourth, this study provides new evidence about radical innovation governance models. DARPA relies on small group decisions by organizational leadership to approve promising new visions, running counter to the dominant literature about stage-gate reviews, peer reviews, and extended consensus-seeking processes.

A Process Model of Radical Innovation

As described in the previous chapter, DARPA follows certain high-level steps in its quest for radical innovation, and this process is reproduced in Figure 5-1. By documenting the process at DARPA, this study helps other researchers and practitioners to understand one organization's formula for sustained radical innovation. Documented processes are the basis for repetition and become the springboard for continuous and measurable performance.

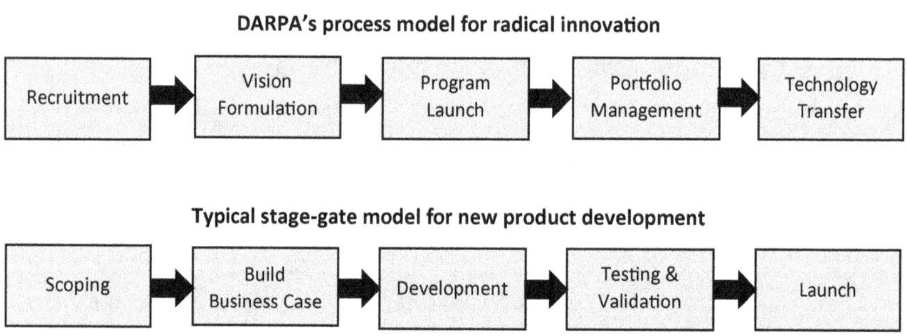

Fig. 5-1 Comparison between DARPA's process model and the stage-gate model. Although DARPA's process model of innovation looks similar to the typical stage-gate model for new product development,[3] the two models differ in terms of objectives, activity, and evaluation mechanics. (Figure prepared by the author.)

3 Cooper, R. G. (2001). *Winning at New Products: Accelerating the Process from Idea to Launch.* 3rd ed. Cambridge, MA: Basic Books.

In addition, some scholars may see a similarity between the depiction of DARPA's process model and the typical stage-gate model for new product development,[4] depicted in Figure 5-1. Both models are comprised of five stages that sequence categories of cross-functional activities, which help to invite a comparison. However, there are at least three key differences between the two models. First, the two models differ in objectives. DARPA's goal is radical innovation, which is intended to produce new technologies that ultimately may lead to new products. In contrast, the stage-gate process is designed to build and launch new products.

Second, the two models differ in their activity timing. DARPA's model is focused on the early stages that precede project scope. The stage-gate model is missing the preliminary or ideation phase, often called Discovery, which occurs before the start of the first stage of scoping.

Third, the two models differ in evaluation mechanisms. DARPA's process is fluid, and although transition arrows are noted between stages, formal decision points are not necessarily required before proceeding onto the next set of activities. In comparison, the stage-gate model is predicated on predefined deliverables and checkpoints with go/no go criteria at the end of each stage (these checkpoints are called gates).

New Dimensions of Vision

Vision plays a central role in DARPA's process of innovation; indeed, DARPA *starts* its process with vision. It matters where and how a vision is started, as does who starts and maintains the vision. DARPA program managers are hired deliberately for their visions of technology, even if partially formed. Then, program managers codify their visions at the start of each new program in a specialized document called a Broad Agency Announcement (BAA), which are used to generate interest in the broader R&D community. Thus, the vision is formulated before groups are funded because DARPA's funding recipients rely on these BAAs to determine potential solutions.

4 *Ibid.*

Visions at the Program Level

By studying the role of vision within DARPA, this study reveals several new dimensions of vision as related to innovation. One dimension is the level at which a vision operates. The dominant business literature has largely studied vision at the organizational level;[5] at the other end of the literature, several studies have investigated technological visions at the product or project level.[6] Within DARPA, work is broken down at three levels: organizational, program, and project, and the data shows that vision is introduced and functions primarily at the *program* level. Figure 5-2 illustrates the multiple levels of visions that could exist within an organization, and shows how visions at DARPA address the gap between the organizational and project/product levels.

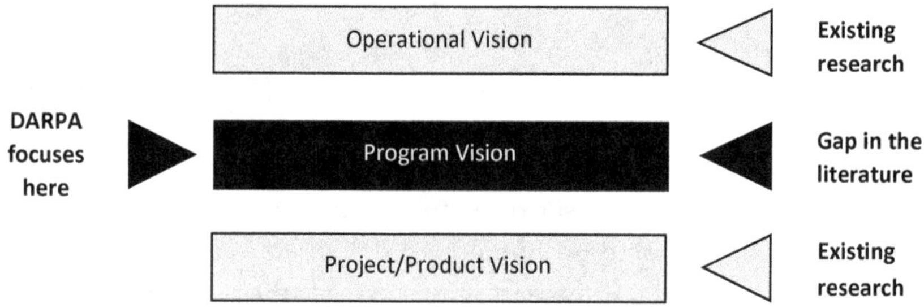

Fig. 5-2 Visions at DARPA operate at the program level. The literature on innovation predominantly discusses technological visions at the organizational level, and several studies have investigated technological visions at the project level. The literature fails to discuss vision at the program level, which is equivalent to the business unit or market level. At DARPA, technological visions function at the program level. (Figure prepared by the author.)

In fact, DARPA lacks a traditional corporate vision, which identifies a set of organizational values and direction for the enterprise. Since its inception in 1958, the agency has not defined (or even reinvented) its long-term goals, aspirations, and values at the organizational level. Instead, DARPA emphasizes visions at the program level, which correlates with a traditional business unit or market focus. Multiple

5 Collins, J. C., and Porras, J. I. (1991). "Organizational Vision and Visionary Organizations", *California Management Review* 34/1: 30–52.

6 Lynn, G. S., and Akgün, A. E. (2001). "Project Visioning: its Components and Impact on New Product Success", *Journal of Product Innovation Management* 18: 374–87.

visions—often totaling over a hundred, depending on the number of program managers actively serving at DARPA—exist in parallel at a given time. Programs serve as new, broad-scale technical initiatives that typically encompass multiple projects, and projects are equivalent to product teams in industry. Again, a DARPA program could be considered equivalent to a business unit or new market category. At DARPA, a program is more of an open-ended question or challenge posed to the R&D community, which might have multiple solutions and product possibilities, and scholars have documented the benefit of a challenge model within an R&D setting.[7]

In addition, visions at the program level allow DARPA program managers to direct multiple projects, multiple teams, and even multiple products over multiple years. Through visions at the program level, DARPA can excite and rally interest across several different technical areas, helping to distribute resources more effectively. Program visions provide a way to organize multiple projects and smaller-scale efforts across a range of funding recipients, who each may interpret the vision differently in application. This approach, in turn, increases the likelihood of a greater diversity of solutions. A program structure also allows for greater flexibility in engendering commitment.

Vision Quality

A second dimension is the quality of the vision. In the literature, few studies focus on technological visions, and most scholars draw on studies of corporate vision. For example, Gary S. Lynn and Ali Akgün describe product visions as a combination of clarity, support, and stability, which are determined relative to the larger organization.[8] While these attributes offer a sense of an ideal vision, they do not provide meaningful guidelines on how to develop a vision, including

7 Bonvillian, W. B. (2006). "Power Play, The DARPA Model and U.S. Energy Policy", *The American Interest* 2/2, November/December, 39–48, https://www.the-american-interest.com/2006/11/01/power-play/; Bonvillian, W. B. (2009). "The Connected Science Model for Innovation—The DARPA Model", in *21st Century Innovation Systems for the U.S. and Japan*, ed. S. Nagaoka, M. Kondo, K. Flamm, and C. Wessner. Washington, DC: National Academies Press. 206–37, https://doi.org/10.17226/12194, http://books.nap.edu/openbook.php?record_id=12194&page=206 (Chapter 4 in this volume).

8 Lynn and Akgün. (2001). "Project Visioning".

the type of vision to create in the technology space. This study shows that technological visions at DARPA have several attributes that are essential to the creation of its visions.

Since its inception, DARPA has socialized a catchphrase known as DARPA Hard. Drawn from the data, a DARPA Hard program vision is characterized as technically challenging, actionable, multidisciplinary, and far-reaching. Taken apart, these attributes can be found discussed in prior studies.

The first attribute—technically challenging—is understood within the operations research and engineering design community as a "wicked problem".[9] A wicked problem is a technically difficult problem that is nearly impossible to solve due to complex interdependencies, a high level of ambiguity, and conflicting interests from stakeholders. Wicked problems cannot be solved through classic experimentation and logic, instead requiring a different and more creative strategy of reasoning. By focusing on these types of problems at DARPA, program managers have ensured that they push the limits of innovation sought, what might be interpreted as "highly radical" innovation according to Abetti's scale.[10] When most definitions of radical innovation argue for market changes, DARPA is pushing for a radical *technology* shift, which then may lead to a radical market shift. Each attempt at creating a new technical solution changes the understanding of the problem in two fundamental ways. First, more information helps to reformulate the initial requirements, and second, every prototype and implementation built advances the state of knowledge overall in the world. In other words, there is no turning back or reverting to the former understanding of the problem. The vision for a DARPA program provides the high-level guidelines to inspire potential funding recipients, and by engaging both more and different groups to respond, DARPA is able to cast a wider net for solutions and likewise accelerate the experimentation process.

This approach helps to drive toward action, and actionable is the second attribute of DARPA Hard. Program visions are intentionally

9 Buchanan, R. (2009). "Thinking about Design: An Historical Perspective", in *Philosophy of Technology and Engineering Sciences*, ed. A. Meijers. Amsterdam, The Netherlands: Elsevier B.V. 409–53, https://doi.org/10.1016/b978-0-444-51667-1.50020-3

10 Abetti, P. A. (2000). "Critical Success Factors for Radical Technological Innovations: A Five Case Study", *Creativity and Innovation Management Journal* 9/4: 208–21, https://doi.org/10.1111/1467-8691.00194

grounded in reality because they are expected to improve and extend the limits of existing technologies. Visions cannot exist as science-fiction fantasy, political rhetoric, or policy scenarios. This attribute is partly captured in earlier research about the reflective practitioner, in which Donald Schön describes how professionals, such as engineers, address problematic situations that are fraught with uncertainty, disorder, and indeterminacy by taking action through real-time cycles of feedback and learning.[11] In DARPA's case, program managers rely on their visions as a way to simulate broader learning in their research networks.

A growing body of research about learning in inter-organizational networks shows that networks facilitate rapid responses. Powell states that, "Whether it is the case that one firm's technological competence has outdistanced the others, or that innovations would be hard to replicate internally, as suggested by the growing reliance on external sources of research and development, network forms of organization represent a fast means of gaining access to know-how that cannot be produced internally".[12]

The third attribute—multidisciplinary—is equally critical to forming the right program visions at DARPA. As many DARPA program managers interviewed for this study noted, they needed to redefine problems outside of usual boundaries, and complex situations required drawing from more than one discipline. Multidisciplinary efforts are not new to government-sponsored R&D and can be evidenced in the rise of systems engineering in the 1950s that supported large scale efforts, such as the Atlas missile program and ARPANET.[13] This type of approach encourages less hierarchical control and more network-based management techniques.

The fourth attribute—far-reaching—is important when creating program visions at DARPA. One part of far-reaching is about having a broad impact in society. Subjects spoke about making a difference in magnitude.

---

11 Schön, D. (1983). *The Reflective Practitioner: How Professionals Think in Action*. New York, NY: Basic Books.

12 Powell, W. W. (1990). "Neither Market nor Hierarchy: Network Forms of Organization", *Organizational Behavior* 12: 295–336, at 316.

13 Hughes, T. P. (1998). *Rescuing Prometheus: Four Monumental Projects that Changed the Modern World*. New York, NY: Pantheon Books.

DARPA program managers stated that they need to think big in order to have big results. Another aspect of far-reaching is the ability to plan long-term. The importance of planning long-term has its roots in World War II, notably the founding of RAND.[14] This idea of planning for the long term made its way into today's management science through thinkers such as Peter F. Drucker.[15]

The real test of a good vision in R&D is whether others will commit resources to action, which will bring results in the future. DARPA deliberately couples action with future intent. However, the conundrum is that traditional R&D results may not be produced or easy to measure because the extent of far-reaching effects take time and are broadly distributed across society. The attribute of far-reaching is consistent with recent work in foresight engineering, which focuses on long-range technology cycles as part of an organization's ongoing search for innovation opportunities.[16]

Together, these four attributes—technically challenging, actionable, multidisciplinary, and far-reaching—that make up a DARPA Hard program provide a metric that can be instrumented and tested. Based on pioneering work in taxonomies,[17] Figure 5-3 presents a sample classification using a 7-point scale that was used for the quantification of human performance variables, specifically describing human ability for side-to-side equilibrium.[18] This type of scale could be adapted in order to classify each of the four attributes characterizing DARPA Hard. Follow-on studies can further define and test the scale values as related to radical innovation. Ultimately, if other organizations seek to recreate

---

14 Campbell, V. (2004). "How RAND Invented the Postwar World", *Invention & Technology* 20/1: 50–59.

15 Drucker, P. F. (1959). "Long-Range Planning: Challenge to Management Science", Management Science 5/3: 238–49; Drucker, P. F. (1973). *Management: Tasks, Responsibilities, Practice.* New York, NY: Harper Colophon.

16 Carleton, T. and Cockayne, W. (2009). "The Power of Prototypes in Foresight Engineering", in *Proceedings of the 17th International Conference on Engineering Design* (ICED'09), ed. M. Norell Bergendahl, M. Grimheden, L. Leifer, P. Skogstad, and U. Lindemann. Stanford, CA: The Design Society. 267–76.

17 Bloom, B. S., ed. (1956). *Taxonomy of Educational Objectives: The Classification of Educational Goals: Handbook I, Cognitive Domain.* New York, NY: Green; Fleishman, E., and Quaintance, M. (1984). *Taxonomies of Human Performance: The Description of Human Tasks.* Orlando, FL: Academic Press.

18 Cockayne W., and Darken, R. (2004). "The Application of Human Ability Requirements to Virtual Environment Interface Design and Evaluation", in *The Handbook of Task Analysis for Human-Computer Interaction*, ed. D. Diaper, and N. Stanton. Mahwah, NJ: Lawrence Erlbaum Associates. 401–21.

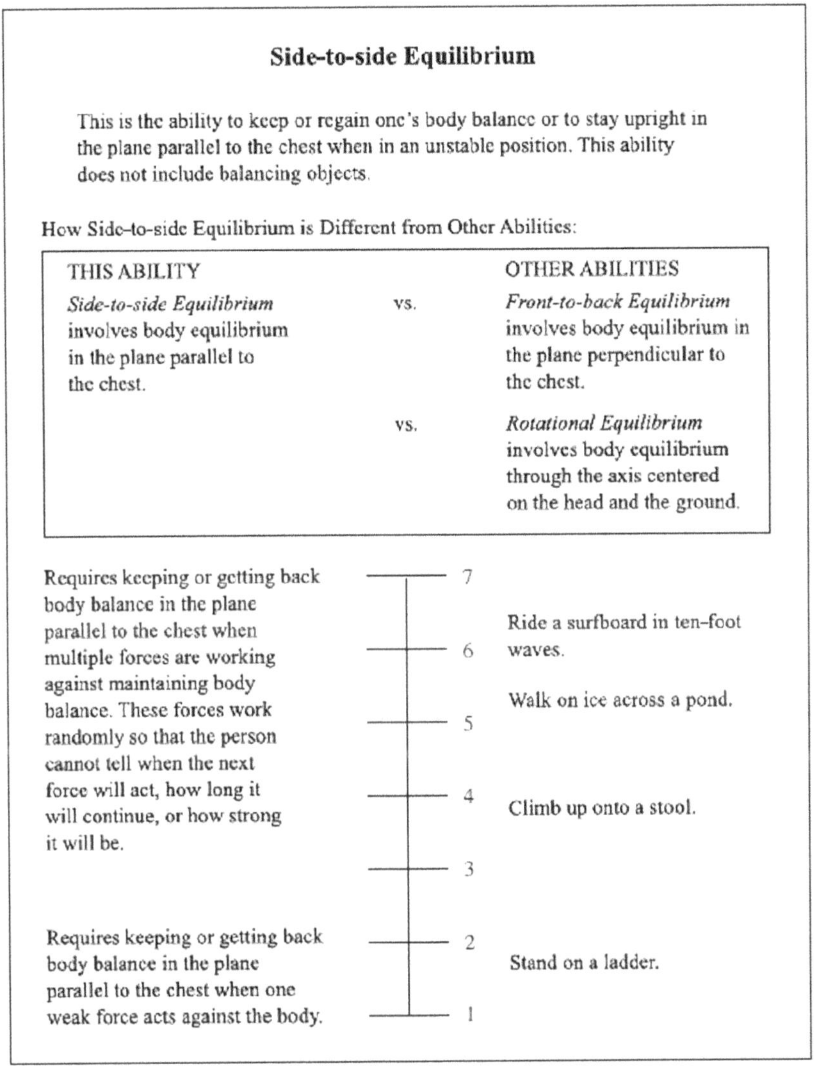

Fig. 5-3 A sample 7-point scale for quantification of human performance variables. A sample 7-point scale, drawn from another study, could be adapted to classify and evaluate each of the four attributes characterizing DARPA Hard. Follow-on studies can define and test the scale values as related to radical innovation. (Figure from William R. Cockayne. (1998). "Two-Handed, Whole-Hand Interaction", Master's thesis, Naval Postgraduate School, Monterey, California. Used here with permission from the author.)

a variant of DARPA Hard, they will benefit from defining and using a clear classification of technological visions.

Visionaries of Technology

A third dimension of vision is the person responsible for fostering it. Visions cannot exist without creators, who must imagine and invent them. Within DARPA, the work on innovation is driven as much by ideas as by individuals. Program managers are hired as technical visionaries, and they are solely responsible for shaping, spearheading, and promoting their respective visions of technology. The project champion is a critically recognized role in innovation, and findings from this study are consistent with literature on this topic.[19] At DARPA, a new program is not confounded with multiple organizational champions; instead, there is a clear relationship in that each program manager builds one vision per program. Figure 5-4 depicts this relationship. However, the DARPA program manager does not operate in isolation. He (or she) is part of a broader ecosystem and network, in which multiple players—both internally and externally to the agency—are engaged to support the formation and execution of a program vision.

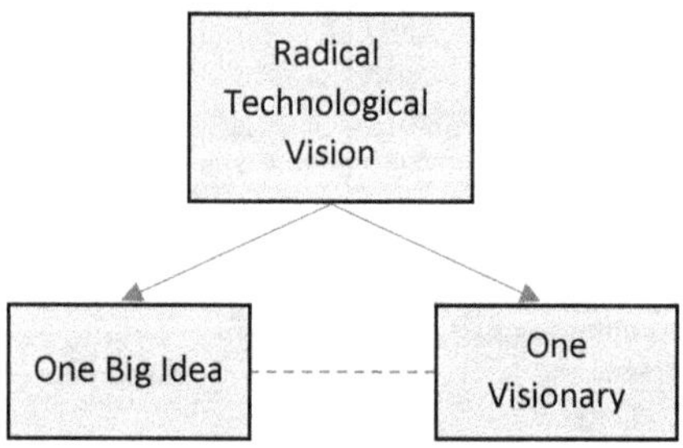

Fig. 5-4 A radical technological vision relies on one big idea and one visionary. At DARPA, a program vision relies on a program manager, who serves as the vision's primary champion internally and externally. Moreover, there is a clear relationship in that each program managers builds one vision per program. (Figure prepared by the author.)

DARPA program managers serve in other innovation roles that have been documented separately in literature. For example, they share

19 Howell, J. M., and Higgins, C. A. (1990). "Champions of Technological Innovation", *Administratively Science Quarterly* 35: 317–41.

some characteristics with *business innovators* because DARPA program managers provide substantial funding, as well as some organizational credibility and access to other resources.[20] Although DARPA program managers do not build and develop their own visions, instead relying on the various funding recipients, they do act as *technical innovators* in other ways.[21] More informed than the usual project champion, DARPA program managers are nearly all technically educated and bring deep expertise from various fields of engineering and science. This background allows them to more effectively understand the given technical problem, as well as advise and guide the technical teams that they sponsor. A growing number of studies discuss the special role of a technical visionary, who combines technical knowledge with project oversight.[22]

DARPA program managers also play the role of technology licenser or technology transfer manager. They are directly responsible for finding potential user groups, typically in the U.S. military services, who might test and ultimately adopt a functional prototype. The final success of DARPA program visions hinges on user adoption.

At DARPA, potential program managers—the champions of new technological visions—are found and recruited through the extended research network. Studies show that as networks mature, they tend to petrify.[23] People prefer to work with familiar connections, which limits network access to new connections. Current program managers will find new program managers based on similar qualities and will continue funding the same relationships. When this happens, an innovation network does not diversify, and the development of new ideas can be potentially severely limited. DARPA has addressed this limitation by deliberately hiring program managers new to the network, who, in turn, bring new visions of technology. Subsequently, the new-to-the-network program manager finds and funds research groups that bring additional new ideas to the network, which helps to refresh institutional thinking and challenge engrained assumptions.

---

20 Howell and Higgins. (1990). "Champions of Technological Innovation".

21 *Ibid.*

22 Hebda, J. M., Vojak, B. A., Griffin, A., and Price, R. L. (2007). "Motivating Technical Visionaries in Large American Companies", *IEEE Transactions on Engineering Management* 54/3: 433–44; Deschamps, J. (2008). *Innovation Leaders: How Senior Executives Stimulate, Steer and Sustain Innovation*. Hoboken, NJ: Jossey-Bass.

23 Powell. (1990). "Neither Market nor Hierarchy".

Lastly, DARPA is now over fifty years old as an organization, and, historically, the agency has relied on its network for internal job referrals. As the people in DARPA's network have aged, they may not be cultivating as many new relationships with other research groups or also with junior engineers and scientists. Age plays a substantial role in creating new fields, and research shows that younger scientists are more likely to be drawn to a new field than older scientists.[24]

Some scholars have studied how large mature organizations must continually reconfigure their systems of power in order to sustain innovation.[25] Recently, DARPA leadership has recognized the need to recruit younger program managers into its mix. For example, the press observed former agency director Tony Tether "has managed to draw younger researchers into an agency whose stalwart backers are growing greyer every year".[26] However, more research is needed to understand the effects of age on DARPA's ability to foster radical innovation.

The Development of Partial Visions

In the key texts that mention vision, few descriptions are provided about how to generate a vision or develop a partial vision into a complete technological vision.[27] Scholars underscore the importance of having a vision, yet they assume a complete vision. Findings from this study demonstrate that multiple steps are consistently taken by DARPA program managers in order to advance their early ideas and thinking before the complete vision is formed. Figure 5-5 illustrates the actions that must occur before a complete vision is achieved. In addition, while the technological idea drives action, the path to the vision itself is emergent.

This study describes the formation of partial visions via two primary mechanisms, specifically expert workshops and proof-of-concepts, which are used consistently throughout DARPA's history to develop partial visions into clear visions. While details may differ, the objective

24 Rappa, M., and Debackere, K. (1993). "Youth and Scientific Innovation: The Role of Young Scientists in the Development of a New Field", *Minerva* 31/1: 1–20.

25 Dougherty, D., and Hardy, C. (1996). "Sustained Product Innovation in Large, Mature Organizations: Overcoming Innovation-to-Organization Problems", *Academy of Management Journal* 39/5: 1120–53.

26 "A Little Less Disneyland", *Nature* 451: 374 (2008), https://doi.org/10.1038/451374a

27 Roberts. (1987). *Generating Technological Innovation.*

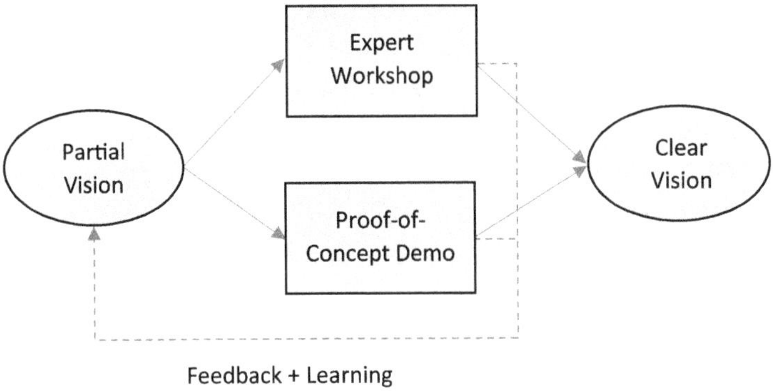

Fig. 5-5 Efforts preceding a complete vision of technology. Earlier actions occur before a complete vision is achieved at DARPA. (Figure prepared by the author.)

is the same between the two mechanisms: to gain more insight into a promising yet incomplete vision. Expert workshops and proof-of-concepts address the people and the idea, respectively. Through expert workshops, each program manager engages his or her network, and the network serves as a way to gain perspective through dialogue among trusted colleagues. In studies about knowledge networks and communities of practice, network members regularly share information through both formal and informal channels,[28] and the DARPA workshops positively exploit the broader knowledge network for the agency. The DARPA workshops are effective because they draw on the collective wisdom for a field, helping DARPA program managers to gain access to the latest knowledge about a particular topic.

If the workshops rely on people, the proof-of-concepts depend on the idea. The objective of the proof-of-concepts is to explore and test the feasibility of an emerging idea. Each proof-of-concept serves as a directed demonstration. Proof-of-concepts are regularly discussed in engineering design research and business studies as a form of prototyping,[29] and specifically, Carleton and Cockayne discuss the

28 Hildreth, P. M., and Kimble, C., eds. (2004). *Knowledge Networks: Innovation Through Communities of Practice.* Hershey, PA: Idea Group Publishing; Powell, W. W., and Grodal, S. (2005). "Networks of Innovators", in *The Oxford Handbook of Innovation*, ed. J. Fagerberg, D. Mowery, and R. R. Nelson. New York, NY: Oxford University Press. 56–85.

29 Schrage, M. (1999). *Serious Play: How the World's Best Companies Simulate to Innovate.* Boston, MA: Harvard Business School Press; Betz, F. (2003). *Managing Technological*

growing role that physical prototypes serve in long-range planning.[30] This study provides new information about the use of proof-of-concepts in vision development as a way to demonstrate feasibility and test early hunches before undertaking a new technical initiative. There is an opportunity to expand on the relationship between prototyping and vision formation.

It is important to note that this combination of expert workshops and proof-of-concepts has provided the primary mechanisms for converting partial visions into full visions at DARPA; no other mechanisms were pursued as long or as reliably, as reported by DARPA program managers and funding recipients. This approach has implications for organizations pursuing radical or disruptive innovation. O'Connor and her colleagues discuss the different experiments that big companies have attempted in order to scout for and generate radical ideas.[31] Some of these experiments resemble the expert workshops at DARPA. IBM has held a large annual R&D event to order to stimulate new ideas internally and identify potential emerging business opportunities. This event has been denoted using multiple names—including idea jams, idea cafes, and deep dives—and while the organizers continually tinker with the process, the event itself remains constant every year. The annual event has led to a high number of opportunities, which in turn have become profitable business lines at IBM.

Learning Radical Innovation Through Socialization

The third finding relates to the culture of innovation at DARPA. Program managers come from a variety of backgrounds. While they have impeccable academic and professional credentials, many lack direct experience with certain innovation skill sets, such as documenting a vision, recruiting and leading others, and technology transfer. Regardless of their background, expectations are high for DARPA program managers to develop and deliver on their program visions quickly.

Innovation: Competitive Advantage from Change. 2nd ed. New York, NY: John Wiley; Moss, L. T. and Atre, S. (2003). *Business Intelligence Roadmap: The Complete Project Lifecycle for Decision-Support Applications*. Boston, MA: Addison-Wesley.

30 Carleton and Cockayne. (2009). "The Power of Prototypes".

31 O'Connor et al. (2008). *Grabbing Lightning*.

In addition, DARPA does not provide formal training in innovation "know how", particularly the skills needed to develop program visions. Is staff training necessary for radical innovation? According to subjects, DARPA has not codified much of its internal procedures historically; so new program managers cannot rely on manuals or similar process guides. Instead, knowing is a matter of participating. At DARPA, subjects reported learning primarily from immersion. From the start, a candidate for a new program manager has to be already embedded in the research community to be considered for recruiting.

Once at DARPA, program managers described learning by doing, particularly by proactively reaching out to colleagues, alumni and other members in the network for advice and resources, as well as by gaining new knowledge from regular field visits.

In many ways, DARPA is a culture of show, not tell. Through a process of socialization, program managers acquire the habits, beliefs, and accumulated knowledge of the organization. In sociology, this period is known as metamorphosis, when a newcomer becomes an established organizational member.[32] How people behave and interact with one another over time shapes an organizational culture, and the data from DARPA is consistent with prior studies about tacit knowledge and informal learning occurring within innovation organizations and communities of practice.

If an organization is to survive, then research shows that stability over time is required, so that one generation of employees transmits the dominant social and cultural patterns to the next generation.[33] In other words, practice is transferred from those who have done it to those who need to do it. At DARPA, this transfer of knowledge occurs through informal conversations, and, given the short contracts of DARPA technical staff, the cycle of generations is rapid. It is remarkable that a knowledge-generating organization over fifty years old, which has resisted lasting knowledge capture, has maintained such a stable set of practices as DARPA has. Based on subject reports, two factors have likely contributed most to the unusual stability of DARPA's culture. First, the

32 Kramer, M. W. (2010). *Organizational Socialization: Joining and Leaving Organizations*. Cambridge, UK: Polity.

33 Alvesson, M. (1995). *Management of Knowledge-Intensive Companies*. New York, NY: Walter de Gruyter.

broader infrastructure supporting DARPA program managers, namely the support staff, provide continuity across leadership turnovers. This support staff functions as an underlying layer of institutional permanence, handling the same routines and project coordination tasks. Second, the agency's network structure supports ongoing learning. For example, even when program managers leave their agency roles officially, they typically stay connected to DARPA in other ways. This connection creates additional channels of knowledge sharing between staff and also ensures that some institutional memory is maintained across staff rotations. New staff rely on the stories and experiences shared within the network in order to prepare themselves at DARPA.

Internal Review of Radical Innovation Ideas

Even with the right person and the right idea, a promising technological vision may not become a new program at DARPA. There is one final test before a Broad Agency Announcement (BAA) is released to the public. A program manager must pitch his vision internally with a small audience for funding approval, and decision-making authority resides namely with the agency director and respective office director. Subject reports demonstrate that DARPA has consistently followed this governance model over the years, actively discouraging larger evaluations in the agency's innovation process. Subjects especially note the benefits of speed, convenience, and flexibility from these small group reviews.

DARPA's model runs counter to the literature and practice of innovation, which discusses consensus-based governance models—such as innovation boards, technology councils, R&D committees, task forces, and stage-gates—as a dominant best practice.[34] These models provide a decision-making framework that help to define evaluation criteria, grant decision-making power, and verify feasibility of a new

34 Bacon, F. R., Jr., and Butler, T. W., Jr. (1973). *Achieving Planned Innovation: A Proven System for Creating Successful New Products and Services.* New York, NY: Simon & Schuster; Hamel, G. (2002). *Leading the Revolution: How to Thrive in Turbulent Times by Making Innovation a Way of Life.* Boston, MA: Harvard Business School Press; Snyder, N. T., and Duarte, D. L. (2003). *Strategic Innovation: Embedding Innovation as a Core Competency in your Organization.* San Francisco, CA: Jossey-Bass; O'Connor et al. (2008). *Grabbing Lightning;* Skarzynski, P., and Gibson, R. (2008). *Innovation to the Core: A Blueprint for Transforming the Way your Company Innovates.* Boston, MA: Harvard Business School Press.

research idea. A growing body of literature has noted that certain models have limitations for radical innovation. Gassmann and von Zedtwitz note:

> In industries or projects where the science or technology push is the dominant driver of innovation, stage-gate processes are too rigid and slow. Innovations that are triggered by a technological invention with unknown market potential need different processes and techniques to succeed.[35]

Overall, innovation studies endorse a strong philosophy that the processes for radical or disruptive innovation must differ from traditional R&D processes in order to be effective within an organization. By deliberately adopting a model of limited, leadership-driven review and following it for over fifty years, DARPA provides empirical support for this belief. Instead of creating large task forces, DARPA relies on its leadership to approve and support the visions. Instead of formally scheduled sessions, DARPA program managers arrange meetings when they feel that their new program visions are ready for funding. Most of corporate R&D, the work of funding agencies, and academic research are actually structured in direct opposition to this approach. Members of the science community, who believe that DARPA provides an enduring and effective model for advancing radical innovation, understand this difference. Penman and Bates write, "Those wishing to emulate the success of DARPA and Bell Labs might consider another important aspect: freedom from the so called 'peer review' that weighs down most National Institutes of Health (NIH) and National Science Foundation efforts".[36]

Conclusion

Four main findings were discussed in relation to the literature review. By describing how visions serve an integral role in DARPA's innovation process, the first finding brings new perspective to innovation studies about the role of visions in radical innovation. In particular, new program

---

35 Gassmann, O., and von Zedtwitz, M. (2003). "Innovation Processes in Transnational Corporations", in *The International Handbook on Innovation*, ed. L. V. Shavinina. Oxford, UK: Elsevier Science. 702–14, at 704.

36 Penman, S., and Bates, C. C. (1999). "DARPA in the Spotlight", *Science* 286/5438: 239.

visions must meet the criteria of being DARPA Hard, and this term of art introduces a working metric for technical breakthroughs that are nearly impossible to achieve based on the current state of knowledge and tools. Second, the discovery that expert workshops and proof-of-concepts have been used repeatedly to convert partial visions into complete visions at DARPA shows that activities exist pre-vision and directly influence the formation of technological visions. Third, the discovery that new program managers receive no formal documentation or training for their roles and instead rely on acculturation is consistent with prior research on innovation networks and communities of practice. Finally, by showing that DARPA has a leadership-driven, decision-making model, in which leadership approves a new program vision, the fourth finding introduces contradictory evidence to the dominant literature. These four findings, supported by empirical evidence, add to the current understanding of technological visions and radical innovation research.

References

"A Little Less Disneyland", *Nature* 451: 374 (2008), https://doi.org/10.1038/451374a

Abetti, P. A. (2000). "Critical Success Factors for Radical Technological Innovations: A Five Case Study", *Creativity and Innovation Management Journal* 9/4: 208–21, https://doi.org/10.1111/1467-8691.00194

Alvesson, M. (1995). *Management of Knowledge-Intensive Companies*. New York, NY: Walter de Gruyter.

Aspray, W., and Norberg, A. (1998). *An Interview with J. C. R. Licklider*, 28 October. University of Minnesota, MN: Charles Babbage Institute, Center for the History of Information Processing.

Bacon, F. R., Jr., and Butler, T. W., Jr. (1973). *Achieving Planned Innovation: A Proven System for Creating Successful New Products and Services*. New York, NY: Simon & Schuster.

Basit, T. N. (2003). "Manual or Electronic? The Role of Coding in Qualitative Data Analysis", *Educational Research* 45/2: 143–54, https://doi.org/10.1080/0013188032000133548

Belfiore, M. (2009). *The Department of Mad Scientists: How DARPA is Remaking our World, from the Internet to Artificial Limbs*. New York, NY: Smithsonian Books/HarperCollins.

Betz, F. (2003). *Managing Technological Innovation: Competitive Advantage from Change*. 2nd ed. New York, NY: John Wiley.

Bloom, B. S., ed. (1956). *Taxonomy of Educational Objectives: The Classification of Educational Goals: Handbook I, Cognitive Domain*. New York, NY: Green.

Bonvillian, W. B. (2006). "Power Play, The DARPA Model and U.S. Energy Policy", *The American Interest* 2/2, November/December, 39–48, https://www.the-american-interest.com/2006/11/01/power-play/

Bonvillian, W. B. (2009). "The Connected Science Model for Innovation—The DARPA Model", in *21st Century Innovation Systems for the U.S. and Japan*, ed. S. Nagaoka, M. Kondo, K. Flamm, and C. Wessner. Washington, DC: National Academies Press. 206–37, https://doi.org/10.17226/12194, http://books.nap.edu/openbook.php?record_id=12194&page=206 (Chapter 4 in this volume).

Bryant, A., and Charmaz, K., eds. (2007). *The SAGE Handbook of Grounded Theory*. Los Angeles, CA: Sage Publications, https://doi.org/10.4135/9781848607941

Buchanan, R. (2009). "Thinking about Design: An Historical Perspective", in *Philosophy of Technology and Engineering Sciences*, ed. A. Meijers. Amsterdam, The Netherlands: Elsevier B.V. 409–53, https://doi.org/10.1016/b978-0-444-51667-1.50020-3

Campbell, V. (2004). "How RAND Invented the Postwar World", *Invention & Technology* 20/1: 50–59.

Carleton, T., and Cockayne, W. (2009). "The Power of Prototypes in Foresight Engineering", in *Proceedings of the 17th International Conference on Engineering Design (ICED'09)*, ed. M. Norell Bergendahl, M. Grimheden, L. Leifer, P. Skogstad, and U. Lindemann. Stanford, CA: The Design Society. 267–76.

Chesbrough, H. (2003). *Open Innovation: The New Imperative for Creating and Profiting from Technology*. Boston, MA: Harvard Business School Press.

Christensen, C. M. (1997). *The Innovator's Dilemma: When New Technologies Cause Great Firms to Fail*. Boston, MA: Harvard Business School Press.

Christensen, C. M., and Raynor, M. E. (2003). *The Innovator's Solution: Creating and Sustaining Successful Growth*. Boston, MA: Harvard Business School Press, https://doi.org/10.5465/ame.2004.12689164

Christiansen, J. A. (2000). *Building the Innovative Organization: Management Systems that Encourage Innovation*. New York, NY: St. Martin's Press.

Chubin, D. E., and Hackett, E. J. (1990). *Peerless Science: Peer Review and U.S. Science Policy*. Albany, NY: State University of New York Press.

Cockayne, W. R. (2004). "A Study of the Formation of Innovation Ideas in Informal Networks", PhD thesis, Stanford University, Palo Alto.

Cockayne, W. R. (1998). "Two-Handed, Whole-Hand Interaction", Master's thesis, Naval Postgraduate School, Monterey, California.

Cockayne W., and Darken, R. (2004). "The Application of Human Ability Requirements to Virtual Environment Interface Design and Evaluation", in

The Handbook of Task Analysis for Human-Computer Interaction, ed. D. Diaper, and N. Stanton. Mahwah, NJ: Lawrence Erlbaum Associates. 401–21.

Collins, J. C., and Porras, J. I. (1991). "Organizational Vision and Visionary Organizations", *California Management Review* 34/1: 30–52.

Collins, J. C., and Porras, J. I. (1996). "Building your Company's Vision", *Harvard Business* Review 74/4: 65–77.

Collins, J. C., and Porras, J. I. (1997). *Built to Last: Successful Habits of Visionary Companies.* New York, NY: HarperBusiness.

Cooper, R. G. (2001). *Winning at New Products: Accelerating the Process from Idea to Launch.* 3rd ed. Cambridge, MA: Basic Books.

Corbin, J. M., and Strauss, A. (2008). *Basics of Qualitative Research: Techniques and Procedures for Developing Grounded Theory.* 3rd ed. Los Angeles, CA: Sage Publications, https://doi.org/10.4135/9781452230153

Cotterman, R., Fusfeld, A., Henderson, P., Leder, J., Loweth, C., and Metoyer, A. (2009). "Aligning Marketing and Technology to Drive Innovation", *Research Technology Management* 52/5: 14–20, https://doi.org/10.1080/08956308.2009.11657585

Daniel, H.-D. (1993). *Guardians of Science: Fairness and Reliability of Peer Review.* Weinheim, Germany: VCH,

https://onlinelibrary.wiley.com/doi/book/10.1002/3527602208

DARPA. (1998). *DARPA Technology Transition,* http://www.darpa.mil/Docs/transition.pdf

DARPA. (2003). *Strategic Plan,* February, https://upload.wikimedia.org/wikipedia/commons/5/57/DARPA_Strategic_Plan_%282003%29.pdf

DARPA. (2007). *Strategic Plan,* February, https://www.hsdl.org/?view&did=769871

DARPA. (2008). *DARPA: 50 Years of Bridging the Gap,* ed. C. Oldham, A. E. Lopez, R. Carpenter, I. Kalhikina, and M. J. Tully. Arlington, VA: DARPA, https://issuu.com/faircountmedia/docs/darpa50

DARPA. (2008). *Human Resources (HR) Recruitment—Sourcing and Placement of DARPA Program Managers (PMs).* Request for Information, reference number SN08-27, 7 April.

DARPA. (2009). *Department of Defense Fiscal Year (FY) 2010 Budget Estimates, Volume 1—Defense Advanced Research Projects Agency,* May, https://www.darpa.mil/attachments/(2G7)%20Global%20Nav%20-%20About%20Us%20-%20Budget%20-%20Budget%20Entries%20-%20FY2010%20(Approved).pdf

DARPA. (2010). *Broad Agency Announcements,* https://www.darpa.mil/work-with-us/office-wide-broad-agency-announcements

DARPA. (2010). *DARPA Mission*, https://www.darpa.mil/about-us/mission

DARPA. (2010). *Director's Biography*, https://www.darpa.mil/about-us/people

DARPA. (2010). *Offices*, https://www.darpa.mil/about-us/offices

DARPA. (2010). *Technology Transition Team*.

DARPA. (2001). *Broad Area Announcement: Augmented Cognition*. Solicitation Number: DARPA-BAA 01–38. Commerce Business Daily, PSA #2902, 27 July, 2001.

DARPA. (2008). *Broad Area Announcement: Legged Squad Support System*. Solicitation Number: DARPA-BAA-08–71, 24 October.

Defense Acquisition, Technology, and Logistics. (2007). *Joining DARPA as a Program Manager Starts With your Idea*, November.

Defense Science Board. (1999). *Investment Strategy for DARPA*, Storming Media Report No. A014763. Washington, DC: Pentagon Reports.

Dekkers, R. (2005). *(R)evolution: Organizations and the Dynamics of the Environment*. New York, NY: Springer Science+Business Media.

Deschamps, J. (2008). *Innovation Leaders: How Senior Executives Stimulate, Steer and Sustain Innovation*. Hoboken, NJ: Jossey-Bass, https://doi.org/10.1002/9781119208761

de Vaus, D. (2001). *Research Design in Social Research*. Thousand Oaks, CA: Sage Publications.

Dougherty, D., and Hardy, C. (1996). "Sustained Product Innovation in Large, Mature Organizations: Overcoming Innovation-to-Organization Problems", *Academy of Management Journal* 39/5: 1120–53.

Drucker, P. F. (1959). "Long-Range Planning: Challenge to Management Science", *Management Science* 5/3: 238–49.

Drucker, P. F. (1973). *Management: Tasks, Responsibilities, Practice*. New York, NY: Harper Colophon.

Dundon, E. (2002). *The Seeds of Innovation: Cultivating the Synergy that Fosters New Ideas*. New York, NY: AMACOM.

Edelheit, L. S. (2004). "The IRI Medalist's Address: Perspective on GE Research & Development", *Research Technology Management* 47/1: 49–54, https://doi.org/10.1080/08956308.2004.11671608

Ettlie, J. E., Bridges, W. P., and O'Keefe, R. D. (1984). "Organization Strategy and Structural Differences for Radical versus Incremental Innovation", *Management Science* 30/6: 682–95.

Fleishman, E., and Quaintance, M. (1984). *Taxonomies of Human Performance: The Description of Human Tasks*. Orlando, FL: Academic Press.

Fleming, L. (2002). "Finding the Organizational Sources of Technological Breakthroughs: The Story of Hewlett-Packard's Thermal Ink-Jet", *Industrial and Corporate Change* 11/5: 1059–84, https://doi.org/10.1093/icc/11.5.1059

Fuchs, E. R. H. (2010). "Rethinking the Role of the State in Technology Development: DARPA and the Case for Embedded Network Governance", *Research Policy* 39/9: 1133–47, https://doi.org/10.1016/j.respol.2010.07.003 (Chapter 7 in this volume).

Gassmann, O., and von Zedtwitz, M. (2003). "Innovation Processes in Transnational Corporations", in *The International Handbook on Innovation,* ed. L. V. Shavinina. Oxford, UK: Elsevier Science, 702–14.

Glaser, B. G. and Strauss, A. L. (1967). *The Discovery of Grounded Theory.* New York, NY: Aldine.

Goulding, C. (2002). *Grounded Theory: A Practical Guide for Management, Business and Market Researchers.* Thousand Oaks, CA: Sage Publications, https://doi.org/10.4135/9781849209236

Hafner, K., and Lyon, M. (1996). *Where Wizards Stay Up Late: The Origins of the Internet.* New York, NY: Touchstone.

Hamel, G. (2002). *Leading the Revolution: How to Thrive in Turbulent Times by Making Innovation a Way of Life.* Boston, MA: Harvard Business School Press.

Harvard Business Essentials. (2003). *Managing Creativity and Innovation.* Boston, MA: Harvard Business School Press.

Hebda, J. M., Vojak, B. A., Griffin, A., and Price, R. L. (2007). "Motivating Technical Visionaries in Large American Companies", *IEEE Transactions on Engineering Management* 54/3: 433–44, https://doi.org/10.1109/tem.2007.900791

Hildreth, P. M., and Kimble, C., eds. (2004). *Knowledge Networks: Innovation Through Communities of Practice.* Hershey, PA: Idea Group Publishing, https://doi.org/10.4018/978-1-59140-200-8

Hiltzik, M. (1999). *Dealers of Lightning: Xerox PARC and the Dawn of the Computer Age.* New York, NY: HarperBusiness.

Howe, H. E. (1952). "'Space Men' Make College Men Think", *Popular Science* 161/4: 124–27.

House, C. H., and Price, R. L. (2009). *The HP Phenomenon: Innovation and Business Transformation.* Stanford, CA: Stanford Business Books.

Howell, J. M., and Higgins, C. A. (1990). "Champions of Technological Innovation", *Administratively Science Quarterly* 35: 317–41.

Hughes, T. P. (1998). *Rescuing Prometheus: Four Monumental Projects that Changed the Modern World.* New York, NY: Pantheon Books.

Johnson, C. L., and Smith, M. (1989). *Kelly: More than my Share of it All.* Washington, DC: Smithsonian.

Johnson, M. W. (2010). *Seizing the White Space: Business Model Innovation for Growth and Renewal*. Boston, MA: Harvard Business School Press.

Junod, T. (2003). "Science & Industry: DARPA", *Esquire*, 1 December.

Kim, W. C., and Mauborgne, R. (2005). *Blue Ocean Strategy: How to Create Uncontested Market Space and Make Competition Irrelevant*. Boston, MA: Harvard Business School Press.

Kostoff, R. N., Boylan, R., and Simon, G. R. (2004). "Disruptive Technology Roadmaps", *Technological Forecasting and Social Change* 71: 141–59, https://doi.org/10.1016/s0040-1625(03)00048-9

Kramer, M. W. (2010). *Organizational Socialization: Joining and Leaving Organizations*. Cambridge, UK: Polity.

Leheny, R. (2007). *What is a DARPA Program Manager?* Speech Presented at DARPATech, DARPA's 25th Systems and Technology Symposium, Anaheim, CA, 7 August.

Leifer, R., McDermott, C. M., O'Connor, G. C., Peters, L. S., Rice, M. P., and Veryzer, R. W. (2000). *Radical Innovation: How Mature Companies Can Outsmart Upstarts*. Boston, MA: Harvard Business Press.

Levy, S. (2000). *Insanely Great: The Life and Times of Macintosh, the Computer that Changed Everything*. New York, NY: Penguin Books.

Lundquist, G. (2004). "One Point of View: The Missing Ingredients in Corporate Innovation", *Research-Technology Management* 47/5: 11–12, https://doi.org/10.1080/08956308.2004.11671646

Lynham, S. A. (2002). "The General Method of Theory-Building Research in Applied Disciplines", Advances *in Developing Human Resources* 4: 221–41 https://doi.org/10.1177/15222302004003002

Lynn, G. S., and Akgün, A. E. (2001). "Project Visioning: its Components and Impact on New Product Success", *Journal of Product Innovation Management* 18: 374–87, https://doi.org/10.2139/ssrn.2151901

Maccoby, M. (2007). "Developing Research/Technology Leaders", *Research-Technology Management* 50/2: 65–67, https://doi.org/10.1080/08956308.2007.11657432

Malakoff, D. (1999). "Pentagon Agency Thrives on In-Your-Face Science", *Science* 285/5433: 1476–80, https://doi.org/10.1126/science.285.5433.1476

Moss, L. T., and Atre, S. (2003). *Business Intelligence Roadmap: The Complete Project Lifecycle for Decision-Support Applications*. Boston, MA: Addison-Wesley.

National Research Council. (2005). *Government/Industry/Academic Relationships for Technology Development: A Workshop Report*. Steering Committee for Workshops on Issues of Technology Development for Human and Robotic Exploration and Development of Space, Aeronautics and Space Engineering

Board, Division on Engineering and Physical Sciences. Washington, DC: The National Academies Press, https://doi.org/10.17226/11206

Nanus, B. (1992). *Visionary Leadership: Creating a Compelling Sense of Direction for your Organization.* San Francisco, CA: Jossey-Bass.

Norling, P. M., and Statz, R. J. (1998). "How Discontinuous Innovation Really Happens", *Research Technology Management* 41/2: 41–44, https://doi.org/10.1080/08956308.1998.11671208

O'Connor, G. C., Leifer, R., Paulson, A. S., and Peters, L. S. (2008). *Grabbing Lightning: Building a Capability for Breakthrough Innovation.* San Francisco, CA: Jossey-Bass.

O'Connor, G. C., and Rice, M. P. (2001). "Opportunity Recognition and Breakthrough Innovation in Large Established Firms", *California Management Review* 43/20: 95–116, https://doi.org/10.2307/41166077

Page, L. (2009). "Spy Chief to Obama: Let DARPA Fix Economy", *The Register,* 21 January, https://www.theregister.co.uk/2009/01/21/darpa_economy_fix_plan/Pappert, S. (2009). *Succeeding with DARPA/MTO: Tips from an ex-Program Manager*. Proceedings from the 2009 Microsystems Technology Office Symposium, March 2–5.

Penman, S., and Bates, C. C. (1999). "DARPA in the Spotlight", *Science* 286/5438: 239, https://doi.org/10.1126/science.286.5438.239b

Perez, C. (2002). *Technological Revolutions and Financial Capital: The Dynamics of Bubbles and Golden Ages*. Northampton, MA: Edward Elgar, https://doi.org/10.4337/9781781005323.00011

Powell, W. W. (1990). "Neither Market nor Hierarchy: Network Forms of Organization", *Organizational Behavior* 12: 295–336.

Powell, W. W., and Grodal, S. (2005). "Networks of Innovators", in *The Oxford Handbook of Innovation,* ed. J. Fagerberg, D. Mowery, and R. R. Nelson. New York, NY: Oxford University Press. 56–85, https://doi.org/10.1093/oxfordhb/9780199286805.001.0001

Quigley, J. V. (1993). *Vision: How Leaders Develop It, Share It, and Sustain It.* New York, NY: McGraw-Hill.

Quigley, J. V. (1994). "Vision: How Leaders Develop It, Share It, and Sustain It", *Business Horizons* 47/5: 37–41.

Rappa, M., and Debackere, K. (1993). "Youth and Scientific Innovation: The Role of Young Scientists in the Development of a New Field", *Minerva* 31/1: 1–20.

Rhea, D. (2003). "Bringing Clarity to the 'Fuzzy Front End': A Predictable Process for Innovation", in *Design Research: Methods and Perspectives,* ed. B. Laurel. Boston. MA: The MIT Press, 145–54.

Richardson, J. J., Bosma, J., Roosild, S., and Larriva, D. (1999). *A Review of the Technology Reinvestment Project.* Arlington, VA: Potomac Institute for Policy Studies.

Richardson, J. J., Larriva, D. L., and Tennyson, S. L. (2001). *Transitioning DARPA Technology*. Arlington, VA: Potomac Institute for Policy Studies.

Rice, M. P., O'Connor, G. C., Peters, L. S., and Morone, J. G. (1998). "Managing Discontinuous Innovation", Research *Technology Management* 41/3: 52–58.

Roberts, E. B., ed. (1987). *Generating Technological Innovation.* New York, NY: Oxford University Press.

Roland, A., and Shiman, P. (2002). *Strategic Computing: DARPA and the Quest for Machine Intelligence, 1983–93.* Cambridge, MA: The MIT Press.

Schilling, M. A. (2005). *Strategic Management of Technological Innovation.* New York, NY: McGraw-Hill/Irwin.

Schön, D. A. (1983). *The Reflective Practitioner: How Professionals Think in Action.* New York, NY: Basic Books.

Schumpeter, J. (1942). *Capitalism, Socialism, and Democracy.* New York, NY: Harper & Row.

Schrage, M. (1999). *Serious Play: How the World's Best Companies Simulate to Innovate.* Boston, MA: Harvard Business School Press.

Siedman, I. (2006). *Interviewing as Qualitative Research: A Guide for Researchers in Education and the Social Sciences.* New York, NY: Teachers College Press.

Sigismund, C. G. (2000). *Champions of Silicon Valley: Visionary Thinking from Today's Technology Pioneers.* New York, NY: John Wiley.

Smith, P. G., and Reinertsen, D. G. (1998). Developing Products in Half the Time: New Rules, New Tools. 2nd ed. New York, NY: John Wiley.

Snyder, N. T., and Duarte, D. L. (2003). *Strategic Innovation: Embedding Innovation as a Core Competency in your Organization.* San Francisco, CA: Jossey-Bass.

Stringer, R. (2000). "How to Manage Radical Innovation", *California Management Review* 42/4: 70–88, https://doi.org/10.2307/41166054

Skarzynski, P., and Gibson, R. (2008). *Innovation to the Core: A Blueprint for Transforming the Way your Company Innovates.* Boston, MA: Harvard Business School Press.

Tennenhouse, D. (2004). "Intel's Open Collaborative Model of Industry-University Research", *Research Technology Management* 47/4: 19–26, https://doi.org/10.1080/08956308.2004.11671637

Tether, T. (2005). *Multidisciplinary Research.* Presentation at the Committee on Science, United States House of Representatives, 12 May.

Tornatzky, L. G., Fleischer, M., and Chakrabarti, A. K. (1990). *The Processes of Technological Innovation.* Lexington, MA: Lexington Books.

Tzeng, C. (2009). "A Review of Contemporary Innovation Literature: A Schumpeterian Perspective", *Innovation: Management, Policy & Practice* 11: 373–94, https://doi.org/10.5172/impp.11.3.373

United States Department of Defense. (2008). *DOD's Financial Management Regulation (Publication No. DOD 7000.14-R), Volume 2B, Chapter 5 (Research, Development and Evaluation Appropriations).* Washington, DC: U.S. Government Printing Office, https://comptroller.defense.gov/portals/45/documents/fmr/archive/02barch/02b_05_dec10.pdf

Urban Dictionary. (2010). *DARPA hard,* https://www.urbandictionary.com/define.php?term=DARPA%20hard

Van Atta, R., Lippitz, M., et al. (2003). *Transformation and Transition, DARPA's Role in Fostering a Revolution in Military Affairs. Volume 1.* Alexandria, VA: Institute for Defense Analyses, https://doi.org/10.21236/ada422835, https://fas.org/irp/agency/dod/idarma.pdf

Van Atta, R. H., Cook, A., et al. (2003). *Transformation and Transition: DARPA's Role in Fostering an Emerging Revolution in Military Affairs. Volume 2.* Washington, DC: U.S. Government Printing Office, https://doi.org/10.21236/ada422835 , https://apps.dtic.mil/dtic/tr/fulltext/u2/a422835.pdf

Van Atta, R. (2008). "Fifty Years of Innovation and Discovery", in *DARPA, 50 Years of Bridging the Gap,* ed. C. Oldham, A. E. Lopez, R. Carpenter, I. Kalhikina, and M. J. Tully. Arlington, VA: DARPA. 20–29, https://issuu.com/faircountmedia/docs/darpa50 (Chapter 2 in this volume).

Van de Ven, A. H., Angles, H. L., and Poole, M. S. (2000). *Research on the Management of Innovation: The Minnesota Studies.* New York, NY: Oxford University Press.

Van de Ven, A. H., Polley, D. E., Garud, R., and Venkataraman, S. (1999). *The Innovation Journey.* New York, NY: Oxford University Press.

Vojak, B. A., Griffin, A., Price, R. L., and Perlov, K. (2006). "Characteristics of Technical Visionaries as Perceived by American and British Industrial Physicists", *R&D Management* 36/1: 17–26, https://doi.org/10.1111/j.1467-9310.2005.00412.x

Waldrop, M. M. (2001). *The Dream Machine: J. C. R. Licklider and the Revolution that Made Computing Personal.* New York, NY: Viking Press.

Watts, R. M. (2001). "Commercializing Discontinuous Innovations", *Research-Technology Management* 44/6: 26–31, https://doi.org/10.1080/08956308.2001.11671461

Weinberger, S. (2006). *Imaginary Weapons: A Journey through the Pentagon's Scientific Underworld.* New York, NY: Nation Books.

6. ARPA Does Windows

The Defense Underpinning of the PC Revolution[1]

Glenn R. Fong

Introduction

> The PC industry is leading our nation's economy into the 21st century… There isn't an industry in America that is more creative, more alive and more competitive. And the amazing thing is all this happened without any government involvement. (Bill Gates, 1998.)[2]

The personal computer revolution, born out of risk-taking corporate ventures and garage-based innovative individualism, is the epitome of the heights than can be achieved by private sector, free-market entrepreneurialism. While this is the conventional story, it is inaccurate. The personal computer (PC) technologies that have revolutionized our everyday lives, whether at the office or at home, have been deeply rooted in public sector initiatives as well. As communities throughout the country and countries around the world rush to clone their own Silicon Valleys, the governmental underpinnings of the original Valley's success should not be overlooked.

1 This chapter originally appeared in *Business and Politics* 3/3 (2001). The editors of this volume gratefully acknowledge the permission to reprint this paper given by Cambridge University Press, the publisher of *Business and Politics*.

2 Microsoft News Release. (1998). "Remarks by Bill Gates", 18 May. Issued on the day the Justice Department launched its anti-trust suit against the company.

 https://doi.org/10.11647/OBP.0184.06

This story parallels the widely-recognized government role in spurring a second revolution in information technology: the Internet. The current-day internet traces its origins back, of course, to the late 1960s ARPANET project of the Defense Department. However, when it comes to our main window on cyberspace—the personal computer—a defense or government link to such a broad-sweeping business and consumer appliance is almost inconceivable. Instead, when it comes to the origins of what makes a PC a PC—its graphical user interface, windows, the desktop metaphor and icons, and the mouse pointing device—the genealogy is usually traced back industrially from Apple and Microsoft, and then back to the Xerox Palo Alto Research Center (Xerox PARC, for short). This accepted history is embodied in the mainstream business literature, general media, and popular culture.

What is less well-known—and serves as the foci of this article—is that Xerox PARC along with other pioneers of PC technology were associated with a significant government-sponsored thrust in desktop computing. The Air Force, Army, Navy, NASA, National Science Foundation, and most notably, the Defense Department's Advanced Research Projects Agency (ARPA or DARPA)[33] aggressively and persistently supported technologies key to the PC revolution.

Uncovering this political-economic link provides an important corrective to the popular lore surrounding the origins of the personal computer. In their emphases on private sector initiative and entrepreneurial risk-taking, conventional PC histories conform to orthodox market-based explanations of technological and economic progress. The role of government in spurring innovation and encouraging risk-taking is downplayed if not outright dismissed. In contradistinction, this article "brings the state" into the PC realm of apparent market purity.[4]

In making this case, we start mid-story with the Xerox-Apple-Microsoft connection. Reflecting a balance in political-economic analysis, this portion of the article is business-centered as it is important to briefly

---

3 The agency was founded in 1958 as ARPA, changed to DARPA ("Defense" added) in 1972, reverted back to ARPA in 1993, and then back to DARPA in 1995. The acronym used in this article will shift according to the time period under discussion.

4 Echoing the statist literature in political science and sociology. See Evans, P., Rueschemeyer, D., and Skocpol, T. (1985). *Bringing the State Back In*. New York, NY: Cambridge University Press.

establish what would come of earlier R&D efforts. The article then jumps back to the pre-Xerox, pre-commercialization story where the government role takes center stage. Before concluding, the penultimate section fast forwards by briefly looking ahead to the government's, particularly DARPA's, continuing influence on personal computing with the onset of the twenty-first century.

PARC and HCI

> Xerox could have been the IBM of the 90's... could have been the Microsoft of the 90's. (Steve Jobs, 1996.)[5]

Microsoft Windows, the Macintosh, the mouse, the desktop metaphor with icons, file directories, and folders—indeed the very notion of computing at the individual, personal level—can all in the first (but not last) instance be traced *directly* back to the Xerox PARC Alto computer. The first two aforementioned systems were introduced in 1985 and 1984, respectively,[6] while the Alto was completed in 1973.

Before tracing this genealogy, it would be appropriate to briefly demarcate what we are tracing—our dependent variable. What the layperson calls a "personal computer" is, of course, an integration of a plethora of different technologies. A core subset of these technologies—and the core focus of this article—is what computer scientists call "human-computer interface", or HCI. HCI is concerned with enhancing the performance of joint tasks by humans and computers. To improve the structure of communication between human and machine, HCI brings together (1) the computer science and engineering fields of computer graphics, operating systems, programming languages, and software development; (2) behavioral science disciplines in communication theory, linguistics, learning theory, and cognitive psychology; and (3) graphic and industrial arts and design, as well as ergonomics. Examples of HCI techniques include keyboard commands; pointing devices; touch

5 *Triumph of the Nerds.* (1996). Public Broadcasting System. 12 June.

6 Windows 1.0 was introduced in 1985, but would not qualify as a fully functional graphical user interface. While version 1.0 and even version 2.0 had windows containing document contents, and while different programs could be open at the same time, the windows could not be overlapped (only tiled) and neither utilized graphical icons. Only with Windows 3.0 in 1990 would Microsoft offer a functional GUI. See Allison, D. (1993). "Bill Gates Interview", *Smithsonian Institution.*

screens and other display technologies; voice, handwriting and gesture recognition; eye movement tracking; biological and psychic sensing; computer speech; graphical user interfaces; user navigation and menu selection tools; windows environments; and desktop metaphors.[7] Ultimately, from the user perspective, HCI technologies result in the user-friendliness and "look and feel"—or lack thereof—of our PCs.

HCI technology provided the crucial linkage between two other developments in the 1970s that brought us the PC.[8] In a top-down development, the processing power of mainframe computers was slowly being brought to individual users through computer time-sharing.[9] The computer was still in the basement, but scores of users could tap into its resources through remote terminals. While representing a disservice to the mainframe's prowess, simple computer games, such as *Spacewar!*, offered a glimpse of real-time interactive computing. Time-sharing, however, could reach only relatively limited numbers of users.

A second, bottom-up development of the 1970s would bring individualized computers to users, but handicapped with primitive features. Here we have the rise of computing devices cobbled together by and offered to electronics hobbyists and enthusiasts. While computers such as the Altair 8800 sat on a desktop, their interfaces were very rudimentary. To program the Altair, users had to flick a series of toggle switches for each program step. Hardly a model of interactivity, these machines had neither displays nor keyboards.

The first major effort to develop a broadly functional individualized computer with HCI-inspired interactivity and user-friendliness took place at the Xerox Palo Alto Research Center. PARC was established in 1970 to provide the technological undergirding for Xerox —the king of paper photocopying—to move into the "paperless" world of office computing. In the process, PARC became the premier draw for the country's best computer scientists—"like Disneyland for seven-year-olds".[10]

---

7 Association for Computing Machinery, Special Interest Group on Computer-Human Interaction. (1992). *Curricula for Human-Computer Interaction*, T. H. Hewitt, et al. New York, NY: ACM

8 These two other developments are covered by Ceruzzi, P. E. (1998). *A Modern History of Computing: 1945–1995*. Cambridge: The MIT Press.

9 Computer time-sharing, like the development of Internet and HCI technologies, was initiated by government program, specifically by ARPA.

10 Hiltzik, M. (1999). *Dealers of Lightning: Xerox PARC and the Dawn of the Computer Age*. New York, NY: Harper Business. 153.

PARC's strategy centered on what it called "distributed interactive computing", and was embodied in the Alto office computer. The Alto was "distributed" in that it was all about getting the computer up from the basement and on to individual desktops. It was "interactive" both in the sense that Altos were to be networked with one another, and in their design for real-time responsiveness and user-friendly approachability for individual users.[11]

The Alto was intended for use by one individual with stand-alone processing power and memory. It was configured much like today's PC. It had a high-resolution monitor that could display a full-sized 8.5 by 11-inch page, a keyboard, a three-button mouse, a removable hard disk cartridge, and ports for printer and Ethernet connections. What today we would call the computer's tower was an Alto cabinet about the size of a portable refrigerator that can be found in today's college dorm rooms.[12]

The Alto's monitor was a key feature of its user interface. Beyond its full-page dimensions, the Alto monitor trumped the standard-of-the-day "character generator" displays—which, in typewriter spirit, would produce fully formed text characters in a preset font and a preset color (usually green). Instead, the Alto could display high-resolution, user-defined fonts and graphics. Using now-standard "bit mapping" technology, the Alto could turn on and off half a million dots across its monitor—essentially turning everything on screen, including text, into pictures. Bit mapping also allowed the computer screen to display exactly what would be output from a printer—a feature that is known as "what you see is what you get" or WYSIWYG.

The Alto's user friendliness is now almost second nature, but was revolutionary in 1973.[13] Xerox designers began with the assumption that computer users were more interested in getting their work done than

11 Consistent with the HCI focus of this article, it does not elaborate on the networking aspects of the Alto.

12 A picture of the Alto can be seen at https://en.wikipedia.org/wiki/Xerox_Alto#/media/File:Xerox_Alto_mit_Rechner.JPG

13 Smith, D. C., Irby, C., Kimball, R., Verplank, B., and Harslem, E. (1982). "Designing the Star User Interface", *Byte* 7/4: 242–82; Johnson, J., Roberts, T. L., Verplank, W., Smith, D. C., Irby, C. H., Beard, M., and Mackey, K. (1989). "The Xerox Star: A Retrospective", *IEEE Computer* 22/9: 11–29; Miller, L. H., and Johnson, J. (1996). "The Xerox Star: An Influential User Interface Design", in *Human-Computer Interface Design: Success Stories, Emerging Methods, and Real-World Context,* ed. M. Rudisill, C. Lewis, and T. D. McKay. San Francisco: Morgan Kaufmann. 70–100

being interested in the computer itself. Therefore, an important Alto design principle was to make the computer as invisible and as intuitive as possible.

They chose a graphical user interface, or GUI, for personal computing.[14] A graphically simulated office served as a working metaphor. Images on screen represented the physical objects of an office—documents, folders, file cabinets, in-baskets, out-baskets, waste baskets, mailboxes, printers—all on an electronic rendition of a desktop. These images or icons could be manipulated with a mouse pointer to simulate the physical actions of opening, moving, filing, saving, deleting, etc. The goal was to make everything needed visible on screen and subject to direct manipulation rather than requiring indirect and memory-taxing (not for the computer, but for humans) keystroke combinations.[15]

More than a decade before the Mac and Microsoft GUIs, the Alto had windows to display document contents. Multiple windows could be open at the same time, overlapped, and resized; documents could integrate text and graphics; and the windows had title bars, mouse-clickable command buttons, and scroll bars. The Alto had a full slate of applications for word processing, graphics (including animation), printing, email, and playing music. The Alto operating system even allowed for task-switching—the capability to easily and quickly switch between programs.

Nearly two thousand Altos were built and used by government, industry and universities. A commercial version of the system, renamed the Xerox Star, was introduced in 1981—a full three and four years ahead of the Mac and Windows, respectively. The Star was marketed as "a new personal computer designed for offices intended for business professionals who create, analyze and distribute information".[16]

By current standards, the Xerox interface did suffer from certain limitations. Commands such as "open", "copy" and "move" required a combination of mouse manipulations and special function key

14 A picture of the Xerox GUI can be seen at https://www.computerhistory.org/revolution/input-output/14/347/1859

15 Ironically, analysts have pointed out that Xerox pushed the physical desktop metaphor too far—requiring cumbersome mouse manipulations where simple keyboard commands would have been sufficient (e.g., requiring that a document icon be moved over a printer icon instead of a simple key command for printing). Miller and Johnson. (1996). "The Xerox Star", 93.

16 Smith, et al. (1982). "Designing", 653.

operations. Resizing windows and moving icons also required mouse and function key combinations. Menu bars were at the top of each window, rather than a single set of menus at the top of the screen as a whole—resulting in the display of multiple and repetitive menu labels.

At the same time, and more significantly, the Alto suffered from being ahead of its time. While it was marketed as the dream machine for the "knowledge worker", such workers hardly existed in any real sense in 1981, let alone in 1973.[17] And even if the market existed, the Alto was far from a marketable product—with each machine costing over $16,000 to build. The resulting commercial demise of the Alto and Star is legend in the business world. A popular recounting of this disaster was titled *Fumbling the Future: How Xerox Invented, then Ignored, the First Personal Computer.*[18]

Alto's Offspring

> When Apple sued Microsoft in 1988 for stealing the "look and feel" of its Macintosh graphical display to use in Windows, Bill Gates' defense was essentially that both companies had stolen it from Xerox.[19]

Xerox "fumbling its future" does not mean that its technologies were commercial failures. Indeed, many of the PARC and Alto technologies were spectacularly commercialized—but just not by Xerox. For instance, outside of the HCI area, notable PARC alumni have made market blockbusters out of their Xerox work:

- Bob Metcalfe brought his Ethernet work to market by founding 3Com.

---

17 Baecker, R. M., and Buxton, W. A. S. (1987). "The Star, the Lisa, and the Macintosh", in *Readings in Human-Computer Interaction: A Multidisciplinary Approach*, ed. R. M. Baecker and W. A. S. Buxton. Los Altos, CA: Morgan Kaufmann. 649–52. Even the Xerox salesforce had difficulty "getting it." Upon the conclusion of an Alto demonstration, one brave soul asked, "Where's the click?" Hiltzik. (1999). *Dealers of Lightning*, 393.

18 Smith, D. K., and Alexander, R. C. (1988). *Fumbling the Future: How Xerox Invented, then Ignored, the First Personal Computer.* New York, NY: W. Morrow. For PARC's commercial fate, see also Hiltzik. (1999). *Dealers of Lightning*.

19 Hiltzik. (1999). *Dealers of Lightning*, xxv. Bill Gates has remarked: "Hey, Steve, just because you broke into Xerox's house before I did and took the TV doesn't mean I can't go in later and take the stereo." *MacWeek*, 14 March 1989, p. 1.

- Charles Geschke and John Warnock have commercialized the computer rendering of graphics for laser printing by co-founding Adobe Systems.
- Edwin Catmull and Alvy Ray Smith took their computer animation work first to Lucasfilm and then co-founded Pixar—making the movies *Star Trek, Toy Story* and *A Bug's Life* along the way.

When it comes to HCI technology, the Xerox legacy and progeny is even greater. In particular, the transfer of technology and, even more importantly, the transfer of people from PARC has been crucial to developments at both Apple and Microsoft.[20]

Apple's Day in the PARC

The flipside of Xerox's fumbling the PC's future is the Macintosh story. These two stories are, in fact, opposite sides of the same coin. The Macintosh story begins when Steve Jobs, Apple's CEO, takes a tour of Xerox PARC in December 1979.

In 1979, Apple was concerned it would soon lose its first mover advantage in the PC industry. Apple employee Jeff Raskin suggested that Xerox PARC held the keys for Apple's future. In the early 1970s, Raskin had spent considerable time at PARC while he was a visiting scholar at Stanford's Artificial Intelligence Laboratory.[21] After Apple arranged for Xerox to purchase $1 million dollars of Apple's skyrocketing shares, PARC agreed to show Apple the Alto.

The Alto team made not one, but two presentations—and not just to Jobs, but to a dozen of Apple's leading executives and programmers. Upon seeing the Alto, Apple software designer Bruce Daniels declared, "That's it—that's what we want to build".[22] While no "blueprints"

20 Other Alto-inspired GUI efforts not covered in this paper include those by Digital Research, IBM, and VisiCalc—efforts that did not match the success of the Mac or Windows.

21 Jeff Raskin, http://www2.h-net.msu.edu/~mac/lore2.html (website no longer active at time of publication); Linzmayer, O. W. (1999). *Apple Confidential: The Real Story of Apple Computer*. San Francisco, CA: No Starch Press. 52.

22 Rogers, M. (1983). "The Birth of the Lisa", *Personal Computing*, February, 89–94; Levy, S. (1994). *Insanely Great: The Life and Times of Macintosh*. New York, NY: Penguin Books (chapter 4).

were transferred, Apple came away from these sessions with a vision of the future of personal computing, and eventually key members of the PARC team.

The Xerox visit first inspired the development of the Lisa computer system—the Apple computer that immediately preceded the Macintosh. The Lisa was in development before the Xerox visit, but it was slated to have a non-graphical user interface and a non-bit mapped character-generator display. It also did not have a mouse. All this changed after the Xerox visit. In the words of Apple executive Larry Tesler, the Lisa was "completely redefined... only the code name, some of the hardware components, and a few of the staff members stayed the same". From the Alto, the Lisa would directly borrow the desktop metaphor, pop-up menus, overlapping windows, and scroll bars. After the 1981 introduction of the Xerox Star, the Lisa team made further changes to their GUI including the incorporation of desktop icons. On Apple's part, the Lisa would be the first to introduce the menu bar at the top to the screen (instead of menus atop each window), the one-button mouse, pull-down menus (point-and-drag mouse movement), and icons that could be dragged with the mouse and double-clicked to open.[23]

Akin to the fate of the Alto, the Lisa was also a commercial failure when it was introduced in January 1983. But its graphical user interface was transferred directly into the Macintosh. Indeed, PARC-savvy Jeff Raskin had begun development of the Mac in Spring 1979. After the Xerox visit, Raskin added the mouse to the Mac.[24] Beginning in January 1982, key members of the Alto-inspired Lisa team were transferred to the Macintosh division. Lisa software programs for word processing and graphics (LisaWrite, LisaDraw) would be converted to the Mac (MacWrite, MacDraw). The two product teams were completely merged in November 1983, and the Mac was introduced January 1984.[25]

Besides inspiration, the Xerox influence on Apple took on a second major form: the transfer of key PARC personnel to Apple. PARC alumni Alan Kay and Larry Tesler were two of the major coups for Apple. Alan Kay was PARC's chief evangelist for personal computing. In his

23 Linzmayer. (1999). *Apple Confidential*, 54–56; Miller and Johnson. (1996). "The Xerox Star", 94; Tesler, L. (1985). "The Legacy of the Lisa", *Macworld*, September, 17–22; Rogers. (1983). "The Birth of the Lisa".

24 Ceruzzi. (1998). *A Modern History of Computing*, 273.

25 Tesler. (1985). "The Legacy of the Lisa"; Linzmayer. (1999). *Apple Confidential*, 57–75.

1969 dissertation, Kay outlined a Dynabook—a computer the size of a notebook with an 8 by 10-inch flat screen, integrated keyboard, all of 2 inches thick, weighing in at two pounds. He had essentially envisioned today's laptop computer.

For the Alto, which he viewed as an "interim Dynabook", Kay led the development of its overlapping windows capability. The Alto not only allowed users to work in and see more than one window at a time, but it was the first system that allowed windows to be resized and moved—including over one another. This overlapping capability was a major advance over the pre-existing standard of tiled multiple windows that were fixed in place, and virtually expanded the working space of a computer monitor. Kay also inspired the Alto's pop-up menus—where the click of one of the mouse's buttons would cause menu options to appear on screen from which a command (e.g., paste) could be selected.[26]

In 1980, Kay became chief scientist at Atari, where he applied his HCI visions to interactive gaming. In 1984 he became an Apple Fellow, and inspired the company's successful PowerBook laptop computer line, and the Newton—the industry's first personal digital assistant (PDA) and forerunner to the Palm Pilot and other handheld computing devices. Since 1996, Kay has been a Disney Fellow and Vice President of Research and Development at the Walt Disney Company.

Larry Tesler preceded Alan Kay in moving from Xerox to Apple. Tesler worked in Kay's section of PARC, where he was dedicated to making computing more intelligible to the average user. For the Alto, Tesler designed Gypsy, a powerful word processing program that employed a graphical user interface with extensive icons and menus. In Gypsy, the mouse could point to and select blocks of text, whereas previous applications only used the mouse to position the cursor and called for keyboard commands for text selection. As an illustration of its user friendliness, Gypsy was the first program to replace commands for deleting a block of text and then placing it elsewhere with the simple labels of "cut" and "paste".[27]

In December 1979, Tesler was one of the two major presenters of the Alto to Steve Jobs and company. In July 1980 he would move to Apple. Tesler first headed up the Lisa user interface team, then helped

26 Hiltzik. (1999). *Dealers of Lightning*, 224–28.
27 *Ibid.*, 201–03, 207–10.

design the Macintosh including its one-button mouse, and then led the Newton PDA development team. He eventually rose to the position of Vice President and Chief Scientist before leaving Apple in 1998 to found a software startup.

Kay and Tesler were not alone in making the move from Palo Alto to Cupertino, where Apple is headquartered. For instance, Dan Ingalls—Kay's right hand man and co-author of one of the Alto's operating system—would follow Kay to Apple. Tom Malloy, who worked on word processing programs for the Alto, would go on to Apple and write the word processor for the Lisa (LisaWrite). Former Xerox PARCers Bruce Horn and Steve Capps would co-write the Macintosh Finder, its graphical file directory. Altogether, some fifteen PARC alumni would make the move to Apple.[28]

Microsoft's Window on Xerox

While the Xerox-Apple story is better known, Microsoft was also a major beneficiary of PARC's work. First, Microsoft Windows drew directly from the Alto-inspired Macintosh. Not unlike Jobs' 1979 visit to Xerox, Microsoft CEO Bill Gates visited Apple in 1981. There he saw a Mac prototype, and immediately thereafter began development of Microsoft's GUI, Windows. In 1982, Mac prototypes were delivered to Microsoft in order for the software company to develop Word and Excel for the new machine. At the same time, the prototypes were used to guide the development of Windows.

This Mac influence would show up even when Gates expressed dissatisfaction at Windows' early development. The Microsoft CEO would complain: "That's not what a Mac does. I want Mac on the PC, I want a Mac on the PC".[29]

To correct the situation, Gates transferred his resident "Macintosh wizard", Neil Konzen, to the Windows team. Having developed Microsoft's initial applications for the Mac, Konzen rewrote much of the

28 *Ibid.*, 214–15, 217–18. 316–17; Miller and Johnson. (1996). "The Xerox Star", 76; Linzmayer. (1999). *Apple Confidential*, 54

29 Campbell-Kelly, M., and Asprey, W. (1996). *Computer: A History of the Information Machine*. New York, NY: Basic Books; and Linzmayer. (1999). Apple Confidential, 136. For an image of the early Windows interface, see https://en.wikipedia.org/wiki/Windows_1.0#/media/File:Windows1.0.png

Windows code by emulating the Mac's internal structure. The results, in Konzen's words, were "Mac knockoffs". Even certain Mac system errors were carried over to the Windows platform.[30]

While the Mac served as a go-between for Xerox's influence on Microsoft, there were direct Xerox-Microsoft connections as well. To begin with, Gates got his tour of PARC and an Alto demonstration in 1980. Soon thereafter, Microsoft purchased a Xerox Star, the commercial version of the Alto. Microsoft did not intend to put the machine to operational use. Instead, in the words of one of Microsoft's leading programmers, "we just wanted everybody in the organization to get used to the desktop and to the mouse… we used it for education of the people".[31]

That programmer was Charles Simonyi, who embodies yet another type of Xerox influence on Microsoft: PARC alumni who moved from Palo Alto to Bellevue and Redmond, Washington, where Microsoft has been headquartered. At PARC, Simonyi co-wrote the Alto's "killer app"—Bravo, its first word processor. Bravo was the first program that could insert text in the middle of a document, display fancy typefaces, number pages, format odd margins, and print almost exactly what was on screen,[32] and it served as the basis for Tesler's Gypsy word processor.

Not unlike Larry Tesler's 1979 presentation to Steve Jobs and subsequent move to Apple, it was Simonyi who demonstrated the Alto to Gates in November 1980, and subsequently moved to Microsoft in February 1981. Joining as Microsoft's fortieth employee, Simonyi essentially "brought Microsoft Word with him".[33] According to Gates, Simonyi was specifically brought "on board to help us write applications that would eventually become very graphical",[34] and Simonyi characterized his mandate as to spread the "PARC virus" in

30 Wallace, J., and Erickson, J. (1992). *Hard Drive: Bill Gates and the Making of the Microsoft Empire*. New York, NY: Harper Business, 221, 273–74. While corporate rivalry has inhibited prominent personnel transfers between the two companies, some members of the Mac team would move on to Microsoft. For instance, Susan Kare did graphic design work for Windows 3.0 after designing the first icons, typefaces, and other graphics for the Macintosh. Linzmayer. (1999). *Apple Confidential*, 73.

31 Brockman, J. (1997). "Intentional Programming: A Talk with Charles Simonyi", *Edge Foundation*, 6 June, https://www.edge.org/conversation/charles_simonyi-intentional-programming

32 Hiltzik. (1999). *Dealers of Lightning*, 198–200, 358–60.

33 *Ibid.*, 395; see also Miller and Johnson. (1996). "The Xerox Star", 76.

34 Allison, D. (1993). "Bill Gates Interview", *Smithsonian Institution*.

Bellevue.[35] As director of advanced product development, Simonyi hired and managed the teams developing the entire suite of Microsoft applications, including Excel and PowerPoint, as well as Word. Simonyi is one of the "seven software samurai" to whom Gates turns for advice, and has been a member of the Executive Committee, the company's most senior-level decision-making team.[36] When Microsoft's Research Division was established in 1991, Simonyi became its Chief Architect.

While Bill Gates hired Simonyi to lead the development of the graphically-oriented Microsoft Office Suite, Gates also tapped a second PARC computer scientist to lead the development of the Windows operating system: Scott MacGregor. At PARC, MacGregor oversaw development of the Xerox Star's windowing system. In summer 1983, Gates recruited MacGregor to became head of the Windows engineering team. In MacGregor's words, "Microsoft was looking for somebody who had done this thing before. They didn't want to reinvent the wheel. That's why they went shopping at Xerox". In that shopping spree, Microsoft would hire others including Dan Lipkie, a Xerox programmer who would work on Word as well as Windows.[37]

Microsoft's Research Division is the site of Xerox's continuing influence on the software company. At the Microsoft labs, Simonyi has been joined by four other of PARC's leading lights: Chuck Thacker, Butler Lampson, Gary Starkweather, and Alvy Ray Smith.[38] Thacker, the lab's Director of Advanced Systems, was none other than the chief designer of the Xerox Alto. He championed the Alto's high-resolution bit-mapped display over the monochrome green monitors of the day, and he designed the Star's first central processor. Lampson, now a Microsoft Distinguished Engineer, first conceived of and started work on Alto's Bravo word processor—work that Simonyi would later pick up on. Lampson also designed the second central processor for the Star. Starkweather developed the Alto's laser printer and, in the process, launched a whole new industry: desktop publishing. Smith, a Microsoft Fellow till 1999, wrote the Alto's graphics program. Before joining Microsoft, Smith would design for Lucasfilm, co-found Pixar, and win

35 Brockman. (1997). "Intentional Programming".
36 Wallace and Erickson. (1992). *Hard Drive*, 369.
37 Wallace and Erickson. (1992). *Hard Drive*, 253–55.
38 Hiltzik. (1999). *Dealers of Lightning*, 397–98.

two technical Academy Awards. For Microsoft, these PARC alumni have worked on advanced programming and graphics, hand-held and wireless computing devices, and computer security.

Xerox's legacy extends, of course, well beyond Apple and Microsoft. Its current-day manifestations are innumerable, but two in particular merit mention here. Akin to Alan Kay's move from Xerox to Atari (before moving on the Apple), HCI advances have been a key driving force behind the interactive gaming industry, with applications ranging from game consoles and joy sticks to virtual reality environments. The World Wide Web, which began with text-based interfaces like Gopher, exploded in popularity only after user-friendly graphical user interfaces were employed by the Mosaic and Netscape web browsers. And members of the original Macintosh development team are about to give the open-source Linux operating system a major shot in the arm by applying a user-friendly GUI to the up-and-coming challenger to the Windows and Mac OS's.[39]

Even without a more comprehensive assessment of Xerox's legacy (a project worthy of an entire piece on its own), its import should not be in doubt. That import sets the proper perspective for considering the R&D that preceded and led into Xerox's effort—a task to which we now turn.

The Rest of the Story

> Silicon Valley. The World Wide Web. Wherever you look in the information age, Vannevar Bush was there first.[40]

> The Alto system grew from a vision of the possibilities inherent in computing: that computers can be used as tools to help people think and communicate. This vision began with Licklider's dream of man-computer symbiosis.[41]

39 Festa, P. (2000). "Apple, AOL Veterans Making Linux Easy", *CNET News.com*, 16 February; Markoff, J. (2000). "Old Apple Macintosh Team Aims to Put Linux on the Desktop", *New York Times*, 21 February; "The New Face of Open Source OS?"; Norr, H. (2000). "A Less Complex Linux", *San Francisco Chronicle*, 21 February.

40 Zachary, G. P. (1997). "The Godfather", *Wired*, November, 152.

41 Lampson, B. W. (1988). "Personal Distributed Computing: The Alto and Ethernet Software", in *A History of Personal Workstations*, ed. A. Goldberg. New York, NY: Addison-Wesley. 291–344, at 293.

> Ivan Sutherland's Sketchpad program is one of the most significant developments in human-computer communication.[42]

While the commercial ramifications of Xerox PARC's work cannot be over emphasized, Xerox was not the sole source of the HCI revolution. Just as Apple and Microsoft drew upon Xerox, so too was Xerox the beneficiary of the prior work of others.

At this point, the political-economic balance of this account shifts. While the narrative thus far has been heavily business-oriented, what follows concerns more of a political dynamic. Most of the innovations and people discussed thus far were in fact influenced by government-sponsored initiatives. Those initiatives began with Vannevar Bush, J. C. R. Licklider, and Ivan Sutherland.

Vannevar Bush

The *Online Encyclopedia Britannica* entry for "graphical user interface" reads as follows: "There was no one inventor of the GUI; it evolved with the help of a series of innovators, each improving on a predecessor's work. The first theorist was Vannevar Bush".[43] The source of this attribution was Bush's vision of a "memex",

> in which an individual stores all his books, records, and communications… It consists of a desk… On the top are slanting translucent screens, on which material can be projected for convenient reading. There is a keyboard, and sets of buttons and levers… if the user inserted 5000 pages of material a day it would take him hundreds of years to fill the repository… If the user wishes to consult a certain book, he taps its code on the keyboard, and the title page of the book promptly appears before him, projected onto one of his viewing positions… [with] one of the levers to the right he runs through the book before him, each page in turn being projected at a speed which just allows a recognizing glance at each. If he deflects it further to the right, he steps through the book 10 pages at a time; still further at 100 pages at a time. Deflection to the left gives him

42 Norberg, A. L., and O'Neill, J. E. (1996). *Transforming Computer Technology: Information Processing for the Pentagon, 1962–86*. Baltimore, MD: Johns Hopkins University Press, 36.

43 Levy, S. (1988). "Graphical User Interface", *Encyclopedia Britannica Online*, https://www.britannica.com/technology/graphical-user-interface. The following quotations are taken from Bush, V. (1945). "As We May Think", *Atlantic Monthly*, July.

> the same control backwards... he can leave one item in position while he calls up another.

Bush went on to consider the memex's applications:

> The lawyer has at his touch the associated opinions and decisions of his whole experience, and of the experience of friends and authorities. The patent attorney has on call the millions of issued patents, with familiar trails to every point of his client's interest. The physician, puzzled by its patient's reactions, strikes the trail established in studying an earlier similar case, and runs rapidly through analogous case histories, with side references to the classics for the pertinent anatomy and histology. The chemist, struggling with the synthesis of an organic compound, has all the chemical literature before him in his laboratory, with trails following the analogies of compounds, and side trails to their physical and chemical behavior.

This vision of the memex is widely recognized in government, industry, and academic circles as the *first* major articulation of the modern personal computer, including hypertext and internet links. Xerox-Apple alumnus Alan Kay observes that "Bush's vision of a hyperlinked 10,000 volume library in a desk had a great impact on the development of personal computing".[44] Tim Berners-Lee, inventor of the World Wide Web, notes that "to a large part we have Memexes on our desks today".[45]

The memex was not the product of a science fiction writer conjuring up visions of the future; nor an entrepreneur toiling away on a garage work bench; nor an industrial researcher supported by a well-financed corporate laboratory. Instead, Vannevar Bush was a government official. More specifically, Vannevar Bush was the Director of Office of Scientific Research and Development—the chief science advisor to the President of the United States. When Bush envisioned the memex, the President was Harry Truman; the date July 1945.

Between 1941 and 1947, Vannevar Bush served as science advisor to both Franklin Roosevelt and Harry Truman. His greatest contribution

44 Kay, A. (1995). *Simex: The Neglected Part of Bush's Vision.* Presentation at "As We May Think—A Celebration of Vannevar Bush's 1945 Vision", MIT Department of Electrical Engineering and Computer Science. 12–13 October, http://dougengelbart.org/content/view/258/000/

45 Berners-Lee, T. (1995). *Hypertext and Our Collective Destiny.* Presentation at "As We May Think—A Celebration of Vannevar Bush's 1945 Vision", MIT Department of Electrical Engineering and Computer Science. 12–13 October, https://www.w3.org/Talks/9510_Bush/Talk.html

in office is highly debatable in both the best and worst of senses. First, he organized the 6000-strong scientific enterprise to help prosecute the U.S. war effort. While he was not physically in the sands of New Mexico, Bush oversaw the Manhattan Project to create the first atomic bomb. Second, he established the structure of the country's postwar science and technology effort—including the prominent roles played by military R&D, the National Science Foundation, and university-based research.[46]

Then there is the memex. Bush's vision inspired R&D efforts throughout government, industry, and academia. The lead player in this R&D was the Advanced Research Projects Agency (ARPA).

J. C. R. Licklider

Since its inception in 1958, ARPA has supported both the development of military-specific weapons technologies, and more generic technologies with the potential for military application. The former includes ballistic missile defense and tactical anti-tank weapons technologies, and even the M-16 rifle. The latter includes R&D in new materials, novel energy sources, and biomedical technologies, as well as computer science.

ARPA began its computer science work in 1962, when it established its Information Processing Techniques Office (IPTO) as one of a half dozen technology-specific offices within the agency.[47] Starting off with a $7 million annual budget, IPTO's funding was larger than the computer research budgets of the rest of the government combined. Over the next eight years, the IPTO budget would more than quadruple.

Most of IPTO's funding went to university research. It is hard to imagine now, but before 1962 no formal university computer science programs existed. ARPA's IPTO grants were essential in establishing the country's first graduate programs in computer science, including those at MIT, Stanford, Berkeley, Utah, and Carnegie Mellon.[48]

These and other ARPA-funded programs will be returned to below. First, however, we turn our attention to the ARPA official who served

46 Zachary, G. P. (1997). *Endless Frontier: Vannevar Bush, Engineer of the American Century*. New York: Free Press.

47 IPTO has undergone a number of name changes over the past 40 years, and is currently named the Information Innovation Office.

48 Norberg and O'Neill. (1996). *Transforming Computer Technology*.

as the guiding light behind this effort—J. C. R. Licklider. As quoted above by Xerox PARCer Butler Lampson, the Alto would grow out of Licklider's vision.

J. C. R. Licklider was IPTO's inaugural director from 1962 to 1964. Earlier, as an MIT professor, Licklider "got fired up about the idea Vannevar Bush had mentioned in 1945, the concept of a new kind of library to fit the world's new knowledge system". Licklider's 1959 book, *Libraries of the Future,* was not only dedicated to Bush but expanded upon the memex concept. When he moved on to ARPA, he brought with him his "religious conversion" to interactive computing.[49]

From ARPA, Licklider galvanized the computing research community around two pathbreaking concepts. Given the first one—"the intergalactic network"—it is almost understandable to overlook the second. The intergalactic network was "the first concrete proposal for establishing a geographically distributed network of computers".[50] As initiated by Licklider, the network would first take the form of computer time-sharing links and later transform into the ARPANET/Internet.

As consequential as this first concept has been, the second—"man-computer symbiosis"—is arguably just as profound. Licklider came to computing not as a computer scientist, but as an academic psychologist. His interest was in how computers could contribute to, rather than replace, human cognitive processes. He was concerned that the rudimentary user interfaces of computers of the 1950s hindered the technology's true potential. To realize that potential, he called for computing advances in real-time processing and interactivity.

He called for advances in the computer's outward face to its user—its display—and in how users input instructions into the computer, including via graphical input and automatic speech recognition. In calling for a "much tighter coupling between man and machine", Licklider sought to realize "interaction with a computer in the same way that you think with a colleague whose competence supplements you own".[51]

---

49 Rheingold, H. (1985). *Tools for Thought: The People and Ideas of the Next Computer Revolution*. New York, NY: Simon & Schuster, http://www.rheingold.com/texts/tft/ (chapter 7).

50 Campbell-Kelly and Aspray. (1996). Computer, 288.

51 Licklider, J. C. R. (1960). "Man-Computer Symbiosis", *IRE Transactions on Human Factors in Electronics* 1: 4–11, https://doi.org/10.1109/thfe2.1960.4503259

These are all matters of human-computer interface, and Licklider defined the HCI agenda for decades to come. ARPA-supported research universities not only took part in building Licklider's "intergalactic network", but they launched major HCI initiatives as well.[52]

Ivan Sutherland

When Licklider prepared to leave ARPA in 1964, he selected Ivan Sutherland to replace him as IPTO director. Sutherland was one of the first researchers to take up Licklider's HCI challenge. His 1962 PhD project at MIT, called Sketchpad, was the first-ever computer graphics program where the user could make drawings on screen interactively.

Sketchpad is widely recognized as the seminal program that started off the entire field of computer graphics.[53] But Sutherland's immediate motivation was to advance human-computer interactivity. Indeed, the subtitle of his project was "A Man-Machine Graphical Communication System".[54] Three features made Sketchpad, as quoted above, "one of the most significant developments in human-computer communication".

First, Sketchpad was one of the first computers with a monitor, and a user's work would immediately be represented on screen. This form of interactivity is now easy to take for granted, but before Sketchpad, users had to wait for a print-out in order to see their work.[55]

Second, Sketchpad was one of the first computers to use a pointing device. A hand-held "light pen" was employed to make drawings. The pen would make physical contact with the screen and its "light" would be picked up by the computer. Moving the pen would draw lines on screen in real-time. The pen could also be used to grab-and-drag images as well as rotate, expand or contract an image. A major user interface break-through, before Sketchpad users had to express object geometry

52 More on Licklider can be found in Waldrop, M. M. (2001). *The Dream Machine: J. C. R. Licklider and the Revolution that Made Computing Personal*. New York, NY: Viking Press.

53 Wolfe, Roaslee, ed. (1998). *Seminal Graphics: Pioneering Efforts that Shaped the Field*. New York, NY: Association for Computing Machinery.

54 Sutherland, I. E. (1963). "Sketchpad: A Man-Machine Graphical Communication System", *Proceedings of the AFIPS Spring Joint Computer Conference* 23: 329–46. See also Norberg and O'Neill. (1996). *Transforming Computer Technology*, 125–28.

55 Wolfe. (1998). *Seminal Graphics*.

by typing coordinates on a keyboard. The light pen would later lead to today's mouse.

Third, Sketchpad was the first system with a rudimentary windowing system. The Sketchpad screen could be split to produce two work areas or windows. One section could, for example, display a close-up view of an object in the other section.[56]

The Sketchpad project was sponsored by the Army, Navy, and Air Force. This funding is a reminder that government agencies other than ARPA have also supported HCI technology. In this particular case, the three military services provided support to Sutherland before IPTO was even established.

Licklider hired Sutherland to explicitly carry on IPTO's HCI work. As IPTO director, Sutherland would fund major university programs in computer graphics. Besides fueling the burgeoning field of computer-generated images, this research would provide the foundation for computers with "graphical" user interfaces, "picture" icons, and high-resolution bit-mapped displays. Such displays, interfaces, and icons—along with Sketchpad-derived windows and pointing devices—would be incorporated into the Xerox Alto.

Xerox's ARPA Brats

> Xerox PARC was set up near the Stanford campus. For the next ten years the ARPA dream took up residence at PARC.[57]

A veritable "ARPA Army"—a phrase coined at PARC—would fill the ranks of computer scientists at the Xerox. This influx into Xerox was led not by a researcher from an ARPA-supported university, but by an official direct from ARPA itself: Robert Taylor.

Robert Taylor

J.C.R. Licklider not only selected Ivan Sutherland to replace him as director of IPTO, but chose Robert Taylor to be associate director. When

56 Perry, T., and Voelcker, J. (1989). "Of Mice and Menus: Designing the User-Friendly Interface", *IEEE Spectrum* 27/9: 46–51, at 48–49.

57 Rose, F. (1989). *West of Eden: The End of Innocence at Apple Computer*. New York, NY: Viking Penguin, 45.

Sutherland finished his term as director in 1966, Taylor took his place, serving through to 1969.

Robert Taylor "heartily subscribed" to Licklider's vision of computing even before joining ARPA.[58] In his first year in office, he advanced Licklider's "intergalactic network", transforming it from a computing time-sharing paradigm to a decentralized packet-switching network, the ARPANET. While ARPANET's construction would begin under Taylor's successor at IPTO, Lawrence Roberts, the network's design was completed under Taylor. Taylor was also a true believer in Licklider's theme of "man-computer symbiosis". Taylor held a NASA research post in HCI just prior to joining ARPA, and distributed interactive computing became his "sacred cause" as director of IPTO.[59] As described in a 1968 paper, co-authored with Licklider, Taylor envisioned a computer for each individual user; each with a large television monitor, a keyboard, and "electronic pointer controllers called 'mice' [that could] control the movements of a tracking pointer on the TV screen".[60] This vision grew directly out of the memex of Vannevar Bush. It also presaged Xerox PARC's Alto.

When Xerox started forming its PARC facility in 1970, one of the first people they tapped was Robert Taylor. As quoted above, "for the next ten years the ARPA dream took up residence at PARC". Taylor has been called "the impresario of computer science at Xerox PARC".[61] Taylor exercised this influence as head of the Computer Science Laboratory (CSL)—the largest of PARC's four internal labs. It was CSL that would become the mecca for fifty of the country's top computer scientists.

In the Spring of 1971, Taylor set CSL's agenda by proposing that it build the machine he had written about in 1968. Two years later, the Alto realized his vision. While his researchers would undertake the Alto's design and development, the general concept and the "Alto" name came from Taylor.[62]

---

58 Norberg and O'Neill. (1996). *Transforming Computer Technology*, 29.

59 Hiltzik. (1999). *Dealers of Lightning*, 19.

60 Licklider, J. C. R., and Taylor, R. (1968). "The Computer as a Communications Device", *Science and Technology* 76: 21–31.

61 Hiltzik. (1999). *Dealers of Lightning*, 3.

62 Smith, D. K., and Alexander, R. C. (1988). *Fumbling the Future: How Xerox Invented, Then Ignored, The First Personal Computer.* New York, NY: W. Morrow, 170.

Besides setting the lab's agenda, Taylor hired its staff. He did so not by merely reading resumes. Instead, he chose his people from ARPA-funded research centers. Indeed, he chose researchers whom he, Licklider, and Sutherland had directly and personally supported through IPTO.

ARPA's Army

Stanford, Berkeley, Utah, and SRI were the major programs that Taylor drew from. Most of these researchers—and their exploits at Xerox, Apple, and/or Microsoft—have already been noted in the first half of this article. Here we reveal their university and ARPA pedigrees. To help keep the names and affiliations straight, Figure 6-1 graphically displays some of these people and places. Stanford's Artificial Intelligence Laboratory was established in 1962 with ARPA funding. Indeed, into the 1970s, most, if not all, of the computing research conducted at Stanford would be supported by ARPA—as would be the case at Berkeley, Carnegie Mellon, Illinois, MIT, UCLA, and Utah.[63] Out of Stanford, Taylor hired Larry Tesler and Charles Simonyi, who would later go on to Apple and Microsoft fame, respectively.

In 1963, IPTO began supporting Project Genie at Berkeley, a small-scale computer time-sharing project. Charles Thacker and Butler Lampson, as well as Simonyi from Stanford, would first come together to work on this project and its commercial Berkeley Computer Corporation spinoff.[64] While burdened by the main-frame paradigm, this experience sparked their pursuit of interactive computing. The three were considered among the country's top programmers, and Taylor hired them as a group to join PARC in 1970. Taylor would hire others from Berkeley including Peter Deutsch, Ed Fiala, Jim Mitchell, and Dick Shoup. Thacker, Lampson, and Simonyi would all end up at Microsoft.

One of the Berkeley faculty members that directed Project Genie, David Evans, would not go to Xerox. Instead, he remained in academia training students, many of whom would make the trek to PARC. This is the Utah connection, where Evans became head of the computer

63 Norberg and O'Neill. (1996). *Transforming Computer Technology*, 290.

64 *Ibid.*, 102–03; Hiltzik. (1999). *Dealers of Lightning*, 18–19, 68–78.

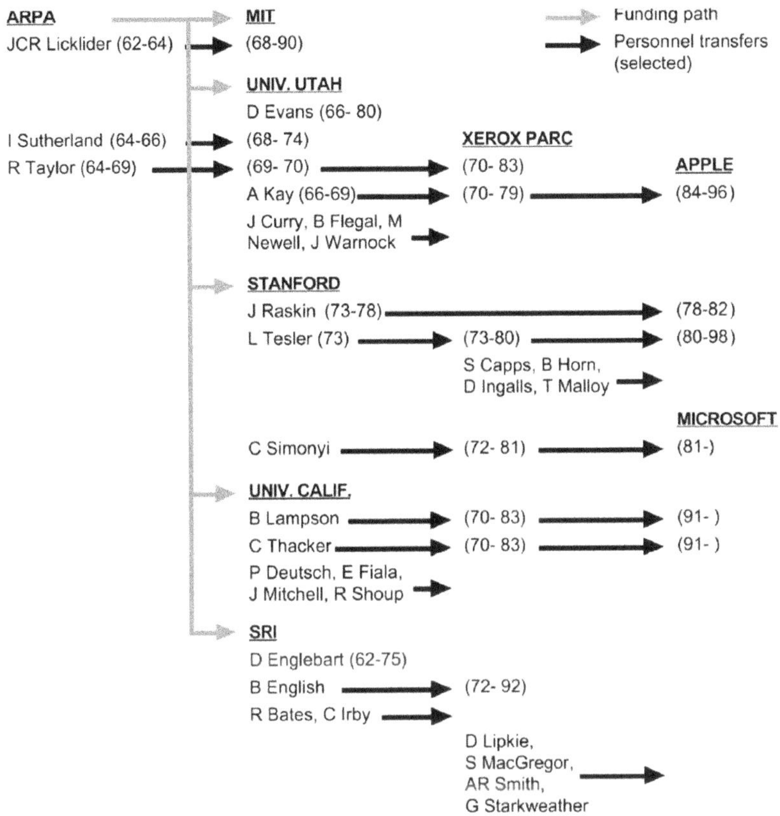

Fig. 6-1 From ARPA to Windows. (Figure prepared by the author.)

science department in 1966. As IPTO director, Taylor would make a $5 million award to Evans to transform Utah into a center of excellence for computer graphics.[65] Ivan Sutherland, Taylor's predecessor and creator of the Sketchpad program, would be on the Utah faculty from 1968 to 1973. Taylor himself would spend a year at Utah between his ARPA and PARC tenures.

Taylor would bring to CSL many Utah students including Jim Curry, Bob Flegal, Martin Newell, and John Warnock. But the key hire for the Alto and HCI at Xerox was Alan Kay in 1972. Kay came to Utah in 1966 as one of Evans' first graduate students. At their very first meeting,

65 Norberg and O'Neill. (1996). *Transforming Computer Technology*, 137–43.

Evans assigned the new student Sutherland's Sketchpad dissertation. In a reaction any professor would die for, Kay has described his reading of Sketchpad as "seeing a glimpse of heaven".[66] Kay would try to capture a bit of that heaven first in his own dissertation, then at PARC, and later at Apple.

One of the major ARPA-supported research centers that has yet to be mentioned, and that has made major contributions to the PC industry, is the think tank Stanford Research Institute (SRI). SRI was the home of computer scientist Douglas Engelbart from 1957 to 1975. Engelbart was inspired by Licklider's notion of augmenting (rather than replacing) human intellect via "man-computer symbiosis". Indeed, Engelbart's lab at SRI was called the Augmentation Research Center.[67]

In designing a system to augment human intelligence, Engelbart used Vannevar Bush's memex concept as an ideal type.[68] Over a two-decade period, Engelbart would develop a computerized personal information storage and retrieval system to replace paper and hardcopy filing systems. Called NLS (for oN Line System), the system was not a personal computer, but rather a networked workstation. It had a large video monitor and input devices to manipulate information on screen, but it was all cabled into a remote mainframe computer.

Still, NLS made two major contributions to "man-computer symbiosis" and HCI. First, it advanced windowing capabilities by being able to divide the display screen into four work areas—an improvement over the split-screen capability of Ivan Sutherland's Sketchpad system. The user could now easily shift work from one window to another.[69] Second, NLS introduced a new pointing device to move a cursor within and between document windows. Engelbart conducted a series of studies comparing various pointing devices including Sketchpad's light pen, track balls, joysticks, and even a knee-switch under the desktop.[70] What he decided upon was a device that "stays put when your hand leaves it do something else (type or move a paper) and reaccessing [it]

66 Hiltzik. (1999). *Dealers of Lightning*, 91.

67 Englebart, C. (1994). "Biographical Sketch: Douglas Carl Engelbart", *Bootstrap Institute*, http://www.dougengelbart.org/content/view/88/45/; Hiltzik. (1999). *Dealers of Lightning*, 63.

68 Rheingold. (1985). *Tools for Thought*, 260.

69 Oerrt and Voelcker, "Mice and Menus", 49.

70 English, W. K., Engelbart, D. C., and Melvyn, A. B. (1967). "Display-Selection Techniques for Text Manipulation", IEEE *Transactions on Human Factors in Electronics* 8/1: 5–15.

proves quick and free from fumbling... and it doesn't require a special and hard-to-move work surface".[71]

This device is, of course, the mouse. Initially the size of a brick and carved out of a block of wood, the underside of Engelbart's mouse had two wheels positioned at right angles to one another that could digitally track and convey its position to the computer.[72] While the wheels would be replaced with a ball, the computer mouse was not invented by Xerox in 1973 let alone Apple in 1984. It was created by Engelbart in 1964.

The system described earlier in Robert Taylor's 1968 paper—a large video screen, keyboard, and a mouse—was Engelbart's NLS. Not only did Taylor properly cite Engelbart in that paper, but Engelbart had three major connections to Taylor and ARPA. To begin with, Taylor—while at NASA—provided initial funding for Engelbart's project. The Air Force did as well. Both NASA and the Air Force were interested in how operators in their command centers could best interface with their computers.[73] As in the case of Sutherland's Sketchpad project, Engelbart received support from these other organizations before IPTO was even established.

Then, with IPTO's establishment in 1962, "Douglas Engelbart was one of the first persons to apply for funding".[74] Not only did he gain IPTO funding, the support would significantly rise during Taylor's tenure. ARPA funding would continue until 1975, and Engelbart's research team would expand from two to nearly fifty. In 1968, ARPA and NASA co-sponsored a major presentation of the NLS to the public that amazed the wider computing research community.

Then there is the Xerox connection. In the words of Butler Lampson, the NLS "made a profound impression on many of the people who later developed the Alto".[75] Both the mouse and windows were directly incorporated from NLS into the Xerox computer.

Moreover, in what became a running theme, Taylor hired key members of the NLS team to come to PARC. Akin to David Evans remaining at Utah, Engelbart would not himself make the move to

71 Levy, S. (1994). *Insanely Great: The Life and Times of Macintosh.* New York, NY: Penguin Books, 41.

72 A picture of the Engelbart mouse can be seen at https://www.computerhistory.org/revolution/input-output/14/350

73 Norberg and O'Neill. (1996). *Transforming Computer Technology*, 131.

74 Ceruzzi. (1998). *A Modern History of Computing*, 260.

75 Lampson. (1988). "Personal Distributed Computing", 294.

Xerox. But Taylor did hire Engelbart's right hand man, Bill English. English was NLS's hardware expert and had done the detailed design work on the mouse. Taylor offered English that chance to "reproduce NLS, or something like it, at PARC".[76]

Another member of the NLS team, Roger Bates, would help develop the Alto's high-resolution bit-mapped display. NLS alumnus Charles Irby would help design the user interface for the Xerox Star. Altogether, a dozen of Engelbart's team would make the move to PARC.[77] Given these hires from SRI and the universities, ARPA-supported research would leave an "indelible stamp on almost every major innovation to emerge from PARC".[78]

Beyond ARPA's influence on Xerox, it is difficult not to mention other major computer scientists that have been supported by IPTO—including Wesley Clark, Lynn Conway, Michael Dertouzos, Edward Feigenbaum, John Hennessy, Daniel Hillis, John McCarthy, Carver Mead, Marvin Minsky, Alan Newell, David Patterson, and Raj Reddy. Then there are those that have left their mark in the commercial world. We have already mentioned Bob Metcalfe of 3COM, John Warnock of Adobe Systems, and Edwin Catmull of Lucasfilm and Pixar—all of whom came out of PARC. We can now note their earlier ARPA-backing at Harvard and Utah (last two). To this list we can add Nolan Bushnell (Utah), founder of Atari; Jim Clark (Utah), co-founder of Silicon Graphics and Netscape; and Bill Joy (Berkeley), co-founder of Sun Microsystems.[79]

But our focus here has been on HCI-specific ARPA-supported researchers who made their way to Xerox PARC *and* then contributed to or influenced developments at Apple or Microsoft. Even with these restrictors, the ARPA reach is substantial. "ARPA does Windows" is more than a catchphrase.

---

76 Hiltzik. (1999). *Dealers of Lightning*, 67; Ceruzzi. (1998). *A Modern History of Computing*, 260.

77 Hiltzik. (1999). *Dealers of Lightning*, 173.

78 *Ibid.*, 67.

79 See Computing Research Association. (1997). *Computing Research: A National Investment for Leadership in the 21st Century*. Washington, DC: Computing Research Association; National Research Council. (1995). *Evolving the High Performance Computing and Communications Initiative to Support the Nation's Information Infrastructure*. Washington, DC: National Academy Press (chapter 1); and Norberg and O'Neill. (1996). *Transforming Computer Technology*.

Windows on the Future

The story has now come full circle. Vannevar Bush's extraordinary vision is followed up by ARPA's Licklider, Sutherland, and Taylor. They sponsor the Stanfords, Berkeleys, Utahs and SRIs. Xerox draws upon this research and the researchers (plus Taylor). Then Apple and Microsoft commercialize Xerox's work. The rest, as they say, is history.

But the PC revolution does not stop with Windows. And ARPA's hand in matters HCI is not confined just to decades past. Indeed, ARPA's and other direct government support for further advances in personal computing continues to this day.

A high-level conference sponsored by Intel in March 2000 illustrates this continuing influence. Five hundred of the world's leading computer scientists came together for Intel's Computing Continuum Conference to "define the next era of computing, communication, and interaction in the digital world".[80] Three dozen "visionaries" made presentations on topics ranging from artificial intelligence to ubiquitous networked computing. Table 6-1 lists the five presentations that were organized for a panel explicitly on HCI.

The primary funding sponsors of this leading-edge HCI research are identified. Seven sponsors are government agencies (including the European Union), and three are industry. Significantly, DARPA is a sponsor in four of the five cases; followed by National Science Foundation (NSF) sponsorship of three.

The DARPA funding is part of its Human Computer Interaction Program. Altogether eleven universities, companies, and government labs have been part of this effort. The NSF funding—under its own Human Computer Interaction Program—went to thirty-four universities by the time this chapter was originally published in 2001. Research being undertaken includes work on three-dimensional graphical user interfaces; intelligent animated life-like computer characters capable of natural face-to-face conversational interaction; and an "intelligent room" embedded with vision, speech understanding, multimedia, and networked interactive computing systems.

80 Intel Computing Continuum Conference. (2000). 15–17 March, https://www.intel.com/pressroom/archive/releases/2000/cn031500a.htm

Table 6-1 Human Interface Panel, March 2000. (Table prepared by the author.)

	Research Group	Major Sponsors[1]
Ronald Cole University of Colorado	Center for spoken Language Research	National Science Foundation Office of Naval Research DARPA Intel
Patrick Hanrahan Stanford University	Computer Graphics Laboratory	Department of Energy Intel
Raj Reddy Carnegie Mellon	School of Computer Science/Speech Group	DARPA
Ben Shneiderman University of Maryland	Human–Computer Interaction Laboratory	National Science Foundation[2] NASA Bureau of the Census European Union DARPA
Victor Zue MIT	Laboratory for Computer Science/Spoken Language Systems Group	DARPA National Science Foundation Information Technology Research Institute NTT

[1] Major sponsors as identified by presenter; listed in order of importance.
[2] Listed in order of projects presented.

The eleven DARPA-sponsored projects include important industry connections. In addition to major co-sponsors such as Intel, NTT, and the Information Technology Research Institute, lower-level funding has come from the likes of Acer, America Online, Apple, Discovery Communications, GE, Hewlett-Packard, Hughes Research, NCR, NEC, Nokia, Philips, Sony, and Toyota. DARPA-sponsored students from these on-going HCI projects have gone on to take positions with these companies as well as with AT&T Research, Bell Labs, Compaq, Dragon Systems, General Magic, IBM Research, Lucent, Microsoft Research, Sarnoff, and Silicon Graphics.

The names of the researchers have changed and the number of funded universities has grown since the 1960s. While the results will be hard to stack up to those of the earlier period, no matter what the results the government influence remains pervasive.

Conclusions

> Government funding of advanced human-computer interaction technologies built the intellectual capital and trained the research teams for pioneer systems that, over a period of 25 years, revolutionized how people interact with computers.[81]

In contrast to the thrust of this argument are sentiments such as that quoted at the top of this article. Bill Gates is not alone in holding this view. His is the mainstream perspective on the development of the PC industry; indeed, of the development of virtually the entire "new economy". Case in point is Tim Draper, who personally provided startup capital for Hotmail (the world's largest email provider), Four11 (internet white pages directory), and *Upside* (one of the most widely read business technology magazines).

In 1997, Draper penned an editorial that articulated much of Silicon Valley's attitude towards the government—an attitude legitimated by the publication in which it appeared, the *Wall Street Journal*. Draper starts by telling us he "earned an MBA from Harvard and an electrical engineering degree from Stanford. I worked at Hewlett-Packard and Alex. Brown before starting a venture capital firm. My favorite periodicals are *Upside* and the *Red Herring*, not the *Washington Post* or the *Weekly Standard*. In my free time I surf the Net; I don't watch Capital Gang or C-SPAN". Writing under the title, "Silicon Valley to Washington—Ignore us, Please", Draper then shares his view of Washington:

> We in the high tech business have reason to feel good... Our industry now accounts for 11 percent of gross domestic product and a quarter of U.S. manufacturing output. We employ more than 4.2 million people, who earn almost double the average salary of manufacturing workers. Our industry is the biggest reason the U.S. has the world's most competitive economy. We ought to count our blessings that most of our industry is 2,500 miles from Washington and that most bureaucrats either fear, don't care about or don't understand technology. And we've done just fine without their help... Washington doesn't understand my business, [and]

81 Card, S. K. (1996). "Pioneers and Settlers: Methods Used in Successful User Interface Design", in *Human-Computer Interface Design: Success Stories, Emerging Methods, and Real-World Context*, ed. M. Rudisill, C. Lewis, P. G. Polson, and T. McKay. San Francisco: Morgan Kaufmann Publishers. 122–69, at 164.

> I'd like it to stay that way. The fact is that politicians and government bureaucrats can't help us; they can only get in the way... If the U.S. wants more good jobs, better lives, and a stronger economy, the best thing lobbyists, bureaucrats and politicians can do is *leave us alone*.[82]

"We've done fine without their help" and "they can only get in the way" are typical of how many "new economy" participants view the development of their own industry. This view permeates coverage in *Fortune* and *Business Week* and the general media. Even the highly regarded six-hour PBS documentary on the history of the PC, *Triumph of the Nerds*, overlooks the government connection.[83] In contrast, we have observations such as those of Stuart Card, quoted at the beginning of this section. Card might be in a position to know. He has been with Xerox PARC for twenty-five years, and currently heads its User Interface Research Group. His comment comes from a fifty-page technical paper he compiled on the historical development of HCI.

Card is not alone. Dr. Brad Myers, Senior Research Scientist at Carnegie Mellon's Human Computer Interaction Institute, warns against "the mistaken impression that much of the important work in Human-Computer Interaction occurred in industry".[84] Instead, as computer historians Martin Campbell-Kelly and William Aspray have written, "almost all the ideas in the modern computer interface emanated from laboratories funded by ARPA's Information Processing Techniques Office".[85] Even one of Silicon Valley's own—Charles Geschke, President and co-founder of Adobe Systems—acknowledges that it was ARPA support that "has allowed the current PC industry to flourish".[86]

Uncovering this political-economic link provides an important corrective to the popular lore surrounding the origins of the personal computer. This article "brings the state" back into the PC realm of apparent market purity. Government support for the development of the PC should take its place on a list that includes the Internet, the

---

82 Draper, T. (1997). "Silicon Valley to Washington—Ignore us, Please", *Wall Street Journal*, 4 March, emphasis in original.

83 Myers, B. A. (1968). "A Brief History of Human Computer Interaction Technology", *ACM Interactions* 5/2, 44–54.

84 Myers. (1968). "A Brief History".

85 Campbell-Kelly and Aspray. (1996). *Computer*, 266.

86 Geschke, C. (1999). "The U.S. Environment for Venture Capital and Technology-Based Start-Ups", in *Harnessing Science and Technology for America's Economic Future: National and Regional Priorities*, ed. National Research Council. Washington, DC: National Academy Press.

computer chip, and the PC's bigger brother, the mainframe.[87] The federal government's role in supporting the development of the Internet is now widely acknowledged. The ARPANET of 1969 was followed by the NSFNET of 1985. This support extends to the government's on-going Next Generation Internet project.

The government's support of the chip industry goes back to military R&D funding in the 1940s and procurements into the 1960s by the Air Force and NASA of 100 percent of the industry's production. Government support of the chip industry would continue into the 1980s and 1990s with the Very High Speed Integrated Circuit Program and SEMATECH consortium.

And, of course, Defense and Energy Department support of the mainframe and supercomputer industry stretches from the ENIAC of 1945, IBM's 1953 Stretch computer, the SAGE computer in 1954, Cray's first supercomputer in 1976, the 1996 Intel teraflop machine, and even IBM's 1997 chess champion Deep Blue. This kind of support continues today with government programs such as the High-Performance Computing and Communication Initiative and the Accelerated Strategic Computing Initiative.

The Internet, the computer chip, the mainframe, and the PC: together, these four innovations define the information technology revolution that has fueled the new economy of the twenty-first century. No doubt university and corporate researchers, as well as private entrepreneurs, have made this revolution possible. But popular mythology, corporate P.R., and political ideology aside, credit also goes to government.

References

Allison, D. (1993). "Bill Gates Interview", *Smithsonian Institution.*

Association for Computing Machinery, Special Interest Group on Computer-Human Interaction. (1992). *Curricula for Human-Computer Interaction*, T. H. Hewett, et al. New York, NY: ACM, https://dl.acm.org/citation.cfm?id=2594128

87 Regarding some of these other areas of government support, see Fong, G. R. 2000. "Breaking New Ground or Breaking the Rules: Strategic Reorientation in U.S. Industrial Policy", *International Security* 25/2: 152–86. The nature of government support across these various cases differs of course, and this article develops a typology that clarifies some of the differences. The Windows case would fall in this typology's "by-product" category.

Baecker, R. M., and Buxton, W. A. S. (1987). "The Star, the Lisa, and the Macintosh", in *Readings in Human-Computer Interaction: A Multidisciplinary Approach,* ed. R. M. Baecker and W. A. S. Buxton. Los Altos, CA: Morgan Kaufmann. 649–52.

Berners-Lee, T. (1995). *Hypertext and Our Collective Destiny*. Presentation at "As We May Think—A Celebration of Vannevar Bush's 1945 Vision", MIT Department of Electrical Engineering and Computer Science, 12–13 October, http://www.w3.org/Talks/9510_Bush/Talk.html

Brockman, J. (1997). "Intentional Programming: A Talk with Charles Simonyi", *Edge Foundation,* 6 June, https://www.edge.org/conversation/charles_simonyi-intentional-programming

Campbell-Kelly, M., and Asprey, W. (1996). *Computer: A History of the Information Machine*. New York, NY: Basic Books.

Card, S. K. (1996). "Pioneers and Settlers: Methods Used in Successful User Interface Design", in *Human-Computer Interface Design: Success Stories, Emerging Methods, and Real-World Context,* ed. M. Rudisill, C. Lewis, P. G. Polson, and T. McKay. San Francisco: Morgan Kaufmann Publishers. 122–69.

Cerf, V. (1993). "How the Internet Came to Be", in *The Online User's Encyclopedia,* ed. B. Aboba. Boston: Addison-Wesley.

Ceruzzi, P. E. (1998). *A Modern History of Computing: 1945–95*. Cambridge, MA: The MIT Press.

Computing Research Association. (1997). *Computing Research: A National Investment for Leadership in the 21st Century*. Washington, DC: Computing Research Association.

Draper, T. (1997). "Silicon Valley to Washington—Ignore us, Please", *Wall Street Journal,* 4 March.

Engelbart, D. C. (1994). "Biographical Sketch: Douglas Carl Engelbart", *Bootstrap Institute,* http://www.dougengelbart.org/content/view/183/153/

English, W. K., Engelbart, D. C., and Melvyn, A. B. (1967). "Display-Selection Techniques for Text Manipulation", *IEEE Transactions on Human Factors in Electronics* 8/1: 5–15.

Evans, P., Rueschemeyer, D., and Skocpol, T. (1985). *Bringing the State Back In*. New York, NY: Cambridge University Press.

Festa, P. (2000). "Apple, AOL veterans making Linux easy", *CNET News.com,* 16 February.

Fong, G. R. 2000. "Breaking New Ground or Breaking the Rules: Strategic Reorientation in U.S. Industrial Policy", *International Security* 25/2: 152–86.

Geschke, C. (1999). "The U.S. Environment for Venture Capital and Technology-Based Startups", in *Harnessing Science and Technology for America's Economic Future: National and Regional Priorities,* ed. National Research Council. Washington, DC: National Academy Press.

Hafner, K., and Lyon, M. (1996). *Where Wizards Stay Up Late: The Origins of the Internet*. New York, NY: Simon & Schuster.

Hart, J. A., Reed, R. R., and Bar, F. (1992). "The Building of the Internet: Implications for the Future of Broadband Networks", *Telecommunications Policy* 16/8: 666–89.

Hiltzik, M. (1999). *Dealers of Lightning: Xerox PARC and the Dawn of the Computer Age*. New York, NY: Harper Business.

Intel Computing Continuum Conference. (2000). 15–17 March, https://www.intel.com/pressroom/archive/releases/2000/cn031500a.htm.

Johnson, J., Roberts, T. L., Verplank, W., Smith, D. C., Irby, C. H., Beard, M., and Mackey, K. (1989). "The Xerox Star: A Retrospective", *IEEE Computer* 22/9: 11–29.

Kay, A. (1995). *Simex: The Neglected Part of Bush's Vision*. Presentation at "As We May Think—A Celebration of Vannevar Bush's 1945 Vision", MIT Department of Electrical Engineering and Computer Science, 12–13 October, http://dougengelbart.org/content/view/258/000/

Lampson, B. W. (1988). "Personal Distributed Computing: The Alto and Ethernet Software", in *A History of Personal Workstations,* ed. A. Goldberg. New York, NY: Addison-Wesley. 291–344.

Leiner, B. M., Cerf, V. G., Clark, D. D., Kahn, R. E., Kleinrock, L., Lynch, D. C., Postel, J., Roberts, L. G., and Wolff, S. (1997). "A Brief History of the Internet", *Internet Society,* https://arxiv.org/html/cs/9901011?

Levy, S. (1988). "Graphical User Interface", *Encyclopedia Britannica Online,* https://www.britannica.com/technology/graphical-user-interface

Levy, S. (1994). *Insanely Great: The Life and Times of Macintosh*. New York, NY: Penguin Books.

Licklider, J. C. R. (1960). "Man-Computer Symbiosis", *IRE Transactions on Human Factors in Electronics* 1: 4–11, https://doi.org/10.1109/thfe2.1960.4503259

Licklider, J. C. R., and Taylor, R. (1968). "The Computer as a Communications Device", *Science and Technology* 76: 21–31.

Linzmayer, O. W. (1999). *Apple Confidential: The Real Story of Apple Computer*. San Francisco, CA: No Starch Press.

Markoff, J. (2000). "Old Apple Macintosh Team Aims to Put Linux on the Desktop", *New York Times,* 21 February.

Miller, L. H., and Johnson, J. (1996). "The Xerox Star: An Influential User Interface Design", in *Human-Computer Interface Design: Success Stories, Emerging Methods, and Real-World Context,* ed. M. Rudisill, C. Lewis, and T. D. McKay. San Francisco: Morgan Kaufmann. 70–100.

Microsoft News Release. (1998). "Remarks by Bill Gates", 18 May.

Myers, B. A. (1968). "A Brief History of Human Computer Interaction Technology", *ACM Interactions* 5/2, 44–54.

National Research Council. (1995). *Evolving the High Performance Computing and Communications Initiative to Support the Nation's Information Infrastructure.* Washington, DC: National Academy Press.

Nerds 2.0.1.: A Brief History of the Internet (1998). Public Broadcasting System. 19 September; 3 October; 25 November.

Norberg, A. L., and O'Neill, J. E. (1996). *Transforming Computer Technology: Information Processing for the Pentagon, 1962–86.* Baltimore, MD: Johns Hopkins University Press.

Norr, H. (2000). "A Less Complex Linux", *San Francisco Chronicle,* 21 February.

Perry, T., and Voelcker, J. (1989). "Of Mice and Menus: Designing the User-Friendly Interface", *IEEE Spectrum* 27/9: 46–51.

Rheingold, H. (1985). *Tools for Thought: The People and Ideas of the Next Computer Revolution.* New York, NY: Simon & Schuster, http://www.rheingold.com/texts/tft/

Rogers, M. (1983). "The Birth of the Lisa", *Personal Computing,* February, 89–94, https://archive.org/details/PersonalComputing198302/page/n89

Rose, F. (1989). *West of Eden: The End of Innocence at Apple Computer.* New York, NY: Viking Penguin.

Smith, D. C., Irby, C., Kimball, R., Verplank, B., and Harslem, E. (1982). "Designing the Star User Interface", *Byte* 7/4: 242–82.

Smith, D. K., and Alexander, R. C. (1988). *Fumbling the Future: How Xerox Invented, Then Ignored, The First Personal Computer.* New York, NY: W. Morrow.

Sutherland, I. E. (1963). "Sketchpad: A Man-Machine Graphical Communication System", *Proceedings of the AFIPS Spring Joint Computer Conference* 23: 329–46.

Tesler, L. (1985). "The Legacy of the Lisa", *Macworld,* September, 17–22.

Triumph of the Nerds. (1996). Public Broadcasting System, 12 June.

Van Atta, R., Deitchman, S., and Reed, S. (1990). *DARPA Technical Accomplishments. Volume I.* Alexandria, VA: Institute for Defense Analyses, https://apps.dtic.mil/dtic/tr/fulltext/u2/a239925.pdf

Waldrop, M. M. (2001). *The Dream Machine: J. C. R. Licklider and the Revolution that Made Computing Personal.* New York, NY: Viking Press.

Wallace, J., and Erickson, J. (1992). *Hard Drive: Bill Gates and the Making of the Microsoft Empire.* New York, NY: HarperBusiness.

Wolfe, Roaslee, ed. (1998). *Seminal Graphics: Pioneering Efforts that Shaped the Field.* New York, NY: Association for Computing Machinery.

Zachary, G. P. (1997). *Endless Frontier: Vannevar Bush, Engineer of the American Century.* New York: Free Press.

7. Rethinking the Role of the State in Technology Development

DARPA and the Case for Embedded Network Governance[1]

Erica R. H. Fuchs

1. Introduction

Debates on the appropriate role for government in technology policy often fall into two camps—proponents of free markets; and proponents of government choosing technology winners. Among those who favor a strong role for government, most view the state's role as limited to facilitating technology investment through tax policy, subsidies, and funding for basic research. A few argue for coordination of technology investment across the many arms of government. In search of this coordination, these few often turn to top-down bureaucracy. What is missing from these debates, however, is an alternative government role that has existed for the past fifty years in the U.S.: the Defense Advanced Research Projects Agency, or DARPA.

Staffed at any moment with little more than one hundred people and $3 billion with which to stimulate U.S. innovation, this small arm of government charged with "preventing technological surprises" has

1 This chapter was originally published in 2010, in *Research Policy Volume* 39/9: 1133–47.

 https://doi.org/10.11647/OBP.0184.07

met with fame and controversy beyond what its size would suggest. Historians have attributed to DARPA creation of everything from the Internet,[2] and the personal computer,[3] to the laser[4] and Microsoft Windows.[5] DARPA has appeared on the pages of Playboy Magazine,[6] and the screen of the popular television show, *The West Wing*.[7] Most pertinently, among those who study national innovation systems, DARPA has come to be seen as the pioneer of the methods now used broadly in what is called the U.S. Developmental Network State.[8] As a consequence, today, agencies ranging from the intelligence community (ARDA—1998, IARPA—2006),[9] to the Department of Homeland Security (HSARPA—2002), to the Department of Energy (ARPA-E—2007), all seem to want their own "ARPA".

Despite such past success, between 2001 and 2008 DARPA underwent tremendous change. This change, initiated by director Tony Tether, brought on an outcry from the computing community—one of the primary benefactors and success stories of DARPA.[10] This criticism suggested that DARPA was no longer "the old DARPA". An in-depth look at history, however, shows that such change, and subsequent criticism, as occurred under Tether were not new. Rather, over the past decades, DARPA has

2 Newman, N. (2002). *Net Loss: Internet Prophets, Private Profits, and the Costs to Community*. University Park, PA: Pennsylvania State University Press.

3 Allan, R. (2001). *A History of the Personal Computer*. Ontario, CA: Allan Publishing.

4 Bromberg, J. (1991). *The Laser in America, 1950–70*. Cambridge, MA: The MIT Press.

5 See Fong's chapter above (Chapter 6).

6 Sedgwick, J. (1991). "The Men from DARPA", *Playboy* 38 (August), at 108, 122, 154–56.

7 Graves, A. (2004). "The Stormy Present", *The West Wing*. National Broadcasting Company. 7 January.

8 Block, F. (2007). "Swimming against the Current: The Hidden Developmental State in the U.S.", *Politics and Society* 36/2: 169–206, https://doi.org/10.1177/0032329208318731; McCray, W. P. (2009). "From Lab to iPod: A Story of Discovery and Commercialization in the post-Cold War Era", *Technology and Culture* 50/1: 58–81, https://doi.org/10.1353/tech.0.0222

9 The Advanced Research and Development Activity (ARDA) was created in 1998, and modeled after DARPA. ARDA's name was changed to the Disruptive Technologies Office (DTO) in 2006. In December 2007, ARDA/DTO was folded into the newly created IARPA.

10 Computing Research Community. (2005). *Joint Statement of the Computing Research Community. House Science Committee Hearing on the Future of Computer Science Research in the U.S.* Washington, DC; Lazowska, E., and Patterson, D. (2005). "An Endless Frontier Postponed", *Science* 308/5723: 757, https://doi.org/10.1126/science.1113963; Markoff, J. (2005). "Pentagon Redirects its Research Dollars", *New York Times*, 2 April.

gone through repeated shifts in its focus and its internal governance structures. This dynamic presents a puzzle—with so much change, what then is the DARPA model that its imitators should be copying?

To answer this question, at least in part, this paper focuses on the period immediately before and after the most recent changes within DARPA. Drawing on over fifty interviews, the paper uses grounded theory-building methods[11] to uncover the processes used by DARPA program managers to influence technology trajectories in the U.S., and how those processes may have changed during Tether's directorship. The paper focuses on the involvement of DARPA's Microsystems Technology Office (MTO) in the development of four semiconductor materials technologies critical to the converging telecom and computing industry and to meeting the performance targets set by Moore's Law.

The telecom and computing industries provide a useful example of industrial sectors traditionally supported by government funding and in particular by DARPA.[12] In addition, the telecom and computing industries are a classic example of sectors that have undergone a recent decline in corporate R&D labs and a shift to a vertically fragmented industry structure, a phenomenon experienced more broadly in the U.S. innovation eco-system.[13] Notably, despite the dramatic differences between DARPA from 1992 to 2001 versus from 2001 to 2008, during both time periods, the telecom and computing industries had already undergone vertical disintegration.

The results of this research suggest that past studies have, by focusing on DARPA's culture and structure, overlooked a set of lasting, informal institutions among DARPA program managers. In the case of DARPA's Microsystems Technology Office, what changed under Tether was not the processes used by the program managers, but rather the situations in which program managers apply these processes. Prior to 2001, DARPA's processes for seeding and encouraging new technology trajectories

---

11 Glasner, B., and Strauss, A. (1967). *The Discovery of Grounded Theory: Strategies of Qualitative Research*. London, UK: Wiedenfeld and Nicholson. Eisenhardt, K. (1989). "Building Theories from Case Study Research", *Academy of Management Review* 14/4: 532–50, https://doi.org/10.5465/amr.1989.4308385

12 Flamm, K. (1988). *Creating the Computer: Government, Industry, and High Technology*. Washington, DC: The Brookings Institute.

13 Mowery, D. C. (1999). *America's Industrial Resurgence? An Overview. U.S. Industry in 2000: Studies in Competitive Performance*. National Academy Press, Washington, DC: National Academy Press, 1–16.

involved (1) bringing star scientists largely from academia together to brain-storm new technology directions, (2) seed funding research themes common across disconnected star researchers, (3) encouraging early knowledge-sharing between these star researchers through required workshops, and (4) providing third-party validation for new technology directions to external funding agencies and industry. These processes support the sources of, knowledge flows around, and development of social networks necessary for initiating new technology directions in the research community. In contrast, since 2001, the DARPA program manager's processes for coordinating technology directions involve (1) orchestrating the involvement of established vendors with academics and startups, (2) supporting knowledge-sharing between industry competitors through invite-only workshops, (3) providing third-party validation of new technology directions, and (4) supporting technology platform leadership at the system level. These new processes support the coordination of technology development within industry across a vertically fragmented industrial ecosystem such that the technology develops in line with longer term commercial and military goals.

These results suggest that, rather than being forced to choose between the extremes of free-markets or the heavy-hand of bureaucratic government, there is a third alternative for government support of cutting-edge technology development. In this third alternative, embedded government agents—who gain knowledge centrality and social capital in their role as DARPA program managers—are able to re-architect social networks[14] among researchers so as to influence new technology directions. In doing so, these embedded agents are in constant contact with the research community, understanding emerging themes, matching these emerging themes to military needs, betting on the right people, bringing together disconnected researchers, standing up competing technologies against each other, and maintaining the systems-level perspective critical to orchestrate these disparate research activities spread throughout our national innovation ecosystem.

14 In this paper, the phrase "re-architect social networks" encompasses all activities in which DARPA program managers bring together disconnected or less connected members of the research community, subsequently building active research communities, and thereby providing validation of technology directions to achieve organizational goals. Section 5. in this chapter ("Results and Discussion") unpacks the full range of activities engaged in by DARPA under the umbrella of this phrase.

2. The Developmental Network State

Debates on the appropriate role for the state in science and technology development have continued for over two hundred years.[15] In the U.S., even when a role for the state is acknowledged, the appropriate government function is often viewed as influencing the volume, not the direction of investment.[16] Under Keynesian thought, "economic policy meant manipulating spending and taxation, money and credit", not coordination of technology development.[17] And yet, a host of literature documents alternative roles for the State in technology development beyond manipulation of spending and regulation.

One categorization of this literature is to split it into two types of theories—those that depict a Weberian-style hierarchy or "developmental bureaucratic state"; and those that argue for "experimental federalism", "flexible developmental state", "developmental network state", or "networked polity".[18] Whereas the "Bureaucratic State" evokes descriptions of "centralized command-and-control" and "top-down policies" leveraging "government-based research and firm subsidies to develop local expertise in targeted industries;" the "networked" alternative is often described as "decentralized and distributed" with "mutual adjustment" and a focus on facilitating "building trust" and "coordination and cooperation among relevant parties".[19] In both governance forms, writers argue that to be successful, public officials must have "embedded autonomy"—i.e. be "embedded in a concrete set of social ties that binds the state to society and provides institutionalized

15 Smith, A. (1776). *An Inquiry into the Nature and Causes of the Wealth of Nations*. London: Metheun & Co.

16 Graham, O. (1992). *Losing Time: The Industrial Policy Debate*. Cambridge, MA: Harvard University Press.

17 *Ibid.*

18 Ansell, C., (2000). "The Networked Polity: Regional Development in Western Europe", *Governance: An International Journal of Policy and Administration* 13/3: 303–33, https://doi.org/10.1111/0952-1895.00136; Block. (2007). "Swimming against the Current"; Breznitz, D. (2007). *Innovation and the State*. New Haven, RI: Yale University Press.

19 Sabel, C. (1993). "Studied Trust: Building New Forms of Cooperation in a Volatile Economy", *Human Relations* 46/9: 1133–70, https://doi.org/10.1177/001872679304600907; Ansell. (2000). "The Networked Polity"; O'Riain, S. (2004). *Politics of High-Tech Growth: Developmental Network States in the Global Economy*. Cambridge, UK: Cambridge University Press; Breznitz. (2007). Innovation and the State.

channels for the continued negotiation and renegotiation of goals and policies".[20]

In describing the networked polity, Chris Ansell proposes "that the state can operate as a liaison or broker in creating networks and empowering nonstate actors, especially when state actors occupy a central position in these networks".[21] The existing network literature helps us understand the emergence and consequences of being a broker. According to Ronald Burt,[22] a broker is an individual who forms the only link between otherwise disconnected actors. Lee Fleming and David Waguespack add to this definition, distinguishing between brokers and boundary spanners.[23] Here, Fleming and Waguespack's boundary spanners are individuals who span different theoretical or organizational areas, but need not be the only individual playing that role. Thus, while all brokers are boundary spanners, not all boundary spanners broker.[24] Notably, neither Burt nor Fleming gives agency to the broker or boundary spanner. While Burt focuses on how the structure of the network puts the broker in a position of power, Fleming and Waguespack focus on how existing human and social capital lead to individuals emerging as leaders in a community.[25] In her qualitative field study, Natalia Levina brings in this agency.[26] Specifically, she suggests that to create a new field, boundary-spanners must produce and use objects that become locally useful to both fields and acquire a common identity.[27] However,

20 Evans, P. (1995). *Embedded Autonomy: States and Industrial Transformation*. Princeton, NJ: Princeton University Press; Ansell. (2000). "The Networked Polity".

21 Ansell. (2000). "The Networked Polity".

22 Burt, R. (1992). *Structural Holes: The Social Structure of Competition*. Cambridge, MA: Harvard University Press.

23 Fleming, L., and Waguespack, D. (2007). "Brokerage, Boundary Spanning, and Leadership in Open Innovation Communities", *Organization Science* 18/2: 165–80, https://doi.org/10.1287/orsc.1060.0242

24 Fleming and Waguespack. (2007). "Brokerage, Boundary Spanning, and Leadership".

25 Fleming and Waguespack. (2007). "Brokerage, Boundary Spanning, and Leadership".

26 Levina, N. (2005). "The Emergence of Boundary Spanning Competence in Practice: Implications for Implementation and Use of Information Systems", *MIS Quarterly* 29/2: 335–63, https://doi.org/10.2307/25148682

27 According to Levina, boundary objects are artifacts such as physical prototypes, engineering sketches, or standardized reporting forms that can span beyond the physical, temporal, or social limitations of an individual boundary spanner. A

while Levina provides practical insights into the boundary-spanning role, her boundary-spanner remains an inside member of the focus community.

In contrast to this earlier work, recent research has begun to explore network plasticity—or the ability of managers to change social networks to achieve organizational objectives.[28] In contrast to structural theories, which focus on how network structures create constraints and opportunities for organizational actors, or naturalistic theories, which focus on how spontaneous forces shape network dynamics, these new agency theories focus on network change agents who sit outside and act upon the community or network of focus.[29] For example, in their study of Levi's jeans, Lester and Piore suggest that in the early, open ended stages of innovation, R&D managers must act as "cocktail hostesses", bringing together the correct parties to the table, and helping facilitate the flow of conversation in order to be successful in their goal of promoting innovative new ideas.[30] Likewise, in his longitudinal study of eight technology collaborations, Jason P. Davis found that managers of successful collaborations prune networks of existing ties that are information bottlenecks in the emerging network collaboration and, rather than rely on social processes, remake these networks with competency pairing, which forms ties between actors with complementary knowledge across organizational boundaries.[31] This new research, in which managers have agency to change the shape of the existing network to achieve organizational objectives, suggests that a different, and more fundamental role may exist for the state in influencing technology development. Describing the role of the state in regional development in Western Europe, Ansell writes, "the state does

---

boundary object as locally useful if it is incorporated into practice in multiple of the fields it spans. A boundary object has a common identity if it is typical enough to be readily recognized in both fields. For example, computer aided design (CAD) software is useful and common to both the photonic and electronic semiconductor communities (Levina. (2005). "The Emergence of Boundary Spanning Competence").

28 Davis, J. P. (2009). "Network Dynamics of Exploration and Exploitation: Pruning and Pairing Processes in Collaborative Innovation", *MIT Working Paper*.

29 *Ibid.*

30 Lester, R. K., and Piore, M. J. (2004). *Innovation: The Missing Dimension*. Cambridge, MA: Harvard University Press.

31 Davis. (2009). "Network Dynamics".

not simply act as a mediator or coordinator, but also actively tries to create relationships between third-party actors".[32]

While the existing network polity literature hints of such activities by the state, the empirical examples of the networked state contained therein are surprisingly similar. The majority of the examples are of the government playing a role in industrial or technology development in industrializing nations in the process of catch-up.[33] In the few examples from developed countries, the role of the state is to connect firms to enable incremental innovation, support collaborative learning among firms, and help smaller firms catch-up, primarily in the context of regional economic development or upgrading in manufacturing.[34] In nearly all examples, the state acts either by linking firms to facilitate increased economic transactions, dissemination of knowledge, and collaborative learning,[35] or by linking individuals to build communities.[36] Throughout these examples the state lacks an active role in identifying and influencing technology directions. Instead, the state, as a central node in the network, helps create network linkages, disseminate knowledge, or act as the breeding ground for communities without influencing the direction or content of discussions. To find an example of the state influencing technology directions one must turn to Japan—a country often characterized as a "bureaucratic"

32 Ansell. (2009). "The Networked Polity".

33 Johnson, C. (1982). *MITI and the Japanese Miracle: The Growth of Industrial Policy, 1925–1975*. Stanford, CA: Stanford University Press; Fransman, M. (1993). *The Market and Beyond: Information Technology in Japan*. Cambridge, UK: Cambridge University Press; Amsden, A., and Chu, W. (2003). *Beyond Late Development: Taiwan's Upgrading Policies*. Cambridge, MA: The MIT Press; O'Riain. (2004). *Politics of High-Tech Growth*; Breznitz. (2007). *Innovation and the State*.

34 Sabel, C. (1996). "A Measure of Federalism: Assessing Manufacturing Technology Centers", *Research Policy* 25/2: 281–307, https://doi.org/10.1016/0048-7333(95)00851-9; McEvily, B., and Zaheer, A. (1999). "Bridging Ties: A Source of Firm Heterogeneity in Competitive Capabilities", *Strategic Management Journal* 20: 1133–56, https://doi.org/10.1002/(sici)1097-0266(199912)20:12%3C1133::aid-smj74%3E3.0.co;2-7; Ansell. (2000). "The Networked Polity"; Whitford, J. (2005). *The New Old Economy: Networks, Institutions, and the Organizational Transformation of American Manufacturing*. Oxford, UK: Oxford University Press.

35 Ansell. (2000). "The Networked Polity"; Amsden and Chu. (2003). *Beyond Late Development*; O'Riain. (2004). *Politics of High-Tech Growth*; Whitford. (2005). *The New Old Economy*.

36 Breznitz, D. (2005). "Collaborative Public Space in a National Innovation System: A Case Study of the Israeli Military's Impact on the Software Industry", *Industry and Innovation* 12/1: 31–64, https://doi.org/10.1080/1366271042000339058

development state that chooses technology winners. And yet, Japan's facilitation of research cooperation between competing firms echoes many of the themes written in the literature on the networked polity.[37] Indeed, the literature on the Japanese government goes farther than what can be found in the networked polity literature on Europe, the U.S., and industrializing nations.

Daniel Okimoto describes the importance of the Japanese government's focus on working with companies on consensus building,[38] and on articulating long-term vision in the development of new technologies.[39] Relatedly, Martin Fransman describes the Japanese government's self-identified role in helping firms overcome the downfalls of "bounded vision" [40]—i.e. the idea that different kinds of organizations (a) receive different kinds of information as the results of their primary activities, and (b) are limited in what they search for and "see" by the overall objectives of the organization. Here, according to Fransman, Japan believes that the limitations in the vision of for-profit firms and the vision of the government can be overcome by bringing the two together.[41] Both of these themes are echoed in the case study presented here on DARPA.

Of course, organizational forms other than the state can also facilitate the connecting of disconnected agents and architect networks. As suggested by Dan Breznitz's example of military training in Israel,[42] education and training—such as being in common graduate programs—can build scientific communities and long-lasting networks. Conferences can act as venues for existing communities to contest and form agreement around the viability of competing technology directions.[43] Firms, such as Intel, can orchestrate the co-development of technologically interdependent platforms across

37 Johnson. (1982). *MITI and the Japanese Miracle*; Fransman. (1993). *The Market and Beyond.*

38 Okimoto, D. (1987). *Between MITI and the Market: Japanese Industrial Technology for High Technology*. Stanford, CA: Stanford University Press.

39 *Ibid.*

40 Fransman. (1993). *The Market and Beyond.*

41 *Ibid.*

42 Breznitz. (2005). "Collaborative Public Space".

43 Garud, R. (2008). "Conferences as Venues for the Configuration of Emerging Organizational Fields: The Case of Cochlear Implants", *Journal of Management Studies* 45/6: 1061–88, https://doi.org/10.1111/j.1467-6486.2008.00783.x

firms, universities, and government labs as is necessary to continue to advance their specific business model.[44] None of the above pieces, however, are by themselves sufficient to seed and develop new technology directions that meet needs beyond the short-term market demands that drive firms. While communities developed through education and training may have common backgrounds, they do not, in and of themselves, have direction. While conferences can act as direction deciders, for a conference to play this role, the community must already exist. Finally, while firms may be able to play many of these roles as platform leaders, they will not have the same incentives as government (having a goal of profits rather than national security, economic growth, and social welfare), and their "vision"[45] will be more short-term.

In this chapter, I leverage extensive empirical data to unpack an active, network-changing role of the state that goes beyond the previous literature on the place and application of a networked polity. First, I focus on cutting-edge, new technology development. In particular, I describe how in the development of new technologies, the state need not stop at merely bringing the appropriate actors together, nor must it go so far as choosing "focus industries" or "technology winners", but rather it can leverage its knowledge centrality and ability to connect disconnected actors to identify and influence new technology directions that achieve its organizational goals.

Further, I describe a state that, in the development of a single new technology, leverages all of the earlier-described roles of network governance—from building new communities to community consensus-making on directions, to platform leadership outside of the constraints of firm incentives—to achieve its goals. Finally, I show that to find such a networked polity influencing technology development we need not look to Japan or to the late industrializing nations, but rather that this networked polity already exists in the U.S. To unpack existing practices, I turn to the pioneer of the U.S. Developmental Network State, DARPA.[46]

44 Gawer, A., and Cusumano, M. A. (2002). *Platform Leadership: How Intel, Microsoft, and Cisco Drive Industry Innovation*. Boston, MA: Harvard Business School Press.

45 Fransman. (1993). *The Market and Beyond.*

46 Block. (2007). "Swimming against the Current".

3. The Changing Faces of DARPA

Long-time defense analyst Richard Van Atta writes, "There is not and should not be a singular answer on 'what is DARPA'—and if someone tells you that [there is], they don't understand DARPA".[47] And yet, with so much success, it has been hard for analysts not to try to pin down the "DARPA model". Van Atta himself summarizes the DARPA organizational environment into three key characteristics: (1) it is independent from service R&D organizations, (2) it is a lean, agile organization with a risk-taking culture, and (3) it is idea-driven and outcome-oriented.[48]

These themes are echoed in DARPA's self-described twelve organizing elements, along with two additional themes—a focus on hiring quality people ("an eclectic, world-class technical staff"), and the importance of DARPA's role in connecting collaborators.[49] Others have suggested that DARPA's "single customer" (the military) and "clear mission" (enhancing U.S. military capabilities) is a critical aspect of the DARPA model.[50] And yet, as shown in the history that follows, the emergence, interpretation and actualization of these organizational features has evolved dramatically over the decades since DARPA's creation in 1958. In many ways, these changes can be grouped into decade-based shifts, as shown in Table 7-1. In this paper, I focus on the shift initiated in 2001 by Tony Tether. To understand this shift, however, it is necessary to look back at the other shifts within DARPA across the previous decades.

The Advanced Research Projects Agency (ARPA) was founded under President Eisenhower in February 1958 by Public Law 85–325 and Department of Defense Directive 5105.41, as a direct consequence

47 Van Atta, R. H. (2007). *Energy Research and the "DARPA Model"*. Subcommittee on Energy and Environment, Committee on Science and Technology. Washington, DC: U.S. House of Representatives, 9.

48 *Ibid.*

49 Bonvillian, W. B. (2006). "Power Play, The DARPA Model and U.S. Energy Policy", *The American Interest* 2/2, November/December, 39–48, https://www.the-american-interest.com/2006/11/01/power-play/

50 Mowery, D. C. (2006). *Lessons from the History of Federal R&D Policy for an "Energy ARPA"*. Washington, DC: Committee on Science.

of the Soviet launching of Sputnik in 1957.[51] Initially, ARPA was charged with preventing technological surprises such as Sputnik.[52] Many blamed the advent of Sputnik on the rivalry at the time between the military services, and ARPA was set up to cut through that rivalry. After its founding, ARPA's first priority was to oversee space activities until NASA was up and running and to screen new technological possibilities, shutting down those without merit.[53] By 1960, all of ARPA's civilian programs were transferred to the National Aeronautics and Space Administration (NASA) and all of its military space programs were transferred to individual Services. At this point, ARPA was forced to face the question of its longer-term role. President Eisenhower had always insisted that the Cold War was fundamentally a contest between two economic systems, and that it would be won or lost economically, not militarily.[54] This perspective, in which the distinction between military and civilian technology was blurred, would stay with ARPA throughout the 1960s.

With space activity oversight behind it, ARPA focused its energies on ballistic missile defense, nuclear test detection, propellants, and materials.[55] It was at this time that ARPA took on the rôle of bringing along military ideas that other segments of the nation would not or could not develop, and carrying them to proof-of-concept.[56] ARPA's goal was then to transition the technology out of the laboratory and into the hands of users or producers who would bring it to full adoption and exploitation.[57]

---

51 National Research Council. (1999). *Funding a Revolution: Government Support for Computing Research*. Computer Science and Telecommunications Board. History, Commission on Physical Sciences Mathematics and Applications. Washington, DC: National Academy Press.

52 *Ibid.*

53 Flamm, K. (1987). *Targeting the Computer: Government Support and International Competition*. Washington, DC: The Brookings Institute; Roland, A. (2002). *Strategic Computing: DARPA and the Quest for Machine Intelligence 1983–93*. Cambridge, MA: The MIT Press.

54 Roland. (2002). *Strategic Computing*.

55 National Research Council. (1999). *Funding a Revolution*.

56 Roland. (2002). *Strategic Computing*.

57 Roland. (2002). *Strategic Computing*.

Table 7-1 The changing face of DARPA: a historical chronology of the organization. (Table prepared by the author.)

The changing face of DARPA: a historical chronology of the organization.

Decade	1958	1960s	1970s	1980s	1990s	2000s
Name	ARPA ('58–72)		DARPA ('72–93)		ARPA ('93–96)	DARPA ('96–08)
Era		Basic Research	Military Missions	Industry Focus	Competitiveness, Internationalization	Industry to Military
President	Eisenhower ('53–61)	Eisenhower ('53–61) Kennedy ('61–63) Johnson ('63–69)	Nixon ('69–74) Ford ('74–77) Carter ('77–81)	Reagan ('81–89)	Bush ('89–93) Clinton ('93–01)	Bush Jr. ('01–08)
Legislative/political environment	Cold War Sputnik ('57)	Cold War Vietnam War ('59–75)	Cold War Vietnam War ('59–75) Mansfield Act ('69)	Cold War Ends Star Wars Noyce - more VC ('78) Concern about competitiveness against Japan; National Cooperative Research Act ('84)	Field forced to leave due to excessive industrial focus ('90); Sematech desires internationalization, weans from public assistance ('95); DARPA criticized for slow transition to military ('97); Increased inter-organizational and international R&D linkages	World Trade Center Attacked (Sept. 11, 2001); Bush Jr. enters Iraq ('03); Increased concerns about U.S. competitiveness, especially against India, China (Rising Above the Gathering Storm, 2005); Criticism of DARPA for not funding basic R&D (Lazowski House Statement, 2005)
DARPA Directors	Johnson ('58–60)	Betts ('60–61) Ruina ('61–63) Sproull ('63–65) Herzfeld ('65–67) Rechtin ('67–70)	Lukasik ('70–75) Heilmeir ('75–77) Fossum ('77–81)	Cooper ('81–85) Duncan ('85–88) Colladay ('88–89) Fields ('89–90)	Reis ('90–92) Denman ('92–95) Lynn ('95–98) Fernandez ('98–01)	Tether ('01–08)
DARPA Environment	Supercede inter-service rivalry; prevent technological surprises	Scientific merit over military; focus on best people - independence, intellectual quality	Mid-term exams, deliverables, success measures	Strategic computing initiative ('83); Sematech ('87); pyramid of technologies; connecting academia and industry	Fernandez priorities: people, competition, outreach, experimentation ('98)	Phases, milestones, accountability; "Transforming Fantasy" (01–03); "Bridging the Gap" ('03–'08)

ARPA's independent status not only insulated it from established service interests, but also tended to foster radical ideas and keep the agency tuned to basic research questions.[58] When the agency-supported work became too much like systems development, it ran the risk of treading on the territory of a specific service.[59] ARPA also established in the 1960s its critical organizational infrastructure and management style: a small, high-quality, managerial staff, supported by scientists and engineers on rotation from industry and academia, successfully employing existing DOD laboratories and contractors (rather that creating its own research facilities), to build solid programs in new, complex fields.[60] Finally, ARPA emerged as an agency extremely sensitive to the personality and vision of its director.[61]

Following Army Brigadier General Austin Betts,[62] Jack Ruina became DARPA's third director in 1961 at the same time as President Kennedy took office. As director, Ruina cemented the agency's reputation as an elite, scientifically respected institution devoted to basic, long-term research projects. Ruina believed that independence and intellectual quality were critical to attracting the best people, both to ARPA as an organization and to ARPA-sponsored projects.[63] A Professor of Electrical Engineering on leave from the University of Illinois, Ruina valued scientific and technical merit above immediate relevance to the military.[64] During his tenure, Ruina decentralized management at ARPA, and began the tradition of relying heavily on independent office directors and program managers to run research programs. To meet his goals for the agency, Ruina encouraged creative use of existing Department of Defense managerial mechanisms including "no-year

58 National Research Council. (1999). *Funding a Revolution.*

59 *Ibid.*

60 Barber Associates, R. (1975). *The Advanced Research Projects Agency, 1958–1974.* Report prepared for the Advanced Projects Research Agency. Springfield, VA: Defense Technical Information Center; National Research Council. (1999). *Funding a Revolution.*

61 National Research Council. (1999). *Funding a Revolution.*

62 Betts, the second ARPA director, had suffered under the perception within the Pentagon that he favored his own service agency. On his recommendation, all subsequent ARPA/DARPA directors have been civilians (Roland. (2002). *Strategic Computing*).

63 Barber Associates. (1975). *Advanced Research Projects Agency;* National Research Council. (1999). *Funding a Revolution.*

64 National Research Council. (1999). *Funding a Revolution;* MIT. (2009). "Research Affiliates: Jack Ruina", MIT Security Studies Program.

money", unsolicited proposals, sole-source procurement, and multi-year forward funding.[65] Through the mid-1960s, DARPA remained committed to supporting basic research with long-term importance, even if there was no immediate military application.[66]

By the 1970s, however, the war in Vietnam had become the driving force at DARPA, tending to redirect research towards military purposes and raising concerns about the effect of defense funding on university research. Under President Richard Nixon, Congress forbade military funding for any research that did not have a "direct or apparent relationship to a specific military function or operations".[67] The legislation, which was enacted into law as the Mansfield Amendment to the Defense Authorization Act of 1970 (Public Law 19–121), was short-lived, but had the longer-term impact of shortening the time horizons for government research support, and in particular defense research.[68] In keeping with the political times, ARPA's name was officially changed to DARPA (the Defense Advanced Research Projects Agency) in 1972. Then, in 1975, George Heilmeier became director of DARPA.[69] Under Heilmeier's directorship, all proposals needed to address six questions: (1) what are the limitations of current practice, (2) what is the current state of technology, (3) what is new about these ideas, (4) what would be the measure of success, (5) what are the milestones and the "mid-term exams," and (6) how will I know you are making progress. In contrast to Ruina, Heilmeier led with a heavy hand, giving all DARPA orders a "wire brushing" to ensure that they had concrete "deliverables" and "milestones".[70] In short, Heilmeier viewed DARPA as a mission agency, whose goal was to fund research that directly supported the mission of the DOD.[71]

In the 1980s, with the Vietnam War over, defense concerns gave way to industrial competitiveness as the primary driver of research policy. The U.S. increasingly feared that the microelectronics and computer industries would go the way of the auto industry—to Japan. These fears were not unfounded. By the end of the 1980s Japanese semiconductor

65 National Research Council. (1999). *Funding a Revolution.*
66 Flamm. (1987). *Targeting the Computer.*
67 National Research Council. (1999). *Funding a Revolution,* 112.
68 *Ibid.*
69 *Ibid.*
70 Roland. (2002). *Strategic Computing.*
71 Roland. (2002). *Strategic Computing.*

manufacturing equipment suppliers were gaining market share at a rate of 3.1 percent a year, and U.S. semiconductor manufacturers planned to purchase the majority of their equipment from Japanese suppliers.[72] Given the heavy-handed role of Japan's Ministry of International Trade and Industry (MITI), later renamed the Ministry of Economy, Trade, and Industry (METI), helping companies cooperate on new markets and technologies, there were increasing cries in the U.S. for government action.[73]

In 1984, the National Cooperative Research Act exempted research consortia from some antitrust laws and further facilitated collaborations. Then, in 1987, fourteen U.S. semiconductor companies joined a not-for-profit venture, SEMATECH, to improve domestic semiconductor manufacturing. The next year, the federal government appropriated $100 million annually for the next five years to match the industrial funding. DARPA had since the late 1970s been supporting the development of "silicon foundry" capabilities to allow cost-effective fabrication of new types of integrated electronic devices by designers lacking easy access to costly production facilities.[74] With semiconductor manufacturing seen as vital to defense technology, the SEMATECH money was channeled through DARPA.[75]

This paper begins its story in the 1990s. During this period, the U.S.'s focus on international competitiveness grew, further distancing DARPA from its role with the military. In 1992, Secretary of Defense Dick Cheney announced "a new, post-Cold War DOD strategy of spending less on procurement of new military systems, while maintaining funding for R&D to develop new technologies for building future systems and for upgrading existing systems".[76] The next year, the Congressional Office of Technology Assessment (OTA) wrote, "Early stages of R&D, in which ARPA is most heavily involved (basic research through technology demonstration), will probably be least affected by reductions in defense spending" (following the cold war). The OTA continued, "Furthermore, based on military interests alone, ARPA will probably

72 National Research Council. (1999). *Funding a Revolution.*

73 *Ibid.*

74 Flamm. (1987). *Targeting the Computer.*

75 National Research Council. (1999). *Funding a Revolution.*

76 Office of Technology Assessment. (1993). *Defense Conversation: Redirecting R&D.* Washington, DC: U.S. Government Printing Office.

Table 7-2 Shift in DARPA funding mechanisms 1992–2008. (Table prepared by the author.)

Shift in DARPA funding mechanisms 1992–2008.

	Pre-Tether (1992–2000)	Post-Tether (2001–2008)
Δ in DARPA Funding Structure	Funding primarily of university-based research	Funding shifted from universities to industry (especially, established vendors)
	Broad area announcements (BAA), few checks and balances on meeting program targets	Multiple phase solicitations: 12–16 month intervals, Funds tied to go/no-go reviews linked to pre-defined deliverables[4]
	Solicitations open to anyone being the prime contractor	Many solicitations preclude universities and small start-ups as prime contractors, instead requiring the formation of teams with the established vendors as the prime contractors

become more involved in the development of dual-use technologies. Despite the apparent divergence of military and commercial systems, many component technologies from which these systems are constructed continue to converge".[77] During the period from 1992 to 2001 DARPA was led by three directors—Gary Denman (1992–1995), Larry Lynn (1995–1998), and Frank Fernandez (1998–2001). During Gary Denman's tenure, DARPA briefly dropped its "D" and returned to its original name of ARPA. Both Lynn and Fernandez continued Denman's focus on basic research. Lynn was part of DARPA's first inclusion of basic biology research into DARPA's budget.[78] Fernandez focused on quality and independence in a manner reminiscent of ARPA's second director, Ruina.[79]

On 20 January 2001, however, George W. Bush took office as the 43rd President of the United States, and DARPA's focus on dual-use technologies came to an end. On 18 June 2001, Tony Tether was appointed as the new Director to head DARPA. Prior to becoming the director of DARPA, Tether had steadily risen in his career through a variety of military and industrial positions. Having served for four years as the director of the DOD's National Intelligence Office (1978–1982), he came to the position of DARPA director under a directive from Secretary of Defense Donald Rumsfeld that the new director must make DARPA "an entrepreneurial hotbed that will give the U.S. military the tools it will need to maintain the nation's access to space and to protect satellites in orbit from attack".[80] Less than three months after Tether was appointed, the U.S.'s post-cold-war peace time landscape began to change. On September 11, 2001, two hijacked planes were flown into the World Trade Center in New York City, a third hijacked plane was flown into the Pentagon, and a fourth hijacked plane attempted an attack on Washington, D.C. In response, on 7 October 2001 the U.S. invaded Afghanistan, and on 21 March 2003, the U.S. began its invasion of Iraq. In his statement to the House of Representatives on 27 March 2003, Tether highlighted DARPA's role in "bridging the gap" between fundamental discoveries and military use.[81] This slogan, "Bridging the Gap", was subsequently added to the official logo for DARPA.

---

77 *Ibid.*

78 Marshall, E. (1997). "Too Radical for NIH? Try DARPA", *Science* 275/5301: 744–46.

79 Fernandez, F. (2000). *Statement by Frank Fernandez Director, Defense Advanced Research Projects Agency*. Given before the U.S. Senate, Washington, DC.

80 Rensselaer. (2002). "DARPA Inside", *Rensselaer Magazine.*

81 Tether, T. (2003). *Statement by Dr. Tony Tether, Director, Defense Advanced Research Projects Agency.* Subcommittee on Terrorism, Unconventional Threats, and

During his time at DARPA, Tether made significant changes to the agency's policies, shown above in Table 7-2, which brought on an outcry from the academic community, especially the computing community.[82] Although overall DARPA funding remained constant, the proportion going to university researchers dropped by nearly half.[83] In contrast to the flexibility and discretion given to researchers in the 1990s, funds under Tether were tied to "go/no-go" reviews linked to pre-defined deliverables—i.e. technical achievements defined either in the solicitation itself or by the researchers as part of responding to the solicitation—that must be achieved within a pre-specified time period (typically six to nine months).[84] This focus on milestones and go/no-go reviews is reminiscent of DARPA policies under Heilmeier. In addition, DARPA raised the classification of research programs and increased restrictions on the participation of non-U.S. citizens.[85] Most significantly, many solicitations precluded universities and small startups from submission as prime contractors, instead requiring the formation of teams and forcing startups and universities to team with large established vendors.[86]

Looking back over the decades since DARPA was founded, it is not immediately clear that the concerns expressed in the 2000s by the academic community with regards to DARPA being "dead" were warranted. Under Tether, DARPA did indeed shift its funding away from academia and, at the same time, shifted its funding model. However, change in DARPA's immediate goals and the director-level rules on how to meet those goals, is common, if not the rule, over the DARPA's history.[87] With so much change, the puzzle is what is the DARPA

Capabilities, House Armed Services Committee, United States House of Representatives. Washington, DC.

82 Computing Research Community. (2005). *Joint Statement*; Lazowska and Patterson. (2005). "An Endless Frontier Postponed"; Markoff. (2005). "Pentagon Redirects".

83 Computing Research Community. (2005). *Joint Statement*; Lazowska and Patterson. (2005). "An Endless Frontier Postponed"; Markoff. (2005). "Pentagon Redirects".

84 The Ultraperformance Nanophotonic Interchip Communications (UNIC) program, discussed in greater detail in Section 5.2. in this chapter ("DARPA under Tony Tether (2001-present)"), provides an example of a proposal under Tether with multiple phases, each with go/no-go deliverables.

85 Computing Research Community. (2005). *Joint Statement*; Lazowska and Patterson. (2005). "An Endless Frontier Postponed".

86 Defense Science Board. (2005). *High Performance Microchip Supply*. Washington, DC: National Academies Press.

87 Mowery and Langlois and others have noted the tension between developing technologies required for highly specialized, low-volume defense applications,

model, and is there something fundamental about DARPA, across the decades, that its imitators should be copying? Past research has focused on DARPA's organizational culture, structure, and goals as the critical and lasting features of the "DARPA-model". In this paper, I argue that beyond these organizational features, there are informal processes used by the program managers to influence technology directions, which have been overlooked in past literature, and have been institutionalized so as to last through changes in directorship and organizational focus.

4. Methods

This paper uses grounded theory-building methods[88] to unpack the processes by which DARPA influences technology development. I conduct a case study[89] of four materials technologies critical to the advancement of Moore's Law. Two of these technologies—SiGe and strained Si—received DARPA funding in the mid-nineties and were subsequently introduced into microprocessor designs and mainstream Si-CMOS production lines. The remaining two materials advances—3D packaging technology and integrated photonics—were funded under Tether and are identified by the ITRS Roadmap and in

and technologies required for civilian applications (Mowery, D. C., and Langlois, R. N. (1996). "Spinning Off and Spinning On (?): The Federal Government Role in the Development of the U.S. Computer Software Industry", *Research Policy* 25: 947–66, https://doi.org/10.1016/0048-7333(96)00888-8). Several items are worth noting on this point. First, as described in the above paragraphs the extent to which DARPA's location within the military narrowed the scope of what science it could fund has varied significantly over the decades—ranging from periods such as those under Heilmeier and Tether, where the immediate needs of military missions figured prominently, to periods such as those under Ruina, or when SEMATECH funding was channeled through DARPA where the needs of the military missions figured less prominently. Second, while DARPA program managers must as part of "selling" any project be able to describe its eventual benefit for the U.S. military, depending on the budget category (basic research, applied), the research can be quite basic and thus far from any application, especially in an office such as the Defense Sciences Office (DSO) and MTO. Finally, due to overlapping needs in the area of microprocessors and commercial demand outpacing that of the military, even under the Tether period DARPA commissioned roadmaps of military versus commercial technical needs to help guide its funding decisions.

88 Glasner, B., and Strauss, A. (1967). *The Discovery of Grounded Theory: Strategies of Qualitative Research*. London, UK: Wiedenfeld & Nicholson; Eisenhardt. (1989). "Building Theories"; Yin, R. K. (1989). *Case Study Research: Design and Methods*. Newbury Park, CA: Sage.

89 Eisenhardt. (1989). "Building Theories"; Yin. (1989). *Case Study Research.*

academic publications as potentially critical to meeting the targets set by Moore's Law in the upcoming decade. All four of these technologies were supported by program managers within DARPA's Microsystems Technology Office (MTO), which, until April 1999, went by the name of the Electronics Technology Office (ETO).[90]

In conducting my research, I triangulated participant observation, qualitative interview data, archival data, and bibliometric data to provide a holistic view of the forces driving technological change.[91] My results draw primarily from fifty semi-structured interviews with DARPA office directors and program managers, industry representatives, and university professors who were involved in the development of SiGe, strained silicon, integrated photonics, and optical interconnects between 1992 and 2008. I identify key scientists and technologists in the "invisible college"[92] in this technical area through a snowball effect based on names mentioned in early interviews and in news documents.[93] I subsequently cross-checked this list using DARPA's online archives for the period and identified additional DARPA program managers involved in funding these technologies. I executed the interviews so as to ensure that they included (1) DARPA MTO office directors and program managers from both before and after Tony Tether took the directorship, and (2) a representative cross-section of scientists and technologists from within academic institutions, startups, and the five established microprocessor vendors—Intel Corporation, Advanced Micro Devices (AMD), International Business Machines (IBM), Hewlett Packard (HP), and Sun Microsystems (Sun). I also asked each respondent to provide an up-to-date biography and curriculum vita (CV), including a list of

90 Reed. (1999). "Defense Advanced Projects Agency's Electronics Technology Office Changes Name", *High Beam Research*. Reed Business Information.

91 Jick, T. D. (1979). "Mixing Qualitative and Quantitative Methods: Triangulation in Action", *Administrative Science Quarterly* 24: 602–11, https://doi.org/10.2307/2392366

92 Price, D. D. S. (1963). *Little Science, Big Science… and Beyond*. New York, NY: Columbia University Press.

93 Derek de Solla Price described the "invisible college" as an informal communication network among elite scholars from different research institutions often within a subject specialty. I use the term "invisible college" a bit loosely since the list is one of researchers identified by each other as "key people in this area" or "key people to talk to", and while communications are documented in the interviews, the exact form or extent of communication is not known. Finally, it is worth noting that in this "invisible college", "research institutions" encompasses everything from universities, to start-ups, to large computing firms, to DARPA itself (Price. (1963). *Little Science*.).

all of their publications and patents to-date in their career. I used these individual CVs to better understand the bibliometric records of each interviewee, as well as their co-patenting and co-publishing records with other scientists. I completed all interviews between September 2006 and October 2008.

I conducted several participant observations throughout the course of the study to gain insights into both the optoelectronics and microelectronics industries and DARPA's role in technology development. Early on, I was able to conduct a three-hour participant observation of a DARPA-funded team in the process of developing its technology so as to acquire Phase II funding. I was also able to attend multiple industry conferences throughout the course of the study, due to my own prior technical activity in the area, through additional connections from my interviews, and through my ongoing professional activities studying the converging telecom and computing industry. These industry conferences included three of the Bi-annual Microphotonics Industry Consortium conferences (Fall 2007, Spring 2007, Fall 2008), Phontics North 2007, the 2007 IEEE Computer Elements Vail Workshop, the Optoelectronics Industry Development Association (OIDA) 2008 Annual Forum, and the OIDA Manufacturing and Innovation in the 20th Century Workshop in Spring 2008.

Finally, I have been able to draw on extensive archival data available through the Carnegie Mellon University libraries, online, and saved within the personal collections of David Hounshell. DARPA provides a wealth of archival data online, as well as through their technical archives. In addition, a host of information about both DARPA and company initiatives can be found in the popular press, congressional hearings, and in industry trade journals. Together, I use these online DARPA archives and available news sources to document DARPA solicitations, workshops, conferences, and press releases as related to the four materials technologies.

5. Results and Discussion

I present my results in three sections. In the first section, I unpack five distinct steps by which DARPA program managers seed and encourage new technology trajectories. This section draws exclusively on archival data and interviews with academics, industry members, and program managers before Tony Tether's period as director, specifically between 1992 and 2001. The second section of the results then explores the changes within DARPA under Tony Tether. Here, I again draw on archival data and interviews with academics, industry members, and program managers but instead from 2001 to 2008. This section again proposes five methods by which DARPA seeded and encouraged new technology trajectories, and compares these methods, and the recipients of their efforts, to those found in the previous period. In the final section, I discuss overarching themes that emerge across the two periods and describe the role of the program manager.

5.1. DARPA in the 1990s (1992–2001)

Based on archives and interviews from academics, industry members, and DARPA program managers active during this period from 1992 to 2001, I identify five processes by which DARPA program managers during this period tap into existing social networks to seed and encourage new technology trajectories. These five processes are (1) identifying directions, (2) seeding common themes, building community, (4) validating new directions and (5) not sustaining the technology. I describe each of these processes in detail, and their significance below.

1) *Identifying directions*: To influence the direction of technology development so as to meet mission goals, a DARPA program manager must first identify the direction in which to go. To do this, DARPA program managers engage in three complementary activities: talking with mission directors to understand the needs of the military, bringing together elite scientists to brainstorm research directions that meet the needs of the military, and talking with existing researchers to understand emerging technology directions within the research community. The first activity DARPA program managers cannot escape. There are military liaisons in the DARPA building, who are senior officers, and have the role of connecting program managers with the needs of the military. In addition, DARPA program managers visit military installations around the country throughout the year to better understand military needs. The second and third activities, however, require greater agency on the part of the DARPA program manager. Below, I discuss the second activity—bringing together elite scientists

to brainstorm research directions—and, in the next section, I discuss the third activity—talking with existing researchers to understand emerging directions.

Table 7-3 Mid-nineties collaborators brought together by a DARPA program manager to brainstorm on carbon nanotubes. (Table prepared by the author.)

Mid-90s collaborators brought together by a DARPA program manager to brainstorm on carbon nanotubes.

	Paul Robinson	Richard Smalley	Charles Lieber
Occupation, mid-90s	President, Sandia Corporation and Laboratories, Director Sandia National Labs	Professor Chemistry, Physics, Astronomy, Rice University	Professor, Chemistry, Harvard University
Total patents	15	>90	>30
Total publications	?	>394	>290
Lifetime achievements	Elected member NAE, Outstanding Public Service Medal from Joint Chiefs of Staff	1996 Nobel Prize for discovery of "buckeyballs"	Elected member NAS
Co-authorships with each other	None	None	None

Over the years, DARPA has developed several formal institutions that enable DARPA program managers to bring together elite scientists to brainstorm research directions that meet the needs of military missions.

Among its formal institutions, most notable is the DARPA-Defense Sciences Research Council. The DARPA-Defense Sciences Research Council holds an annual summer conference that brings together "a group of the country's leading scientists and engineers for an extended period, to permit them to apply their combined talents in studying and reviewing future research areas in defense sciences".[94] At this summer conference, top scientific and technical researchers in the country are exposed to major problems facing the U.S. military, and asked to identify technological directions to solve these challenges.

In addition to the Council's annual summer conference, DARPA leverages several smaller task forces and technology groups. Each year following the Council's summer meeting, smaller groups of Council members meet for Council workshops and program reviews, whose reports are made directly to DARPA.[95] Other formal advisory activities include Department of Defense's Defense Science Board (DSB) task forces, and Information Sciences and Technology Study Groups (ISAT).[96] Like the Council's workshops and program reviews, DSB[97] and ISAT task forces can be called to address specific topics or challenges.

DARPA is not limited to holding these brainstorming sessions to identify directions within formal committees. Brainstorming sessions can also be called together by individual DARPA program managers, and can be much more informal. One DARPA program manager describes his role in bringing scientific leaders together around a common theme.

---

94 Defense Sciences Research Council. (1997). *Defense Sciences Research Council Summer Conference Summary Report*. Defense Science Research Council Summer Conference, LaJolla, California, Defense Advanced Research Projects Agency.

95 *Ibid.*

96 ISAT has similar workings to the Defense Science Board task forces, but are focused on military challenges associated with information technology.

97 The DSB was established in 1956, in response to recommendations of the Hoover Commission. Today, the DSB's authorized size is thirty-two members selected for the pre-eminence in science and technology and its application to military operations, and seven ex-officio members. The task force consists of DSB board members, and other selected consultants or experts (Defense Science Board. (2008). *Defense Science Board: History*. Washington, DC: Department of Defense, Office of Research and Engineering, https://dsb.cto.mil/history.htm).

> We were talking with Paul Robinson about the notion of building very high volume carbon nanotubes that were functionally matched... And I said, gee, Rick's always been working in that area, let's just call him in. Rick's a Nobel Prize chemist. So we called him. He was there in two days. And so Lieber came over from Harvard. We sat around. And it was a great discussion.

The above-described interaction occurred in the mid-90s. Here, in supporting innovation DARPA program managers are the cocktail hosts described by Lester and Piore as necessary for the early-stage brainstorming of new ideas.[98] The DARPA program managers select the members of the party, and help start the conversation necessary to brainstorm and identify the necessary new directions.

It is important, however, to look closer at the above quotation. As shown in Table 7-3, all of the people at the above-mentioned gathering, with the exception of the DARPA program manager, could be characterized as Lynne Zucker's and Michale Darby's "star scientists".[99] None of them, however, have bibliometric or other paper trails of intellectual ties with each other. These results are in striking contrast with the majority of social networks research, which focuses on documenting collaborations through patent co-authorships. These early-stage, informal, roundtable technical conversations are the type of conversations that cannot be found in bibliometric studies. Further, it is in precisely these formative conversations where the state's involvement in bringing together the right parties may be particularly influential in determining future directions.

2) *Seeding common research themes*: DARPA program managers do not stop at a series of brainstorming session with elite scientists. In addition, DARPA program managers are continually returning to the field to find emerging projects and capabilities within the research community. In this role, they not only identify additional research directions, but also encourage research in those directions by funding researchers working on common themes that have the potential to contribute to military

98 Lester and Piore. (2004). *Innovation*.

99 Zucker, L., and Darby, M. (1996). "Star Scientists and Institutional Transformation: Patterns of Invention and Innovation in the Formation of the Biotechnology Industry", *Proceedings of the National Academy of Sciences* 93/23: 709–12, https://doi.org/10.1073/pnas.93.23.12709

needs. Further, in contrast to the brainstorming sessions, in this field-based activity of identifying emerging directions and encouraging research in those directions, the DARPA program manager need not necessarily, or at least immediately, bring everyone into the same geographic space.

One DARPA program manager explains,

> So I'll tell you the SiGe story… So, the first guy to show me this, actually two guys,… was the guy who founded Amberwave. He showed me this is possible. And then Jason Woo and UCLA,… he showed me a plot of bandgap as a function of percent Ge. And he had two plots. He came to DARPA. And he said, look, there is a dependency, here it is, it follows band gap theory… And I said, "Jason, two dots don't make a program… I need a third dot". And he faxed me a chart the next day… So I sent him a small seeding.
>
> At the same time I called Bernie (a fellow at IBM), and I said, "Bernie, have you ever seen this bandgap dependency in SiGe? You know, do you think it's something we can exploit?" He said, "Funny you should ask. We've been looking at the same thing, and we've got some ideas as well". So I funded him $2 million or whatever it was.

In this function, the DARPA program manager is neither acting as a broker—connecting otherwise disparate actors; nor as a boundary-spanner—identifying, translating, and relaying information across firm, cultural, or technical boundaries; in the traditional sense.[100] Instead, the DARPA program manager is using his connections with researchers to identify emerging directions and capabilities within the research community, and seed-fund common themes across these disparate researchers. While the program manager is perhaps relaying some knowledge about the one researcher to the other or about general activities in the technical community, at first, he may be the only connection between them.

Upon closer scrutiny of the above quote, these results also have a second significance. Specifically, similar to the results in section (1), background research on the technologists referenced by the DARPA program manager in the above quote, show both Eugene Fitzgerald ("the guy who founded of Amberwave") and Bernard Meyerson ("Bernie")

100 Fleming and Waguespack. (2007). "Brokerage, Boundary Spanning, and Leadership".

Table 7-4 Technologists funded by a DARPA program manager to gain momentum around Si Ge and strained Si technology. (Table prepared by the author.)

Technologists funded by a DARPA program manager to gain momentum around Si Ge and strained Si technology.

	Eugene Fitzgerald	Bernard Meyerson	Jason Woo
Occupation, mid-90s	Associate Professor, Materials Science and Engineering	IBM Fellow, Group Director	Professor, Electrical Engineering, UCLA
Occupation, 2008	Professor, Materials Science and Engineering	IBM Fellow, V.P. and Chief Technologist, Systems and Technology Group	Professor, Electrical Engineering, UCLA
Total patents	>15	>40	>1
Total publications	>186	>180	>100
First paper in SiGe technology	1986	1986	1991
Evidence of co-authorships with each other, or other cooperation	None	None	IBM Faculty Award, 1998

again to be what Zucker and Darby would classify as star scientists[101] (see Table 7-4). Thus, this DARPA program manager is describing his contact with three star scientists, working in the same area. These results are significant given Zucker and Darby's findings that star scientists are very protective of their techniques, ideas, and discoveries in their early years, tending to collaborate most within their own institution, which slows diffusion to other scientists.[102] Assuming Zucker and Darby's findings are correct, here, the sole connecting person, who is aware of all three of the star scientists' activities, may be the DARPA program manager (Table 7-4).

Finally, it is worth noting that in playing out this role of seeding common research themes across disparate researchers, the DARPA program manager does not always fund the same technologies. At times, DARPA program managers fund competing technologies aimed at solving the same problem. The same program manager explains such an example in a different funding situation,

> Take the case of thin-film technologies. In that case I funded two parallel programs. I funded IBM, because they were convinced that the parallel junction for thin-film SOI wasn't going to go on forever, and they wanted more thick-film SOIs for the company manufacturing purposes. And then I funded Lincoln Labs to do thin-film SOI... I pitted Lincoln against IBM... So, they both succeeded, and IBM is still manufacturing thick-film SOI today.

3) *Building community*: increasing information flows, growing the base: DARPA's role in seeding disparate researchers working on common research themes (whether the same or competing technologies) has a second significance. In receiving funding from DARPA, researchers are required to present to each other in workshops, thus further increasing the flow of knowledge between star scientists during early-stage research. Fitting with their classification as star scientists, neither Fitzgerald nor Meyerson—who are at different institutions—have ever co-patented or co-published. Yet, through DARPA, Fitzgerald and Meyerson were brought together in workshops to present to each other their research. What would otherwise have been knowledge kept within their organization was forced at some level (with the exception of some

101 Zucker and Darby. (1996). "Star Scientists".
102 *Ibid.*

company-proprietary details which are presented solely to the program managers) to flow between the two. In funding disparate researchers, DARPA program managers promote the sharing of knowledge between star scientists, who left to their own devices would, according to the literature, tend to be very protective of their knowledge. In some cases, these workshops may even lead to new collaborations. Jason Woo, for example, started in the field somewhat later than Fitzgerald or Meyerson (1991), and, as the 1998 IBM Faculty Award he received suggests, may have even developed a relationship with IBM through his funding from DARPA.

4) *Providing third-party validation of new technology directions*: in addition to DARPA program managers' roles in bringing researchers together to brainstorm new technology directions, seeding disparate researchers to gain momentum around those directions, and bringing those researchers together to share their results, DARPA program managers play a fourth role in technology development. Specifically, DARPA program managers' funding actions act to provide external validation for new directions. One program manager explains, "So the DARPA piece, while large, was the validation for IBM to spend their own money". He continues, "The same way for the Intel piece. You know, Intel certainly looked at that project, and then Intel ended up funding it internally, but the fact that DARPA went back to them three and four times and said, this is an important thing, this is an important thing, you know, it got to the board of directors, and it got high enough that they set up a division to do this". A university professor makes the same point with respect to DARPA's role with other funding agencies, in this case NSF. The professor explains, "See, once you've gotten funding from DARPA, you have an issue resolved, and so on, then you go right ahead and submit an NSF proposal. By which time your ideas are known out there, people know you, you've published a paper or two. And then guys at NSF say, yeah, yeah, this is a good thing". He continues, distinguishing DARPA's place within the broader U.S. government system, "NSF funding usually comes in a second wave. DARPA provides initial funding". As a consequence, he concludes, "DARPA plays a huge role in selecting key ideas" (from among the broader set of ideas present in the research community).

5) *Avoiding reliance on the state*: Finally, despite DARPA's role in validating new technology directions both to other funding agencies and in industry, DARPA program managers from the 1992–2001 period take note to point out that DARPA is not the "sustaining piece" in commercializing a new technology. As one DARPA program manager explains, "So we ran all of these design- of-experiment concepts, and you know,... we were doing great stuff, really good science. But the tipping point,... is the fact that IBM saw the value in this to the point that they started investing in it".

This emphasis on the state not sustaining technology is an important final piece. Past research has warned of the tendencies for companies to become reliant on support from the state.[103] History suggests that DARPA has had many successes transitioning subsequent development and production of its early-stage technologies to commercial (e.g. laser,[104] the Internet,[105] and the personal computer)[106] and military (e.g. F-117A, Predator, Global Hawk)[107] organizations. Future research should explore DARPA program manager's mechanisms for transitioning technology development, and how they handle technologies that do not transition.

5.2. DARPA under Tony Tether (2001-present)

Tony Tether was appointed director of DARPA on 18 June 2001. As discussed above, Tether made many changes within DARPA, which were poorly received from the academic, and particularly the

103 Allen, T., Utterback, J., Sirbu, M., Ashford, N., and Hollomon, J., (1978). "Government Influence on the Process of Innovation in Europe and Japan", *Research Policy* 7/2: 124–49; Sirbu, M. (1978). "Government Aid for the Development of Innovative Technology: Lessons from the French", *Research Policy* 7/2: 176–96, https://doi.org/10.1016/0048-7333(78)90004-5; Zysman, J. (1983). *Governments, Markets, and Growth: Financial Systems and the Politics of Industrial Change*. Ithaca, NY: Cornell University Press.

104 Bromberg. (1991). *The Laser in America*.

105 Newman, N. (2002). *Net Loss: Internet Prophets, Private Profits, and the Costs to Community*. University Park, PA: Pennsylvania State University Press.

106 Allan. (2001). *History of the Personal Computer*.

107 Van Atta, R., Lippitz, M., et al. (2003). *Transformation and Transition, DARPA's Role in Fostering a Revolution in Military Affairs. Volume 1*. Alexandria, VA: Institute for Defense Analyses, https://doi.org/10.21236/ada422835, https://fas.org/irp/agency/dod/idarma.pdf

computing, community. These changes included shifting funding from universities to industry (especially, established vendors); changing funding solicitations from broad agency announcements with few checks and balances to announcements with go/no-go reviews linked to pre-defined deliverables; and precluding universities and startups as prime contractors on many solicitations, instead requiring the formation of teams with established vendors as the prime contractors.

These changes in the framework of funding at DARPA can best be understood by looking at a program during this period.[108] One such program, DARPA's Ultraperformance Nanophotonic Intrachip Communications (UNIC) program,[109] is outlined in Table 7-5 above. As shown in the table, the UNIC program consisted of three phases. The first phase lasted nine months. To pass this phase the program required the "development, fabrication, and demonstration, of silicon nanophotonic devices". The second phase was two years. This phase was focused on designing and validating photonic networks between the devices developed in phase I, and "established the credibility of the technology within the microprocessor community". Program submissions were required to establish "interim milestones every six months", associated with "demonstrable, quantitative measures of performance". As shown in Table 7-5, with the exception of one team at the Massachusetts Institute of Technology (MIT I), established companies, like HP, IBM, and Sun Microsystems, were placed in the position of prime contractors, while

---

108 One predominant type of DARPA solicitation is called a Broad Agency Announcement, or BAA. BAAs occurred regularly during both the pre-Tether and the Tether period. The nature of many BAAs changed, however, under Tether. An example of the phases and pre-defined deliverables associated with a Tether-period BAA is provided in this paragraph. For the purpose of comparison, a Very Large Scale Integrated (VLSI) Photonics solicitation from the pre-Tether period (i.e. with fewer checks and balances) reads as follows: "(DARPA/ETO) is soliciting innovative research proposals to develop VLSI-level microfluidic analysis and synthesis systems (MicroFlumes) and to develop the design tools for the implementation of mixed technology systems that include microfluidic, electrical, kinematic, optical, and electromagnetic domains (Composite CAD). Of particular interest in Area 1 (MicroFlumes) are technology developments... that integrate multiple analysis & synthesis programs (or sequences of microfluidic processing steps) in one system... Of particular but not sole interest in Area 2 (Composite CAD) are design support tools, models and methods that include, but are not limited to, [list of possible interest areas]."

109 DARPA Ultraperformance-Nanophotonic-Intrachip-Communication Program, April 24, 2007, DARPA BAA 07–35, https://fbo.gov.surf/FBO/Solicitation/BAA07–35

Table 7-5 DARPA Microsystems Technology Office (MTO) Ultraperformance Nanophotonic Intrachip Communications (UNIC) program. (Table prepared by the author.)

DARPA microsystems technology office (MTO) ultraperformance nanophotonic intrachip communications (UNIC) program.

	Phase I	Phase II	Phase II
Award date	February 2006	November 2006	March 2008
Description	Super-seedling, validity demonstration		$44M
Timeline	9 Months	2 Years	5½ Years
Primary contractor awardees	1. HP 2. IBM 3. Sun Microsystems 4. MIT I[a] 5. Analog Devices?	1. HP 2. IBM 3. Sun Microsystems 4. MIT I[a]	1. Sun Microsystems
Additional team members	1. ? 2. Luxtera 3. Luxtera 4. BAE Systems 5. MIT II[a]	1. Intel 2. Luxtera 3. Luxtera 4. BAE Systems	1. Luxtera, Kotura, Stanford, UCLA

[a] Here, MIT I and MIT II are abbreviations for two teams out of the Massachusetts Institute of Technology. MIT I was led by M.I.T. Materials Science and Engineering Professor Lionel Kimerling and M.I.T. Electrical Engineering Professor Anant Agarwal. MIT II was led by M.I.T. Electrical Engineering Professor Rajeev Ram.

[a] Here, MIT I and MIT II are abbreviations for two teams out of the Massachusetts Institute of Technology. MIT I was led by M.I.T. Materials Science and Engineering Professor Lionel Kimerling and M.I.T. Electrical Engineering Professor Anant Agarwal. MIT II was led by M.I.T. Electrical Engineering Professor Rajeev Ram.

universities (MIT II, Stanford, UCLA) and startups (Luxtera, Kotura) were members of the contractor-led team.

And yet, despite these dramatic changes under Tether in the framework of funding at DARPA, as shown in the upcoming section, the five processes by which DARPA program managers influence technology directions have remarkably remained the same. The recipients of these processes, however, and as a consequence, the implications, have changed significantly (Table 7-5).

1) *Identifying directions*: As in the 1992–2001 period, to identify new technology directions that meet military needs, DARPA program managers in the 2001–2008 period engaged in three complementary activities: talking with mission directors to understand the needs of the military, bringing together elite scientists to brainstorm research directions that meet military needs, and talking with existing researchers to understand emerging technology directions within the research community. As there are no changes in their activities talking with mission directors, I skip that discussion here. I discuss the program managers' activities bringing together elite scientists to brainstorm research directions that meet military needs briefly below. I discuss program managers' activities talking with existing researchers to understand emerging technology directions within the research community in the next section.

Based on the empirical data to which I had access, nothing changed within the formal institutions used by DARPA program managers for bringing together elite technology leaders to brainstorm new technology directions. The same institutions as were used during the 1992–2008 period, existed and were used throughout the 2001–2008 period. For example, a February 2005 DSB task force focused on High Performance Microchip Supply, a topic of great interest to DARPA, and around which the Microsystems Technology Office had several solicitations. What I could not tell from my empirical data, was whether the composition of these brainstorming sessions may have changed after Tether took on the directorship. In particular, while I was able to access nearly half of the DARPA-Defense Sciences Research Council summaries for the pre-Tether period (1992, 1993, 1996, and, 1997), I was not able to gain access to any of these summaries from the period after Tether took office. While this lack of public access to these reports could be representative of increased classification of research programs during this period, it also could be that the 2001–2008 period is more recent, and these summaries have simply not yet been released.

2) *Seeding common themes: orchestrating the involvement of established vendors with academics and startup companies*: As from 1992 to 2001, DARPA program managers during 2001–2008 did not stop at a series of brainstorming sessions with elite scientists. Instead, program managers continually return to the field to find out emerging directions and capabilities within the research community. As described in Section 5.1. of this chapter ("DARPA in the 1990s (1992–2001)"), the DARPA program managers need "vision", but not necessarily the original ideas.

One program manager explains, "This is an opportunity that people will actually tell me their best ideas and we can see what we can do with those. It's really amazing in that sense". Another program manager clarifies, "I was not working in a vacuum, right?" He continues, "[I would ask people], 'Can you provide this functionality? Can you provide that functionality?'" This probing and testing of the research community to explore what is possible in a given technology—here silicon photonics, mimics the same probing being done by the program managers in the 1992–2001 period, in the case quoted, in SiGe.

As discussed in Section 5.1., at times DARPA program managers fund disparate researchers doing similar research for achieving a particular end-goal, and at other times, DARPA program managers fund competing technologies for achieving a particular end goal. One DARPA program manager suggests, "I think our best [programs] are the ones where there's multiple solutions to a common problem". He explains that in one program, "I have six performers and the reason I have six is because I was able to convince the Director that this is an extremely high-risk effort. I don't know which technology or which architecture is going to win, if any… [But], if you give me four and they all fail, maybe you left the wrong two out". This theme is echoed in the first program manager's comments, "I wanted to have three or four ideas that I could say, 'Look… here are paths we could go along. I don't know which if any of them will be successful.'… if I didn't have those, then I cannot go and sell the program".

In their continual connection with the field, DARPA program managers not only identify additional research directions, but also encourage research in those directions by funding researchers working on common themes that have the potential to address military needs. In seeding disparate researchers around common themes, the DARPA program manager is neither a broker nor a boundary-spanner. Rather, he takes in ideas from the existing research community, identifies directions, and then funds disparate researchers working on common themes that hold potential in contributing to achieving an end-goal. He synthesizes emerging ideas into common themes. He integrates common themes into directions to meet military goals. Finally, he directs researchers along these directions through carefully crafted funding solicitations.

The disparate researchers in the 2001–2008 time period are, however, very different than those funded in the period from 1992 to 2001. Where in the first time period the disparate researchers were star scientists, in the latter period, the disparate researchers are teams of startups, universities, and prime contractors. A startup company founder described his interactions with DARPA's program managers, and the role the program managers played in encouraging research in the academic and industrial communities around their ideas: "So DARPA has program managers, and we were talking to them, and they got excited about this project, and they said, let's try to get a program out. So we worked with... the DARPA program manager, and they got interested in the field, and they got a program out of this. They got a bunch of other people involved in the program". Here, the "other people" are the companies and universities for the UNIC program shown in Table 7-5.

Unlike in 1992–2001, when startup companies would have been funded directly, in 2001–2008, startup companies were frequently not able to be the primary contractor on a proposal. In the case of the above startup company, the company needed to team up with an established vendor to receive funding for the project. Describing this process, the program manager clarifies, "I have never... said, 'I want you to work with these two.'" He clarifies, "You have to structure the solicitation in such a way that... they would do that on their own". The program manager goes on to describe this system-level goal, "There was one... condition imposed on [the teams], and that was that these things must be developed in a... foundry compatible process". He explains, "I don't want people to go out and do something in the basement, and say that, 'Ah, I produced the best results in the world,' in a process that is totally incompatible with anything else that the industry does. Because the whole idea here was to leverage the industry's path down the road of smaller and smaller devices". While at first glance, this requirement for established vendors to be the primary contractors may seem limiting, it may also have an important purpose. In particular, recent research has shown that with the decline of corporate R&D labs and the vertical fragmentation of industries, firms today face new challenges coordinating across firms when advancing technology platforms,[110] aligning incentive structures across these interdependent firms,[111] and supporting long-term research within such ecosystems.[112] By leveraging his birds-eye view of research in the community, the DARPA program manager can help ensure that technical activities being engaged by disparate entities, such as startups, in the vertically disintegrated framework fit in the broader industry picture.[113]

3) *Building community: supporting knowledge flows between competitors and enabling technology platform leadership at the systems level*: As seen in Section 5.1 of this chapter, DARPA's role in seeding disparate researchers working on common themes has a second significance. In receiving funding from DARPA, researchers are required to present to each other in workshops, thus further increasing the flow of knowledge between researchers working on common themes. Under

110 Gawer, A., and Cusumano, M. A. (2002). *Platform Leadership: How Intel, Microsoft, and Cisco Drive Industry Innovation.* Boston, MA: Harvard Business School Press; Iansati and Levien. (2004). *The Keystone Advantage.*

111 Casadesus-Masanell, R., and Yoffie, D. B. (2005). "Wintel: Cooperation and Conflict", *Management Science* 53/5: 584–98, https://doi.org/10.1287/mnsc.1060.0672

112 *Ibid.*

113 Multiple academic informants in areas outside the technical scope of this paper described situations under the Tether administration in which they shared a new idea with a program manager, expecting to subsequently be funded under the BAA, only to find later that a large contract manufacturer had been funded to do their idea, and they had received no funding. While these stories could not be validated, they highlight the importance of trust between the program managers and the researchers they fund in DARPA's system of technology development. If this trust is lost, the DARPA program managers lose their position of knowledge centrality, and are no longer able to successfully identify and influence new technology directions within the research community.

Tether, DARPA funding recipients are required to attend and present to each other in workshops at the end of each go/no-go program phase. However, in contrast to the 1992–2001 period, where institutionally isolated start scientists were brought together, in the 2001–2008 period, these researchers are established vendors and their teams of startups and university professors. In response to a presentation of an early proposal for this work, which I gave at an industry conference, one university professor angrily responded, "I can tell you what you'll find. I was there (at the DARPA workshop), and they're (the companies) all presenting to each other what they're going to do. They're all talking to each other. And they're all doing the same thing". And yet, in the case of established vendors, DARPA workshops may provide them with a critical opportunity to share new ideas and agree (implicitly or explicitly) on technology directions. One industry respondent explained the importance of such an opportunity to coordinate in today's industry environment, "You just can't make anything happen in industry (today) on your own, because it's completely impossible. You have to find a partner, you have to convince your competition this is the right thing to do". He continued, "You're guiding people [your competitors]... and they ask, 'Why are you helping me with this?' and the fact is you give them information so the suppliers are in the right place to help you".

DARPA is not only supporting the coordination of technology directions across competitors. By encouraging teams of startups, universities, and prime contractors, DARPA may also be helping coordinate technology directions in a vertically fragmented industry in a second way. One established vendor emphasizes both the importance of DARPA's systems perspective and of DARPA giving the established vendors power by making them the primary contractors. He explains, "Here, the technology is being driven by the systems companies. Very few companies have the resources to do system-level exploration without DARPA funding. DARPA funding is enabling the system players to determine the direction of this technology. If you don't get the system guys involved, you end up getting widgets that don't work in the bigger picture".

This system-level goal is already hinted at in Section 2 ("The Developmental Network") by both the startup company—which notes that DARPA "got a bunch of other people involved", and by the DARPA program manager—who emphasizes the importance of developing new technologies compatible with the established industry platform. Finally, another established vendor emphasized the importance of DARPA's longer-term vision in supporting technology trajectories across the vertically disintegrated industry, saying, "You need someone with a longer-term horizon. Ten years from now, we want a teraflop of computing. But we don't have more than a six-month time horizon".

With the decline of corporate R&D labs and the vertical fragmentation of industries, firms today face new challenges in establishing appropriate sources of new inventions and in coordinating subsequent technology development across the myriad of affected firms. Recent research has documented challenges in the coordination across firms in advancing technology platforms,[114] in aligning incentive structures across interdependent firms,[115] and, in particular, in supporting long-term research within such ecosystems.

Within DARPA between 1992 and 2001, the mandate to present early-stage research in DARPA workshops encouraged star scientists to divulge information that they might otherwise have kept confidential within their institution, and thereby helped align them on similar trajectories. In contrast, in the case of DARPA under Tether, the teams DARPA forms between universities, startups, and established vendors, and its subsequent mandatory workshops are supporting the coordination of technology trajectories across a vertically fragmented industry and the alignment of long-term technology trajectories.

4) *Providing third party validation for new technology directions*: As during 1992–2001, DARPA also played a fourth role in technology development from 2001 to 2008. Specifically, it provided external validation for new directions. Under Tether, instead of DARPA's funding providing validation to industry and NSF for latter-stage funding and commercialization, it instead validates technology directions within the vertically fragmented industrial ecosystem. This validation of a new technology can be particularly helpful for startups. The CEO and founder of one startup described the challenge of breaking into the broader industry knowledge network, saying, "[In contrast to a large company or M.I.T.],… as a small company, you have to develop

114 Gawer and Cusumano. (2002). *Platform Leadership*; Iansati and Levien. (2004). *The Keystone Advantage*.

115 Casadesus-Masanell and Yoffie. (2005). "Wintel".

a contact. Headhunters... [can] also bring information to you. We are starting to discuss with (large systems vendor)... They're trying to keep us developing pieces of technology they need". Another startup founder emphasizes the importance of DARPA's validation. He explains, "[Venture capital] investors are highly motivated to see the company succeed. As a consequence, they will lie through their teeth about what the company can do. DARPA funding and ATP funding [funding from the Commerce Department's former Advanced Technology Program] have the added benefit of communicating to a third party a validation of the technology".

5) *Breeding reliance on the state?* Finally, like the DARPA program managers from 1992 to 2001, DARPA program managers from 2001 to 2008 were concerned to not become the sustaining force for any technology. Under Tether, DARPA program managers were particularly encouraged to focus on the last step of transitioning the technology to the military and (or) to industry. As one program manager explained, "The third phase is a very important phase usually... it's the last phase... [It] defines how you will transition the technology in this office, say, to somewhere else". He continues, "Dr. Tether pays extra attention to your plan for Phase III".

And yet, some members of the industrial community whose positions involved shorter term time horizons and the pressing realities of commercialization expressed caution about participating in DARPA-based activities. One established computing vendor explained, "So, <my company> as a whole has just shied away from government funding... <Our company> labs, or whatever, they'll get a little DARPA funding, but most of that is, has never, produced anything of value, from a... commercial perspective. That wasn't saying it wasn't of value within the industry, but just trying to delineate". A startup company CEO and founder expressed similar concerns, "Sometimes I'm very nervous about getting too much focus on defense money. I don't want to lose track of the fact that I'm developing products, not technology". He continues, "DARPA is funding the industry so far ahead. If you're developing for 10 years from now, DARPA is great. But how do you manage not to lose revenue unless the market is starting in now... Some of the technology developed for the next generation—I don't know if it is applicable that well to (now). I'm not sure DARPA's direction is the direction to go". He concludes, "I think... <my company> is ideally placed for (today's technology). But, admittedly, not necessarily for the long term".

These results do not conflict with the supportive comments made by established computing vendors in Section 3. ("The Changing Faces of DARPA") above. Rather, they help underscore DARPA's role in coordinating longer term technology trajectories, while not being accepted by industry for coordinating technologies required in the shorter term. Notably, while the interviewee was not participating in any DARPA-funded projects, the labs at the same established computing vendor were participating in DARPA contracts from the Microsystems Technology Office at the time of the interview.

Further, the concerns expressed by the above established vendor and startup founder may not be unwarranted. A recent study on awards from the Small Business Innovation Research Program (SBIR) by the National Academy of Sciences shows that while small businesses receiving government funding are good at achieving mission goals, they are frequently not successful at surviving in the long-term or at technology commercialization.[116] Since the time of the interview, the above-described startup has joined an established vendor's team, and acquired DARPA funding for developing the longer-term technology. Most recently, as part of the UNIC program described in Table 7-5, Sun Microsystems received a $44 million contract for the next five years to continue to develop the photonic system-on-a-chip technology. Whether or not some startups and established vendors who were involved in DARPA funding during the Tether period end up developing a reliance on the State, will remain to be seen.

6. Discussion: the DARPA Program Manager—Embedded Network Agent

Key to understanding DARPA's role in influencing technology directions is understanding the role of the program manager, not as someone who "opens windows" to which researchers can bring funding

116 Wessner, C. W., ed. (2007). *SBIR and the Phase III Challenge of Commercialization. Policy and Global Affairs Division.* Washington, DC: National Academies Press, https://doi.org/10.17226/11851

ideas,[117] nor merely as a "boundary spanner" (and possibly a "broker") who connects different communities,[118] but rather in a more active role. The nature of this role makes it difficult to describe, as it comes through in the seemingly conflicting descriptions of the DARPA program manager role by members of the research community. One former office director explains, "It really comes down to the program manager. A program manager that has a passion for an idea, that understands the technical elements of an idea, and has some vision for where it might go". On the other hand, industry and academic researchers consistently describe themselves as the people with the ideas, and DARPA program managers as the people who funded them, provided legitimacy, and helped provide the funding and community support to bring the vision to fruition. In the words of one university professor, a DARPA program manager would "touch" on "people like [professor's name] and others he knew well, and [say] 'hey, help me, give me the ideas.'"

This seeming inconsistency, however, can be resolved through the DARPA program managers' own description of their role. As a former member of the research community who suddenly rises in status and holds the promise of money, the DARPA program manager becomes a central node to which information from the larger research community flows. In this role, the DARPA program managers are in constant contact with the research community, bringing people together to brainstorm new directions, understanding emerging research themes, matching these emerging themes to military needs, "betting on the right people", connecting disconnected communities, standing-up competing technology solutions against each other, and maintaining the system-integrating view. In executing these tasks, they must, indeed, have "vision", but this vision does not necessarily involve themselves having the ideas. In the words of one program manager, "There were people around who I could go [to] and talk to [and] see what their ideas were… What they could do". Program managers from both periods, 1992–2001 and 2001–2008, describe this same idea-seeking behavior.

Most importantly, DARPA program managers conduct all of these activities, without explicitly choosing the technology winners. At times

117 Block. (2007). "Swimming against the Current".

118 Ansell. (2000). "The Networked Polity"; Block. (2007). "Swimming against the Current".

they seed disconnected researchers working on common themes—whether with the same or with competing technologies—that hold potential to meet military needs. As the DARPA program manager from the preceding paragraph explains, "So obviously… I would not propose a program if there were no ideas [among researchers] that would address the challenges that we [at DARPA] had to address. I just didn't know what… particular idea would work". He continues, "But I wanted to have three or four ideas that I could say, 'Look,… here are paths we could go along. I don't know which, if any of them, will be successful.'… if I didn't have those, then I cannot go and sell the program". In other cases, they bring together disconnected researchers, whether to brainstorm directions, to work together (on teams), or to learn from each other (in workshops). As described by a program manager from the 1992–2001 period, "You get communities together that don't naturally talk and you give them some latitude and some life, and you push them forward and see what comes out of it". In this situation, "Conversations were often… one-upmanship… You know, sort of realizing what other people were doing and you'd reset your goals, and you'd kinda all move. And the role of the program manager was kind of to keep the band marching down the street". Finally, throughout these activities, whether bringing together members of research communities that may not normally talk, or funding an entire suite of technologies necessary to meet an integrated outcome, DARPA program managers contribute a system-level perspective to organizing national R&D. As one program manager from the 1992–2001 period explained, "… we were able to broaden it out, do the VLSI, do the hardware, acceleration, do all the stuff [necessary to advance Moore's Law] and sure enough we stayed on that ops curve and we were pulling the industry along". The same systems-level view is seen in the 2001–2008 period.

Thus, while the DARPA program manager is, indeed, sometimes a broker—acting as the only connection between disconnected researchers or communities—and sometimes a boundary spanner—connecting communities to support the development of a new field—his role is much more active than that prescribed to these positions in previous literature. The DARPA program manager is not only a connector, but also a conductor and a systems integrator. He comes to his position through his prior social capital and position in the network. Once in this position, he holds and leverages particular powers. Yet, what is most significant, is the deliberate role the DARPA program manager plays in changing the shape of the network once in this position, so as to identify and influence new directions for technology development.

7. Conclusions

Several years after Tony Tether took office, popular press articles began suggesting that the U.S.'s great engine of technology change—DARPA—was "dead".[119] Drawing on a case study of DARPA's Microsystem's Technology Office from 1992 to 2008, I argue that this perceived death is because past analyses have, by focusing on the organization's culture and structure, overlooked a set of lasting, informal institutions among DARPA program managers. In the case of DARPA's Microsystems technology office before and during the directorship of Tony Tether, what changed is not the processes used by the program managers, but rather the situations to which program managers apply these processes. Prior to 2001, DARPA's processes for seeding and encouraging new technology trajectories involved (1) bringing star scientists largely from academia together to brainstorm new ideas, (2) seeding disparate researchers around common themes, (3) encourage early knowledge-sharing between these star researchers through workshops, and (4) providing third-party validation for new technology directions to external funding agencies and industry. By identifying ideas across, bringing together, and seed funding star scientists (who may otherwise institutionally isolate their knowledge) around common themes, DARPA was able to support the sources of, knowledge flows around, and development of social networks necessary for initiating new technology directions in early-stage research. In contrast, since 2001, the DARPA program manager's processes for gaining momentum around new ideas involve (1) orchestrating the involvement of established vendors with academics and startups, (2) supporting knowledge-sharing between industry competitors through invite-only workshops,

119 CRC. (2005). *Joint Statement*; Lazowska and Patterson. (2005). "An Endless Frontier Postponed"; Markoff. (2005). "Pentagon Redirects"; Shachtman, N. (2008). "Darpa Budget Cut $130 Million for 'Poor Execution'", *Wired*, 25 September, https://www.wired.com/2008/09/darpa-budget-sl/

(3) providing third-party validation of new technology directions to a vertically fragmented industry, and (4) supporting technology platform leadership at the system level. Here, DARPA is supporting the coordination of technology development across a vertically fragmented industry in whose direction the military has interest and in which long-term coordination of technology platforms is particularly challenging.

These results suggest a new form of technology policy, in which embedded government agents re-architect social networks among researchers so as to identify and influence new technology directions in the U.S. to achieve an organizational goal. In this role, these agents do not give way to the invisible hand of markets, nor do they step in with top-down bureaucracy to "pick technology winners". Instead, they are in constant contact with the research community, understanding emerging themes, matching these emerging themes to military needs, betting on the right people, connecting disconnected communities, standing up competing technologies against each other, and maintaining that birds-eye perspective critical to integrating disparate activities across our national innovation ecosystem.

Acknowledgements

My sincerest thanks to John Alic, David Andersen, Fred Block, Glen Fong, David Hounshell, Matthew Keller, Michael Piore, Ray Reagans, Alex Roland, Sean Safford, Shane Greenstein, Josh Whitford, participants in the 2008 Sloan Industry Studies Conference, and participants in the Society for the and Advancement of Socio-Economic Research Conference for their insights at various stages of this paper. Thanks to the M.I.T. Microphotonics Center and the Chemical Heritage Foundation for their generous funding of parts of this research. Many thanks, also, to Woody Powell, Martin Kenney, and the four anonymous reviewers at Research Policy for their excellent feedback. Finally, thanks to the countless government, academia, and industry interview participants, who shared their own personal experience and insights in making this document possible, and remain unnamed. All mistakes in the document remain my own.

References

Allan, R. (2001). *A History of the Personal Computer*. Ontario, CA: Allan Publishing.

Allen, T., Utterback, J., Sirbu, M., Ashford, N., and Hollomon, J., (1978). "Government Influence on the Process of Innovation in Europe and Japan", *Research Policy* 7/2: 124–49.

Amsden, A., and Chu, W. (2003). *Beyond Late Development: Taiwan's Upgrading Policies*. Cambridge, MA: The MIT Press.

Ansell, C., (2000). "The Networked Polity: Regional Development in Western Europe", *Governance: An International Journal of Policy and Administration* 13/3: 303–33, https://doi.org/10.1111/0952-1895.00136

Barber Associates, R. (1975). *The Advanced Research Projects Agency, 1958–1974*. Report prepared for the Advanced Projects Research Agency. Springfield, VA: Defense Technical Information Center.

Block, F. (2007). "Swimming against the Current: The Hidden Developmental State in the U.S.", *Politics and Society* 36/2: 169–206, https://doi.org/10.1177/0032329208318731

Bonvillian, W. B. (2006). "Power Play, The DARPA Model and U.S. Energy Policy", *The American Interest* 2/2, November/December, 39–48, https://www.the-american-interest.com/2006/11/01/power-play/

Breznitz, D. (2005). "Collaborative Public Space in a National Innovation System: A Case Study of the Israeli Military's Impact on the Software Industry", *Industry and Innovation* 12/1: 31–64, https://doi.org/10.1080/1366271042000339058

Breznitz, D. (2007). *Innovation and the State*. New Haven, RI: Yale University Press.

Bromberg, J. (1991). *The Laser in America, 1950–1970*. Cambridge, MA: The MIT Press.

Burt, R., (1992). *Structural Holes: The Social Structure of Competition*. Cambridge, MA: Harvard University Press.

Casadesus-Masanell, R., and Yoffie, D. B. (2005). "Wintel: Cooperation and Conflict", *Management Science* 53/5: 584–98, https://doi.org/10.1287/mnsc.1060.0672

Computing Research Community. (2005). *Joint Statement of the Computing Research Community*. House Science Committee Hearing on the Future of Computer Science Research in the U.S. Washington, DC.

Davis, J. P. (2009). "Network Dynamics of Exploration and Exploitation: Pruning and Pairing Processes in Collaborative Innovation", *MIT Working Paper*.

Defense Science Board. (2008). *Defense Science Board: History*. Washington, DC: Department of Defense, Office of Research and Engineering, https://dsb.cto.mil/history.htm

Defense Science Board. (2005). *High Performance Microchip Supply*. Washington, DC: National Academies Press, https://www.hsdl.org/?view&did=454591

Defense Sciences Research Council. (1997). *Defense Sciences Research Council Summer Conference Summary Report*. Defense Science Research Council Summer Conference, LaJolla, California, Defense Advanced Research Projects Agency.

Eisenhardt, K. (1989). "Building Theories from Case Study Research", *Academy of Management Review* 14/4: 532–50, https://doi.org/10.5465/amr.1989.4308385

Evans, P. (1995). *Embedded Autonomy: States and Industrial Transformation*. Princeton, NJ: Princeton University Press.

Fernandez, F. (2000). *Statement by Frank Fernandez Director, Defense Advanced Research Projects Agency*. Given before the U.S. Senate, Washington, DC.

Flamm, K. (1988). *Creating the Computer: Government, Industry, and High Technology*. Washington, DC: The Brookings Institute.

Flamm, K. (1987). *Targeting the Computer: Government Support and International Competition*. Washington, DC: The Brookings Institute.

Fleming, L., and Waguespack, D. (2007). "Brokerage, Boundary Spanning, and Leadership in Open Innovation Communities", *Organization Science* 18/2: 165–80, https://doi.org/10.1287/orsc.1060.0242

Fong, G. R. (2001). "ARPA Does Windows; the Defense Underpinning of the PC Revolution", *Business and Politics* 3/3: 213–37, https://doi.org/10.2202/1469-3569.1025 (Chapter 6 in this volume).

Fransman, M. (1993). *The Market and Beyond: Information Technology in Japan*. Cambridge, UK: Cambridge University Press.

Garud, R. (2008). "Conferences as Venues for the Configuration of Emerging Organizational Fields: The Case of Cochlear Implants", *Journal of Management Studies* 45/6: 1061–88, https://doi.org/10.1111/j.1467-6486.2008.00783.x

Gawer, A., and Cusumano, M. A. (2002). *Platform Leadership: How Intel, Microsoft, and Cisco Drive Industry Innovation*. Boston, MA: Harvard Business School Press.

Glasner, B., and Strauss, A. (1967). *The Discovery of Grounded Theory: Strategies of Qualitative Research*. London, UK: Wiedenfeld & Nicholson.

Graham, O. (1992). *Losing Time: The Industrial Policy Debate*. Cambridge, MA: Harvard University Press.

Graves, A. (2004). "The Stormy Present", *The West Wing*. National Broadcasting Company. 7 January.

Iansati, M., and Levien, R. (2004). *The Keystone Advantage: What the New Dynamics of Business Ecosystems Mean for Strategy, Innovation, and Sustainability*. Boston, MA: Harvard Business School Press.

Jick, T. D. (1979). "Mixing Qualitative and Quantitative Methods: Triangulation in Action", *Administrative Science Quarterly* 24: 602–11, https://doi.org/10.2307/2392366

Johnson, C. (1982). *MITI and the Japanese Miracle: The Growth of Industrial Policy, 1925–75*. Stanford, CA: Stanford University Press.

Lazowska, E., and Patterson, D. (2005). "An Endless Frontier Postponed", *Science* 308/5723: 757, https://doi.org/10.1126/science.1113963

Lester, R. K., and Piore, M. J. (2004). *Innovation: The Missing Dimension*. Cambridge, MA: Harvard University Press.

Levina, N. (2005). "The Emergence of Boundary Spanning Competence in Practice: Implications for Implementation and Use of Information Systems", *MIS Quarterly* 29/2: 335–63, https://doi.org/10.2307/25148682

Macher, J., and Mowery, D. (2004). "Vertical Specialization and Industry Structure in High Technology Industries. Business Strategy over the Industry Life Cycle", *Advances in Strategic Management: A Research Annual* 21: 31–55, https://doi.org/10.1016/s0742-3322(04)21011-7

Markoff, J. (2005). "Pentagon Redirects its Research Dollars", *New York Times*, 2 April, https://www.nytimes.com/2005/04/02/technology/pentagon-redirects-its-research-dollars.html?searchResultPosition=1

Marshall, E. (1997). "Too Radical for NIH? Try DARPA", *Science* 275/5301: 744–46, https://doi.org/10.1126/science.275.5301.744

McCray, W. P. (2009). "From Lab to iPod: A Story of Discovery and Commercialization in the post-Cold War Era", *Technology and Culture* 50/1: 58–81, https://doi.org/10.1353/tech.0.0222

McEvily, B., and Zaheer, A. (1999). "Bridging Ties: A Source of Firm Heterogeneity in Competitive Capabilities", *Strategic Management Journal* 20: 1133–56, https://doi.org/10.1002/(sici)1097-0266(199912)20:12%3C1133::aid-smj74%3E3.0.co;2-7

MIT. (2009). "Research Affiliates: Jack Ruina", *MIT Security Studies Program*.

Mowery, D. C. (2006). *Lessons from the History of Federal R&D Policy for an "Energy ARPA"*. Washington, DC: Committee on Science.

Mowery, D. C. (1999). *America's Industrial Resurgence? An Overview. U.S. Industry in 2000: Studies in Competitive Performance*. National Academies Press, Washington, DC: National Academy Press.

Mowery, D. C., and Langlois, R. N. (1996). "Spinning Off and Spinning On (?): The Federal Government Role in the Development of the U.S. Computer

Software Industry", *Research Policy* 25: 947–66, https://doi.org/10.1016/0048-7333(96)00888-8

Newman, N. (2002). *Net Loss: Internet Prophets, Private Profits, and the Costs to Community.* University Park, PA: Pennsylvania State University Press.

National Research Council. (1999). *Funding a Revolution: Government Support for Computing Research. Computer Science and Telecommunications Board.* History, Commission on Physical Sciences Mathematics and Applications. Washington, DC: National Academies Press.

O'Riain, S. (2004). *Politics of High-Tech Growth: Developmental Network States in the Global Economy*. Cambridge, UK: Cambridge University Press.

Okimoto, D. (1987). *Between MITI and the Market: Japanese Industrial Technology for High Technology*. Stanford, CA: Stanford University Press

Office of Technology Assessment. (1993). *Defense Conversation: Redirecting R&D.* Washington, DC: U.S. Government Printing Office.

Price, D. D. S. (1963). *Little Science, Big Science… and Beyond.* New York, NY: Columbia University Press.

Reed. (1999). "Defense Advanced Projects Agency's Electronics Technology Office Changes Name", *High Beam Research.* Reed Business Information.

Rensselaer. (2002). "DARPA Inside", *Rensselaer Magazine.*

Roland, A. (2002). *Strategic Computing: DARPA and the Quest for Machine Intelligence 1983–1993*. Cambridge, MA: The MIT Press.

Sabel, C. (1993). "Studied Trust: Building New Forms of Cooperation in a Volatile Economy", *Human Relations* 46/9: 1133–70, https://doi.org/10.1177/001872679304600907

Sabel, C. (1996). "A Measure of Federalism: Assessing Manufacturing Technology Centers", *Research Policy* 25/2: 281–307, https://doi.org/10.1016/0048-7333(95)00851-9

Sedgwick, J. (1991). "The Men from DARPA", *Playboy* 38 (August), 108, 122, 154–56.

Shachtman, N. (2008). "Darpa Budget Cut $130 Million for 'Poor Execution'", *Wired*, 25 September, https://www.wired.com/2008/09/darpa-budget-sl/

Sirbu, M. (1978). "Government Aid for the Development of Innovative Technology: Lessons from the French", *Research Policy* 7/2: 176–96, https://doi.org/10.1016/0048-7333(78)90004-5

Smith, A. (1776). *An Inquiry into the Nature and Causes of the Wealth of Nations.* London: Metheun & Co.

Tether, T. (2003). *Statement by Dr. Tony Tether, Director, Defense Advanced Research Projects Agency*. Subcommittee on Terrorism, Unconventional Threats, and

Capabilities, House Armed Services Committee, United States House of Representatives, Washington, DC.

Van Atta, R. (2007). *Energy Research and the "DARPA Model"*. Subcommittee on Energy and Environment, Committee on Science and Technology. Washington, DC: U.S. House of Representatives.

Van Atta, R., Lippitz, M., et al. (2003). *Transformation and Transition, DARPA's Role in Fostering a Revolution in Military Affairs. Volume 1.* Alexandria, VA: Institute for Defense Analyses, https://doi.org/10.21236/ada422835, https://fas.org/irp/agency/dod/idarma.pdf

Wessner, C. W., ed. (2007). *SBIR and the Phase III Challenge of Commercialization.* Policy and Global Affairs Division. Washington, DC: National Academies Press, https://doi.org/10.17226/11851.

Whitford, J. (2005). *The New Old Economy: Networks, Institutions, and the Organizational Transformation of American Manufacturing*. Oxford, UK: Oxford University Press.

Yin, R. K. (1989). *Case Study Research: Design and Methods*. Newbury Park, CA: Sage.

Zucker, L., and Darby, M. (1996). "Star Scientists and Institutional Transformation: Patterns of Invention and Innovation in the Formation of the Biotechnology Industry", *Proceedings of the National Academy of Sciences* 93/23: 709–12, https://doi.org/10.1073/pnas.93.23.12709

Zysman, J. (1983). *Governments, Markets, and Growth: Financial Systems and the Politics of Industrial Change*. Ithaca, NY: Cornell University Press.

8. DARPA's Process for Creating New Programs[1]

David W. Cheney and Richard Van Atta

Introduction

The U.S. Defense Advanced Research Projects Agency (DARPA) is widely recognized to be a highly successful R&D agency. It has been credited with making investments that have led to a large number of innovations and important advances in electronics, computing, and robotics, as well military advances such as stealth aircraft, smart weapons, and autonomous vehicles. In light of its success, there has been interest in learning from DARPA and adopting its methods. In the United States, there have been several attempts to apply the DARPA model to other agencies, including the Intelligence Advanced Research Projects Agency (IARPA), the Homeland Security Advanced Research Projects Agency (HSARPA), and the Advanced Research Projects Agency-Energy (ARPA-E). Other countries have also been interested in learning from the DARPA model. Most notably, in Japan, the Cabinet Office's Council on Science, Technology and Innovation has sponsored the ImPACT program, which was in part inspired by DARPA and is intended to support high impact, high risk R&D.

1 This paper was written for Japan's New Energy and Industrial Technology Development Organization (NEDO) and was completed in March 2016. The authors gratefully acknowledge NEDO's support.

 https://doi.org/10.11647/OBP.0184.08

A key aspect of any successful R&D program is to pick the right problems to work on—problems that are both important and also addressable within the time and resources of the program. This typically is one of the greatest challenges in creating a successful R&D program. DARPA appears to be very successful at picking good problems to address, and it has a remarkable record of supporting timely and ground breaking projects. DARPA programs often appear to be unconventional and represent different choices than normal government or private R&D investment. How does DARPA identify and decide on these unconventional topics?

In recent years, literature on DARPA's management practices has emphasized:[2]

- DARPA's non-hierarchical and non-bureaucratic organization
- The role of highly talented, entrepreneurial program managers (PMs) who serve for limited (three- to five-year) duration
- That research is performed entirely under contract with outside organization
- The use of short-term funding for seed efforts to test promising concepts, and a clear willingness to terminate non-performing projects

With respect to the selection of focus areas, the literature has noted:

- DARPA's emphasis on "high-risk/high-payoff" projects, selected and evaluated based on the impact they could make to achieve a new capability or meeting a defense need.
- The key role that its program managers play in developing programs, gathering ideas from the technical community, making funding decisions and in managing programs, and working DARPA's technical community as well as the defense community.[3]

2 Bonvillian, W. B., and Van Atta, R. (2011). "ARPA-E and DARPA: Applying the DARPA Model to Energy Innovation", *The Journal of Technology Transfer* 36: 469–513, https://doi.org/10.1007/s10961-011-9223-x (Chapter 13 in this volume).

3 Fuchs, E. R. H. (2009). "The Road to a New Energy System: Cloning DARPA Successfully", *Issues in Science and Technology* 26/1, http://issues.org/26-1/fuchs/

Most studies have not focused specifically on where program ideas come from, and many studies have drawn their conclusions from one part of the agency, or at one time.

Against this research backdrop, NEDO Washington asked us to do a study of specific cases that illustrate how DARPA chooses its program areas. These cases focus on the selection of programs, not on the individual projects that make up programs (although the distinction is not always so clear, leading us to discuss a few major projects). Moreover, the focus is on the formation, and not the execution of programs.

We were asked to have the cases cover:

- Some well-known and easily understandable technologies
- A range of DARPA offices
- Programs that generated technologies for different military services
- A variety of time periods, with a preference for relatively recent projects.

Our study has several important limitations. First, the study was limited in scope, time, and resources, and is not comprehensive. While any R&D agency with more than fifty years of history cannot be fully characterized by a handful of case studies, a particular challenge in studying DARPA is that DARPA has changed over time and that its processes differ in different parts of the organization. DARPA is often recognized to be relatively free of bureaucracy, but the lack of rules and structure also leads to a lack of consistency throughout the organization and over time. As a result, while our study describes how DARPA has operated at different times and in different parts of the organization, it cannot be considered a complete description of how DARPA develops new programs.

Our selection of cases studies may also have several biases. Due to limitations in time and resources, we focused on programs for which information was more readily available. These included cases for which the authors personally knew key individuals who could discuss the cases, as well as cases that had already been well described, either by us or by others.

Most of our cases took place in the late 1980s and early 2000s, and many of our cases are concentrated in periods that are often considered somewhat atypical of DARPA. First, the period from 1988 to about 1996 was characterized by a very high interest in dual-use technology. The loss of industrial competitiveness in key industries and technologies, combined with changing defense needs with the end of the cold war, led to an expansion of programs that were outside of DARPA's traditional mission and were intended to help support the competitiveness of key industries. During this period, Congressional and Administration priorities exerted an unusual influence in creating new programs.

Second, the period from 2001 through 2008 was characterized by unusually strong top-down direction, due to the management style of the director during this period, Dr. Anthony (Tony) Tether. Programs that were started in this period tended to have more influence from the DARPA director than in most other periods. Thus, while there is no single period of DARPA's history than can be described as completely typical, the period in which many of our cases are concentrated are notably atypical.

There are several other sources of potential bias in the selection of cases. One is that it is easier to get information about programs that DARPA chooses to publicize. Like most organizations, DARPA highlights its successes more than its failures. When DARPA makes information available on a program, program managers are less inhibited in discussing it, and journalists or analysts are more likely to write about it, all of which increases the information on the program available in the public domain. DARPA programs that are well-known may differ systematically from less visible programs.

Because we did not do a random sample of DARPA programs, we cannot generalize our findings to all of DARPA. Other analysts, looking at different parts of DARPA at different times, may come to different conclusions. Several of our interviewees reported that they viewed their program as an atypical DARPA program. Indeed, one of the findings of the report is that atypical programs are common.

A further limitation is that each case is not comprehensive. In most cases we relied on one interview supported by background materials; it is quite possible that other participants would have different perspectives on each case.

General Framework and Typical Patterns of Program Development

Figure 8-1 illustrates the influences on the development of new programs at DARPA. As will be discussed in the case descriptions, not all of the influences are present in every case, and the relative strength of the influences from the various sources differs significantly among the cases.

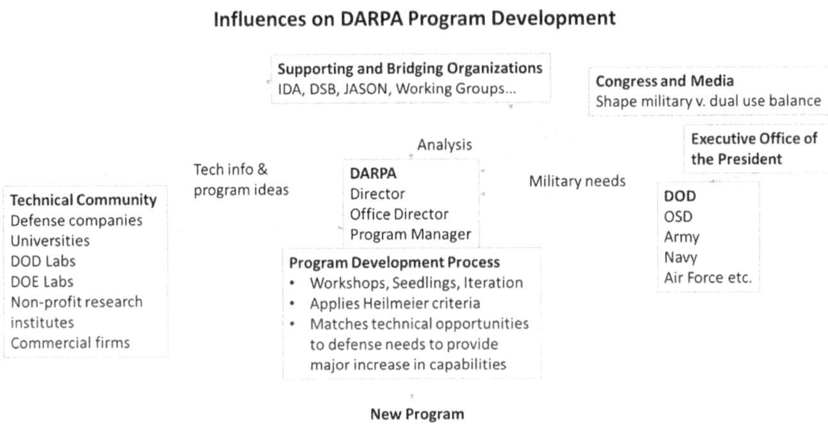

Fig. 8-1 Influences on DARPA's Program Development. Source: TPI. Notes: IDA is Institute for Defense Analyses; DSB is Defense Science Board; JASON is a group of high-level government science and technology advisors; DOD is Department of Defense; DOE is Department of Energy; OSD is Office of the Secretary of Defense. (Figure prepared by the authors.)

In the archetypal DARPA program development process, information concerning useful new capabilities comes from the Department of Defense, while information concerning what is technically possible, and what areas might be ripe for advancement, comes from the technical community. Information and analysis may come from the community of think tanks and advisory committees that advise the Department of Defense and DARPA. The DARPA program manager has the responsibility for taking this input and constructing a program, usually made up of a set of projects, with defined technical goals that are aggressive but can potentially be met within a defined time frame and within a budget. The PM (Project Manager) must put together a program that is sufficiently challenging, important, and doable, to be approved by the office director and ultimately the DARPA director. The "Heilmeier Catechism" (see Table 8-1) provides a set of questions the DARPA program managers should be able to answer to get approval for their program.

Table 8-1 Heilmeier's Catechism.
Source: https://www.darpa.mil/work-with-us/heilmeier-catechism

George H. Heilmeier (DARPA director 1975-1977) developed a set of questions known as the "Heilmeier Catechism" to help Agency officials think through and evaluate proposed research programs:

- What are you trying to do? Articulate your objectives using absolutely no jargon.
- How is it done today, and what are the limits of current practice?
- What is new in your approach and why do you think it will be successful?
- Who cares? If you succeed, what difference will it make?
- What are the risks?
- How much will it cost?
- How long will it take?
- What are the mid-term and final "exams" to check for success?

Within the broad categories of groups that provide information into the program development process, there are many subcategories. Within the Department of Defense, there may be input from the military services (Army, Navy, and Air Force) as well as the Office of the Secretary of Defense (OSD), and these may all have different views on the importance of new technology. Within the technical community, there are universities, defense laboratories, defense contractors, and others, each of which bring different viewpoints. Of special influence are the parts of technical community that have had long-term interactions with DARPA, as contractors and as sources of program managers.

There are many variations in the influences on programs. In some cases, the military need drives the process, and the program is set up to develop a prototype that may not require fundamental advances in technology. In other cases, the DARPA director or a DARPA office director may drive the process. They may have particular interests that they believe DARPA should pursue, and they will recruit a program manager to execute a program built around those interests. In other cases, the drive may come from DARPA's technical community, which may make DARPA aware of the potential that advances in science and technology may have for the military. In some cases, a general need may come from the defense community, but the key ideas that form the basis for a program may come from the technical community, in workshops or in response to a Request for Information (RFI) or a Broad Agency Announcement (BAA). In some cases, outside advisors, the Congress, and/or the Executive Office of the President (including the President's key science, economic, and national security advisors) have played important roles in shaping DARPA programs.

The different DARPA offices can vary in their processes for program development. In general, the Defense Sciences Office (DSO) can be expected to interact more with the research community while the Strategic Technology Office (STO) and Tactical Technology Office (TTO) tend to interact more with the military services. The other three technical offices—the Biological Technologies Office (BTO), Microsystem Technology Office (MTO) and Information Innovation Office (I2O)—are somewhere in between.

The influences on DARPA program development have also changed over time. In DARPA's early days, much of its work was driven by large defense projects in space, missile and satellite development (especially before NASA was established) and nuclear test detection. In the late 1960s and early 1970s, the needs of the Vietnam War were a major influence. In the late 1970s into the mid-1980s, DARPA initiated major thrusts in radically new weapons concepts, such as stealth aviation and standoff precision strike. In the late 1980s and early 1990s, DARPA was given new dual-use roles by the Congress and the Administration, and funds for industrial consortia in semiconductors, optoelectronics and other areas were administered by DARPA. In much of the 2000s there was a refocusing on defense applications, as well a strong top-down influence from the Director, Anthony Tether.

These various influences on DARPA program development are illustrated in the next section through case studies.

Case Studies of the Development of DARPA Programs

We focused our study on the nine cases of the development of DARPA programs as described in Table 8-2.

Table 8-2 Case studies in this chapter. (Table prepared by the authors.)

Case[3]	Period	DARPA Office	Idea Origin
Have Blue (Stealth)	1974-1981	TTO	OSD
Assault Breaker	1978-1984	TTO	External DARPA-funded studies
Amber (High Altitude, Long Endurance Unmanned Aerial Vehicles)	1978-1986	TTO	OSD
Optoelectronics (Wavelength Division Multiplexing)	1989-1992	DSO and MTO	PM and community
High Definition Systems	1989-1993	DMO	DARPA Director
Magnetic Materials / Spintronics	1993-2005	DSO	PM
Personalized Assistant that Learns (PAL) led to SIRI)	2002-2009	IPTO	DARPA director and IPTO Office Director
Topological Data Analysis	2004-2008	DSO	PM
Revolutionizing Prosthetics	2005 to present	DSO to BTO	Need driven

In each case, we have tried to focus on a specific program, but we discuss related activities that preceded and followed the program. In some cases, the identification of a program for analysis, and when it started, is not so clear, as the agency may have funded small projects before the main program began, so the precursors to a program may have begun well before the DARPA program was created.

For each program, we characterize the program by name, goal, DARPA office, time period, and main results. Then we examine the history of the program and where the idea for the program came from. Whose idea was it? How advanced was the idea when DARPA took it on? Were there antecedent ideas and programs? Was the program part of a broader and long-standing set of DARPA activities? How long was the proposal in development? Were there small projects, termed "seedlings", to test key concepts before the main program was established? Was the program significantly modified in goals or approach?

We discuss the background of the program managers and their role. What were they hired to do? Did the PMs have the idea, were they given the idea, or did they find the idea? Then we discuss other key roles in the formation of the programs including the role of the technical community, the DARPA senior management, and other elements of the Department of Defense.

Finally, we discuss the lessons learned from the case and what the case illustrates about DARPA's process of program formation.

Have Blue (Stealth)[4]

Overview

The "Have Blue" program was the DARPA program that produced the original prototype "stealth" aircraft that is much less visible to radar and other detection methods. It was managed in the Tactical Technology Office.

Planning studies began in 1974, and the program to develop the prototype plane took place in 1976 to 1978, with subsequent follow-on support to the Air Force through 1981. The program was highly successful, and led to a new generation of aircraft, starting with the F-117A, that represented a major increase in military capabilities.

4 Van Atta, R., Lippitz, M., et al. (2003). *Transformation and Transition, DARPA's Role in Fostering a Revolution in Military Affairs. Volume 1.* Alexandria, VA: Institute for Defense Analyses, 11–15, https://doi.org/10.21236/ada422835, https://fas.org/irp/agency/dod/idarma.pdf

Context and history of the program development

Origins

By the early 1970s it was clear to the U.S.'s strategic defense planners that the Soviet Union had achieved air defense capabilities that would have made penetrating Soviet airspace difficult. This presented the U.S. with a fundamental strategic challenge, requiring the development of new alternatives if the U.S. and NATO were to deter or combat the Soviet Bloc without having to resort to nuclear war. A central party to address this threat was the Director of Defense Research and Engineering (DDR&E) of the Office of the Secretary of Defense, who at that time was Dr. Malcolm Currie. Currie assumed this position in 1974 and, based on guidance from Secretary of Defense Schlesinger, sought greater innovation from the defense research community to develop emerging technologies to address the Soviet military buildup.

It was in this larger context in 1974 that Chuck Myers, Director of Air Warfare Programs in the Office of the DDR&E, mentioned to Robert Moore, then Deputy Director of DARPA's Tactical Technology Office (TTO),[5] an idea he called the "Harvey concept".[6] The concept was to create a tactical combat aircraft that was much less detectable by radar or infrared, acoustic, or visual means.

A primary objective was to use only passive measures (coatings and shaping) rather than depending on support aircraft carrying jammers. Such a plane would allow for new types of deep air attacks, replacing the "air armada" tactics using a large number of aircraft that had become the norm in Air Force and Navy aviation.

The Harvey idea was not entirely new, as some techniques to make aircraft less visible had been used in highly classified reconnaissance aircraft (both manned and unmanned). However, there were no serious efforts to employ such capabilities on a weapons platform. To do this, significant advances in radar cross-section reduction were needed to overcome Soviet integrated anti-aircraft systems. Myers wanted to fund aircraft companies to propose conceptual designs. Coincidentally, shortly after the Myers-Moore discussion, DDR&E Malcolm Currie sent

5 Moore became TTO Director in 1975.

6 "Harvey" was the name of an invisible rabbit in a popular play and 1950 movie of the same name.

out a memo stating that he was not satisfied with the innovation he saw coming out of DOD research. The memo also invited organizations to propose radical new ideas. Representing the TTO Office, Moore nominated the "Harvey" idea, renaming it "High Stealth Aircraft".

Ken Perko from the Air Force Systems Command at Wright Patterson Air Force Base had recently been recruited as a program manager to build up a tactical air program within DARPA's TTO. Perko had worked in the Air Force on DARPA-sponsored work on "low-observable" research for drones and remotely-piloted vehicles, and had some knowledge of this field. DARPA's Moore therefore assigned Perko the task of contacting U.S. defense aviation contractors directly to solicit their ideas on approaches to achieve extremely low radar cross-section. Moore recalled that most of the vendors submitted slightly improved radar cross-section reduction, but nothing that would reach the order-of-magnitude goals that DARPA was seeking. Based on these initial submissions, DARPA ultimately funded small preliminary studies at Grumman, McDonnell-Douglas, and Northrop. Three formal study contracts followed, awarded to McDonnell-Douglas, Northrop, and Hughes (for its radar expertise). While these studies were under way, Lockheed became aware of the project (Lockheed had not been invited to participate initially because it was not considered to be active in tactical aircraft) and contacted DARPA requesting permission to participate in the first phase concept development, without compensation. This request went to DARPA Director George Heilmeier, who granted Lockheed permission.

DARPA Have Blue Prototype

By the summer of 1975, it was clear that only Lockheed and Northrop had credible, near-term concepts for making aircraft radically less visible to enemy antiaircraft radar. Perko, Moore and Heilmeier met to develop a strategy and decided that a full-scale flight demonstration would be needed to make the results convincing. However, Heilmeier insisted that the program should not go forward without Air Force backing. Air Force support was highly uncertain, as the Air Force saw limited value in a stealthy strike aircraft, given the severe performance compromises that they assumed would be required to achieve a very low radar cross-section. There were also competing Air Force R&D priorities, most notably the Advanced Combat Fighter program (which eventually became the F-16).

DDR&E Currie discussed the problem directly with General David Jones, the Air Force Chief of Staff, and General Alton Slay, the Air Force R&D Director. Although the Air Force remained skeptical as to a stealth strike fighter's value, Currie and Jones brokered a deal to obtain active Air Force support for the DARPA stealth program, provided that funding for the stealth development would not come out of existing Air Force programs, especially the F-16. With that agreement, Phase II of DARPA's stealth aircraft program—Have Blue—began in 1976. Lockheed won the sole Phase II award, in part due to the record of its "Skunk Works"[7] for on-schedule accomplishment of high-risk, high-classification projects, especially the SR-71 Blackbird.

Have Blue was a quarter-scale proof-of-concept aircraft designed to evaluate Lockheed's concept for "very low-observable" capabilities while meeting a set of realistic operational requirements. The development program at Lockheed's Skunk Works was highly classified (a Special Access Program or SAP), but managed in an environment open to experimentation and flexible problem solving, with a high degree of communication among scientists, developers, managers, and users. Shortly after the program began, its management was transferred to the Air Force, due to its being highly classified. Importantly, only a total of a dozen or so people in OSD, DARPA and the Air Force knew of the Have Blue program. OSD leadership under Currie and Myers kept the program focused and moving forward in the face of many fundamental uncertainties.

Transition to Air Force—Senior Trend

Successful flights of Have Blue planes in 1977 made it clear that a stealthy aircraft could be built and flown. Based on these results—and guided by the high priority of countering Soviet numerical superiority with U.S. technology—Currie's replacement in the Carter

7 "Skunk Works" was the name given to Lockheed's Advanced Development Programs (ADP), which was famous for rapidly developing new airplanes in an un-bureaucratic environment, https://en.wikipedia.org/wiki/Skunk_Works.

administration that took office in January 1977, Under Secretary of Defense (Research and Engineering) (USD(R&E)) William Perry sought accelerated development of a real weapons system. The DARPA stealth program was then immediately transitioned to an Air Force acquisition program—called "Senior Trend"— with an aggressive schedule to have operational planes in only four years, forgoing the normal development and prototyping stage. The objective was to build and deploy a wing of stealth tactical fighter-bombers (seventy-five planes) as rapidly as possible. Furthermore, in order to obtain the largest possible technical lead, it was deemed necessary to hide the acquisition by making Senior Trend a highly secret program. The resulting operational aircraft was dubbed the F-117A.

Impact

The first F-117A "stealth fighter" was delivered in 1981, and fifty-nine were deployed by 1990. In 1991, the F-117A was an outstanding success in the Gulf War. It helped the U.S. achieve early air superiority critical for defeating heavily defended targets. It did so in the face of the same type of Soviet anti-aircraft systems that had been effective against U.S. aircraft in Vietnam and other wars. In championing stealth, DARPA harnessed ideas from industry and the military service laboratories to pursue a radical new warfighting capability. Stealth combat systems had not been pursued because the Services lacked a strong interest in such a nontraditional concept. With high-level support from civilian leadership across presidential administrations, DARPA overcame that resistance, set out priorities, and obtained funding for the considerable engineering work to develop a proof-of-concept aircraft demonstration system. This demonstration enabled top civilian and Service leadership to proceed with confidence. OSD and Service leadership, once persuaded, rose to the challenge, and provided funding and support to implement a full-scale weapons program.

From the outset Have Blue was a "crash" program, designed to develop and deploy a breakthrough capability in as short a time as possible. Achieving this required a highly focused technology development, prototyping and acquisition approach. The approach was driven by a national-level strategic imperative that was initiated out of the Office of the Secretary of Defense and developed by DARPA. The subsequent implementation was through a highly classified Air Force Program with direct and close oversight of the Under Secretary of Defense for Research and Engineering. Throughout this process the focus was delivering an operationally capable stealth strike aircraft in four years. The imperative of offsetting the Soviet air defense capabilities drove decisions on the structure of the program, the selection of the performer, the oversight mechanisms. The program had ambitious but clear objectives that helped focus the contractor and the government on working together pragmatically to achieve the outcome.

Background and Role of the Program Manager

Ken Perko, the program manager for Have Blue, worked closely with TTO Director Robert Moore in (1) getting industry inputs, (2) assessing the competing approaches, and (3) selecting the eventual contractor, Lockheed, to produce the Have Blue prototype. While Perko had earlier experience in related DARPA programs in low observables when working for the Air Force, the idea to actively pursue such a radically different aircraft came from the top down, led by Myers and supported at DARPA by Moore.

Other Key Roles in Program Formation

Myers (Director of Air Warfare Programs in the Office of the DDR&E) was the true instigator of a "stealthy" tactical aircraft—initially called "Project Harvey". Indeed, Myers was a driver of new aviation concepts more broadly, including the notion of a mini-fighter that would be intrinsically low-observable. In essence he was OSD's aviation leader and engaged the Services and DARPA actively to pursue new ideas.

DARPA Director George Heilmeier was both a champion and a skeptic. He was an advocate of pursuing radical new concepts, and especially in scaling these up as proof-of-concept demonstrations. However, he also realized that only the Air Force could actually produce a successful aircraft weapon system. Therefore, he insisted that Air Force backing be obtained, which required intervention by Dr. Currie, the DDR&E. Heilmeier was actively involved with Moore and Perko in strategizing how the program should be scoped and conducted. His involvement was predicated on Have Blue being such a high-priority program with such high-level interest (as well as being a very high-cost program relative to most DARPA programs).

Director of Defense Research and Engineering (DDR&E) Currie had sent out a memo stating that he was not satisfied with innovation he saw coming out of DOD research. The memo also invited organizations to propose radical new ideas. Representing the TTO Office, Moore nominated the "Harvey" idea, renaming it "High Stealth Aircraft". Currie subsequently used his office to leverage Air Force participation in Have Blue and subsequently the Senior Trend program that led to the F-117A.

Moore focused DARPA's involvement in the Have Blue program. He took on Myers' challenge to see whether an "invisible" combat aircraft was possible and worked with program manager Perko to determine the options and develop the approach.

Key Insights

Have Blue shows that DARPA could be extremely responsive to high-level priorities of OSD and indeed the White House. DARPA saw itself as the organization that could and should take on high-risk programs that could fundamentally improve the national security position of the United States. This was exactly what it did in response to DDR&E Currie's (and Defense Secretary Schlesinger's) call for greater defense innovation to meet the Soviet threat. OSD articulated the challenge—can a stealthy aircraft be made? DARPA organized and funded the research to discern what could be done and then developed the prototype that demonstrated this.

DARPA conducted Have Blue as a "black program"—classified above Top Secret. This was done to keep the Soviet Union from knowing what was being done. Importantly, such programs are known within the DOD to very few, and also very few individuals outside (including only a handful in Congress). This permits them to proceed with less scrutiny than is the norm. However, such classification places a great deal of extra burden on the project management.

Have Blue shows the role of civilian leadership in pushing concepts that the military services resist. Stealth combat systems had not been pursued by the Air Force because they conflicted with their priorities and concepts for combat aviation. The Air Force lacked interest in such a nontraditional concept that compromised performance—especially speed, maneuverability, and self-defense. However, with high-level support from civilian leadership across administrations, DARPA overcame that resistance, set out priorities, and obtained funding for the considerable engineering work to develop a proof-of-concept aircraft demonstration system. Have Blue is also an example of where an OSD-identified need led DARPA to fund several conceptual studies, and then DARPA developed the most promising of these into a program. Such conceptual studies can be a key part of program development.

Assault Breaker (Standoff Precision Strike)

Overview

Assault Breaker was the demonstration of a concept for finding, hitting and destroying targets on a battlefield from a distance—known as "standoff precision strike"—by employing a "system of systems". The program combined airborne radar, long-range tactical ground-based missiles and terminally-guided submunitions, linked to a rapid, all-source targeting system. The Assault Breaker program began in 1978 and concluded in 1983, and was run through DARPA's Tactical Technology Office (TTO). It is generally recognized that the result of this program was a joint operational concept that would revolutionize the battlefield.[8]

Context and History of Program Development

Assault Breaker had its origins in a DARPA study jointly funded with the Defense Nuclear Agency (DNA) to define alternatives to allow the United States "to respond flexibly to a military threat from an aggressor nation". This was a large, multi-participant study comprised of strategic thinkers and technologists who were drawn together as the "New Alternatives Panels", organized under DARPA and DNA to respond to Presidential, National Security Advisor, and Secretary of Defense concerns that there was a need to "broaden

8 Van Atta, et al. (2003). *Transformation and Transition. Volume 1*, 15–16.

the spectrum of strategic alternatives" available (other than nuclear strike) to "limit Soviet aggression".[9] The classified work of these panels was simply titled the *Long Range Research and Development Plan*. These deliberations converged around new defense concepts that emphasized standoff precision strike. It was understood that to actually combine capabilities to do this would require unproven and unprecedented integration of a wide variety of technologies that dictated a unified development, integration and employment of both targeting and weapons systems.[10]

DARPA was given the task of implementing the precision strike concept based on the integration of inputs from (1) the Long-Range Research and Development Planning Program; (2) ideas from DARPA program manager Leland Strom for using Moving Target Indicator (MTI) radar to guide a missile to a target area and then use terminally guided submunitions to destroy the targets; and (3) briefings from industry on using tactical missiles with submunitions with electro-optical seekers. The Director of DARPA's Tactical Technology Office, Moore, drew upon these ideas to propose the Integrated Target Acquisition and Strike System (ITASS) as a DARPA program to develop and demonstrate such capabilities. Moore asked MIT's Lincoln Laboratory to flesh out this concept, including potential systems that could be incorporated, and the feasibility of enabling technologies that would be needed.[11] When DARPA Director Robert Fossum approved the program in 1978 it was renamed Assault Breaker.

Establishing the Assault Breaker Program

There had been several rather disparate R&D efforts of the military services on parts of the technology underpinnings of what became Assault Breaker, such as the newly deployed E-3 Sentry (AWACS) aircraft, which led to the DARPA-Air Force Tactical Air Weapons Direction System Program (TAWDS), which then was renamed Pave Mover.

9 *Ibid.*, 16, quoting *ARPA/DNA* Long Range Research and Development Plan, *Final Report of the Advanced Technology Panel* (1975), vi.

10 Van Atta, et al. (2003). *Transformation and Transition. Volume 1*, 8.

11 *Ibid.*, 18.

Pave Mover was then merged into the Assault Breaker program, and subsequently became JSTARS (Joint Surveillance Target Attack Radar System).[12] Similarly the Air Force and Army were both working on various programs to develop new munitions for attacking ground targets and ways to deliver these from a distance including an array of submunitions that could be directed to individual targets, including the Air Force's Wide Area Anti-Armor Munitions (WAAM) and the Army's Terminally Guided Sub-Munition (TGSM). These new individual weapons technology concepts were all inputs to Moore in DARPA's Tactical Technology Office, and all influenced DARPA PM Leland Strom in formulating a concept that integrated such capabilities, which he presented to Moore. These separate developments in sensing, missiles, submunitions, as well as command and control, were inputs into an *integrated* capability (system of systems) in a DARPA-funded project (ITASS) conducted by Lincoln Laboratory.

While these concepts were developed by 1976, the actual Assault Breaker Program to develop and demonstrate these integrated capabilities did not start until 1978. This was the result of several factors: (1) the change of Administrations in 1976, bringing in new leadership; and (2) concerns by new DARPA Director Fossum that the Assault Breaker was "fragile" in combining multiple capabilities that were unproven both individually and together in a combat environment. Moreover, Assault Breaker was itself different from "normal" DARPA military programs in that it was more about integration of several relatively near-term technologies, rather than a leap in technology itself. Thus, DARPA Director Fossum and his immediate superior, Under Secretary of Defense for Research and Engineering (USDR&E), William Perry, both new to the Pentagon in 1976, had to evaluate the complex proposals for standoff precision strike and determine whether and how to proceed. It should be noted that both Fossum and Perry were well versed in the earlier developments through their industry backgrounds and as advisors to DOD. Moreover, Perry was an enthusiastic advocate for the overall concept of standoff precision-guided weapons, as articulated in his testimony in 1978 upon becoming USDR&E.

12 Van Atta, R., Deitchman, S., and Reed, S. (1991). *DARPA Technical Accomplishments. Volume II*. Alexandria, VA: Institute for Defense Analyses, 5–6, https://apps.dtic.mil/dtic/tr/fulltext/u2/a241725.pdf

Background and Role of the Program Manager

The Assault Breaker program was the result of higher-level inputs above the DARPA program manager. The key individual in developing the program was Moore, who was Director of the Tactical Technology office. Assault Breaker was driven by a high-level strategic imperative from the White House (President Nixon and Security Advisor Kissinger) to address Soviet military capabilities threatening Western Europe. This translated into a DARPA-DNA sponsored study group that identified the general concept of standoff precision strike using conventional weapons as a way to "offset" Soviet-Warsaw Pact armor. However, it was Moore who harnessed inputs from a TTO program manager, Leland Strom, and inputs from industry, into an initial study by Lincoln Laboratory and then used that to formulate the Assault Breaker Program.

Other Key Roles in Program Formation

Assault Breaker was in fact a multi-project four-phase program, with these sub-projects managed by a set of DARPA TTO PMs. For example, the Pave Mover airborne reconnaissance aircraft (which subsequently became JSTARS) was under PM Nicholas Willis. In its first phase, the program supported continued development of individual component technologies, such as the sensors, radars, and automatic target recognition—most of which were being pursued within DARPA under various PMs. The second phase was then testing in parallel different contractor approaches for systems level capabilities. In the third phase, more complex integration of systems-of-systems was demonstrated in competition. Finally, the fourth phase linked together the integrated system into a large, complex demonstration.[13]

Assault Breaker was managed under a unique approach under DARPA with an actively involved steering group that included the Director of DARPA, Fossum, as well as Lt. Generals from both the Army and the Air Force. Notably DARPA reported directly to Under Secretary of Defense (Research and Engineering), Perry for this project. Moreover, Moore was elevated from DARPA to the position of Deputy

13 These phases are described in specifics in Van Atta, et al. (1991). *DARPA Technical Accomplishments. Volume II*, V-9, V-10.

Under Secretary of Defense for Tactical Warfare Programs, to provide continued oversight of this and related programs.[14] DARPA was thus given direct responsibility for managing what became a combined set of projects conducted mainly under the Army and Air Force.

Key Insights

The Assault Breaker program is an example of a very large-scale systems integration project, driven by highest-level military priorities, with the DARPA office director playing a key role in orchestrating the development of the implementing concepts.

- DARPA first supported a conceptual study (with the DNA) to determine an overall concept to meet a high-level security problem.
- DARPA then funded under the Office Director's initiative a detailed technical assessment of options and approaches for the integrated system-of-systems.
- Assault Breaker was an integration of multiple projects that were being individually pursued and managed by a set of DARPA PMs mostly being supported by individual military services. DARPA fostered the demonstration of these as an integrated system, which was largely counter to the culture and priorities of the separate military services.
- A unique management structure reporting to the Under Secretary of Defense (Research and Engineering) was established with the DARPA Tactical Technologies Office conducting day-to-day management.
- Program managers played primarily a management oversight role over very large individual sub-programs and their overall integration into a proof-of-concept demonstration.

14 Assault Breaker was in fact one of several large-scale DARPA programs for developing an integrated response to the Soviet Bloc. Another one was the Stealth aircraft program reported upon here as well. Moore moved to his position in OSD to provide broad oversight of all these programs as they matured and transitioned.

Amber/Predator (High Altitude Long Endurance UAVs)

Overview[15]

Amber, out of which grew the Predator unmanned aerial vehicle (UAV), was a specific program that developed from the Teal Rain program for advancing technologies for High Altitude Long Endurance (HALE) UAVs. The UAV that became Amber was proposed to DARPA in 1978 by its developer, Abraham Karem, who owned a firm called Leading Systems, Inc. The PM whom he briefed did not pursue the idea, but DARPA Director Fossum heard the presentation, overruled this rejection, and funded it out of his own office's funds. Based upon this support, Karem successfully developed and demonstrated a UAV called Albatross. DARPA then in 1984 began a program for Amber, a scaled-up version of Albatross. Amber was a classified reconnaissance UAV, which was flown in 1986—just two years after the initial DARPA contract. However, Amber was used only in small numbers (by the CIA), and, with no subsequent DOD business, Karem's firm, Leading Systems, Inc., went into bankruptcy and was sold to General Atomics. After a decade of delay, OSD pushed renewed interest in HALE UAVs and Amber was modified under a DARPA program to become Predator, an extremely successful intelligence, surveillance and reconnaissance (ISR) system that has been used extensively by U.S. and allied forces in conflicts in Iraq and Afghanistan.[16]

Context and History of the Program Development

The concept of an unmanned aerial vehicle can be traced as far back as World War I with a British radio-controlled "guided explosive laden unmanned air vehicle [intended] to glide into German ships". During World War II Germany further developed radio-controlled rockets,

15 Van Atta, R. H., Cook, A., et al. (2003). *Transformation and Transition: DARPA's Role in Fostering an Emerging Revolution in Military Affairs.* Volume 2. Washington, DC: U.S. Government Printing Office, VI-2, VI-5, https://doi.org/10.21236/ada422835, https://apps.dtic.mil/dtic/tr/fulltext/u2/a422835.pdf

16 Predator was subsequently fitted with a missile that allowed it to become an attack weapon itself.

including the V-1. During World War II the United States converted B-17s into BQ-7 radio controlled "flying bombs", and then after the war modified additional B-17s as the QB-17G for such purposes as collecting atmospheric samples from nuclear tests, and later as target drones. The Air Force in the 1960s worked with Ryan Aerospace to develop an unmanned reconnaissance aircraft called the Firebee, which was used to conduct reconnaissance over North Vietnam and southern China, particularly to substitute for the manned U-2 spy plane in heavily defended areas. The Firebees were air-launched from a C-130 aircraft. With the termination of the Vietnam conflict, and subsequent drawdown of forces, Air Force interest in UAVs waned.

DARPA and UAVs

DARPA's initial involvement with UAVs was with remotely piloted vehicles (RPVs) used first in support of tactical reconnaissance in Vietnam.[17] However, by the early 1970s the expense and complexity of these earlier systems led to their demise, and the Director of Defense Research and Engineering, John Foster, urged that DARPA should focus instead on using lightweight, rugged, inexpensive model airplane technology, which became DARPA's Mini-RPV program. That program led to the successful development and testing of relatively small, fixed-wing UAVs, but these did not transition into any operational UAVs, as the Army's Aquila program which was based on these ultimately failed when requirements, weight and costs spiraled out of control. Thus, DARPA's first foray into UAVs ended with little actual deployed capabilities.[18]

DARPA High Altitude Long Endurance UAVs

In 1978, DARPA funded the aircraft developer Abraham Karem to develop a very-long endurance very high altitude (90,000 feet and 5-day

17 Van Atta, et al. (1990). *DARPA Technical Accomplishments. Volume I*, 28–23, 28–25.

18 It should be noted that the technologies did become further developed and deployed as combat systems by Israel as the Mastiff, Scout and Pioneer UAVs. Ironically, the U.S. Navy and Army acquired the Pioneer from Israel and eventually this led to the development of the Shadow tactical UAV by AAI, which is now part of Textron. See Hirschberg, M. J. (2010). "To Boldly Go Where No Unmanned Aircraft Has Gone Before: A Half-Century of DARPA's Contributions to Unmanned Aircraft", *American Institute of Aeronautics and Astronautics* (January): 11–13.

flight endurance) UAV under the Teal Rain program. Teal Rain was a classified DARPA program to explore technology for long endurance UAVs driven by the problem that prior efforts, largely by the Air Force, had resulted in very large and expensive aircraft. Teal Rain projects were expressly "unfettered, technology-push studies to generate new ideas".[19] Based on this initial support Karem, using his own funds, and under his own firm, Leading Systems, Inc., built prototypes of a new UAV, the Albatross, for which DARPA then supported flight tests. DARPA then began a program in 1984 for Amber, a scaled-up version of Albatross. Amber was a classified reconnaissance UAV, which was flown in 1986—just two years after the initial DARPA contract.

From a technical standpoint Amber was highly successful and Leading Systems invested in considerable technology development for improving performance and operational capabilities. However, in 1987, when the program was transferred to the Navy, Amber became a victim of Navy funding priorities. Moreover, Congress established within the DOD a Joint Program Office for UAVs consolidating all the military efforts.

With existing UAVs meeting then current Service requirements, the more advanced Amber was not selected to continue into acquisition. Leading Systems could not survive this misfortune and was sold first to Hughes and then to General Atomics. Karem, now associated with General Atomics, kept the Amber concept alive by developing a lower performance version called the Gnat 750, which was aimed at the international market. A few were sold to Turkey. Others were acquired by the CIA, which supported further development.

Predator

In 1990 the Joint Requirements Oversight Council (JROC) of the Joint Chiefs of Staff established a requirement for Long Range Endurance Reconnaissance, Surveillance, and Target Acquisition. The JROC put forward a three-tier approach for this.[20] Tier I was a quick reaction

19 Van Atta, et al. (2003). *Transformation and Transition. Volume 1*, VI-15, quoting DARPA Program Manager Charles Heber.

20 The three-tier concept was articulated in a memo by Deputy Secretary of Defense John Deutch. Tier III was to be a very high altitude, long endurance stealthy UAV. After considerable machinations, Tier III devolved into two alternative

capability that could be satisfied by the General Atomics Gnat 750. Tier II was labeled "Medium Altitude Endurance" and a scaled-up version of the Gnat 750 was seen as the best approach for this. This became the Predator. Predator was initially an incremental modification of the Gnat 750—essentially a stretched airframe and longer wings with additional ISR sensors—with linkage to satellite system for communications. Subsequent developments added substantial new operational capabilities for target acquisition and strike. The initial system comprised an aircraft, sensors, communications capabilities, and a ground station for aircraft control. Subsequently, laser target designator capabilities and then Hellfire missile launch capabilities were added.

As an aircraft, Predator is not highly complex. Primary complexities were involved in the control software and in the satellite communications linkage. The operational linkage through the Ground Control Station was a complicating factor. The technologies were generally mature. Most of the technology had been developed under DARPA, although with limitations and iterative developments. A major new development was use of satellite communications. Predator used GPS satellites for navigation, being the first UAV to overcome line-of-sight range limitations through use of satellite technology. Predator used commercial satellite data links for control and imagery transmission.

While much of the technology in the Predator system was in place, the implementation of a tactical intelligence, surveillance, and reconnaissance (ISR) UAV in the field was largely untried. The implementation of this system became an urgent priority of Office of the Secretary of Defense (OSD)-driven due to a need to have ISR capabilities to support efforts in Bosnia and later in Iraq. Consequently, Predator was developed as an urgent program, although not based on formal military-service derived requirements. The Gulf War in 1991 highlighted serious deficiencies in airborne tactical-level ISR, particularly for wide-area coverage. The Predator arose out of high-level (Secretary of Defense, Under Secretary of Defense, Joint Chiefs of Staff, Director CIA) concerns that these ISR capabilities needed to be kept affordable.

---

platforms—Tier II+, which became the Global Hawk UAV, and Tier III-, which was called Dark Star, a smaller, stealthy system. Dark Star was cancelled after two crashes and costs that escalated excessively. Global Hawk subsequently became very successful in operations in Iraq, Afghanistan and elsewhere.

Predator was put into service using a "non-standard" accelerated process known as the Advanced Concept Technology Demonstration (ACTD) process. Use of the ACTD "allowed use of a streamlined management and oversight process, provided for early participation of the user community, and bound the schedule length. The goal of the ACTD was to demonstrate military utility in a relatively short timeframe. The use of mature technology was intended to limit risk".[21] Under the ACTD process, Predator was delivered for user experimentation in just six months. Predator was successfully employed in Bosnia (just a year after its first flight), Kosovo, and the no-fly zone in Iraq. Predator was later used in Afghanistan, becoming a weapons platform, firing Hellfire missiles.

Predator provides a clear example of a successful demonstration of innovative new capabilities prior to their being identified as military requirements. With this demonstration the operational community championed the novel HALE UAV capabilities for use in combat. Through this demonstration "technology push" became "demand pull" and the Predator went from demonstration to an accelerated acquisition. Of paramount importance was the fact that Predator met a compelling need for which there was no existing system, and that it was able to evolve to meet additional needs as these were identified.

Background and Role of the Program Manager

The program manager for the HALE Program was Charles Heber who served as director of the High Altitude Endurance Unmanned Air Vehicle Joint Program Office at the Defense Advanced Research Projects Agency (DARPA). Previously, he had served as Deputy Director of DARPA's Tactical Technology Office, where he oversaw UAV programs. Prior to that he was deputy director of technology for the Office of Naval Research's (ONR's) Low Observables Technology Office. Heber was the manager of this set of programs, not the initiator of the ideas for it. The ideas were brought to DARPA from the outside (primarily by Abraham Karem for Amber and then Predator).

21 Drezner, G., et al. (1999). *Innovative Management in the DARPA HAE UAV Program*, MR-1054-DARPA. Santa Monica, CA: RAND Corporation.

Other Key Roles in Program Formation

Dr. Robert Fossum, the Director of DARPA, played a crucial role in formulating a program around the notion of a high altitude, long endurance (HALE) UAV. He initiated DARPA's Teal Rain program that investigated advanced technology concepts for HALE—essentially technology push programs to generate new ideas. One of these programs was Karem's Amber. According to Fossum, he personally supported Amber when the cognizant PM was uninterested in pursuing it.

While DARPA, particularly under Fossum, supported HALE developments, these developments foundered with military service lack of interest until a decade later. In the 1990s DARPA became reengaged with the high-level of interest of OSD and the Joint Chiefs of Staff in implementing UAV-based long endurance ISR capabilities. Notably, many of those pushing for this were experienced in the prior HALE UAV efforts through DARPA in the 1980s. This included Secretary of Defense Perry, and Larry Lynn, who had been Deputy Director of DARPA in the early 1980s and was now Deputy Under Secretary of Defense for Advanced Systems and Concepts. Lynn's position was, in fact, created expressly to achieve a breakthrough in ISR technologies. He and others were convinced that only DARPA could effectively manage the ambitious HALE UAV implementation that would lead to both Predator and Global Hawk being fielded.

Key Insights

Some key lessons from the HALE UAV evolution and development include:

- The concept of UAVs did not originate with DARPA—there had been prior efforts to develop and deploy them. However, military service interests in UAVs were generally short-lived and at critical junctures DARPA was critical in promulgating and refocusing UAV developments.
- DARPA's work in support of UAVs has spanned several decades, starting in the 1970s, but was not continuous. The

programs that eventually led to implemented systems were built upon previous efforts.

- The initial impetus for smaller mini-RPVs came directly from Director of Defense Research and Engineering (DDR&E), John Foster, who encouraged DARPA to take on this new direction.
- DARPA's focus on RPVs corresponded with an OSD level focus on addressing Soviet Bloc (and Chinese) threats by being able to see and hit deep targets and quickly destroy their forces before they could mass for strike.
- DARPA supported development of several enabling technologies essential to overall UAV capabilities including sensors, command and control, structures, which contributed to UAV communication, navigation, targeting.
- DARPA determined at the highest level (DARPA Director) to move away from smaller tactical UAVs (RPVs) to High Altitude, Long Endurance UAVs. This refocusing was supported by inputs from high-level advisory organizations (Defense Science Board) and OSD leadership.
- DARPA leadership generally supported the concept of High Altitude Long Endurance (HALE) UAVs despite the lack of interest of the Military Services. However, the specific HALE concepts were brought to DARPA by individuals (Karem) and firms (Boeing, Ryan Aeronautical, General Atomics).
- DARPA helped to develop a novel, non-standard approaches for development and initial acquisition (the ACTD mechanism) to speed implementation of UAVs.
- Strong high-level (OSD) support for the development, demonstration and deployment of novel HALE UAV defense capabilities outside of standard Service processes were crucial for these new capabilities to gain traction.

Optoelectronics Program[22]

Overview

Optoelectronics at DARPA is generally considered to have started as the optronics program that began in 1984, under John Neff. The program was stimulated by requirements of the Strategic Computing Initiative that DARPA launched in 1983, which required advances in networking and signal processing. In the 1989–1992 period the program was expanded and renamed the Optoelectronics Program, taking advantage of congressionally provided funds for university-industry consortia and university optoelectronic centers. The program started in the Defense Science Office (DSO), but moved to the Microsystems Technology Office (MTO) when that office was established in 1991. The program led to major advances in optical communication, including networks that use "wavelength division multiplexing" (WDM). The program was followed by several additional DARPA programs that made further advances at the component and system level, and in the integration of optical and electronic technologies. The program and its successors are credited with accelerating the development and demonstration of WDM components and systems, encouraging the adoption of technical standards that helped the industry grow rapidly, and creating community of experts who helped North American companies move quickly in WDM.

Context and History of the Program Development

Several influences came together to shape the optoelectronics program. One influence was the increasing importance of high-performance computing and networking. By the 1980s, the U.S. military relied increasingly on advanced information technology and communications for intelligence, battlefield intelligence, and logistics. DARPA had long supported computing and networking technology, including the foundation of the ARPANET and Internet. In 1983 DARPA launched

22 Sources for this section include: interview of Dr. Andrew Yang, by authors March, 2016; Optoelectronics Industry Development Association. (2001). *Creating Bandwidth for the Internet Age*. Washington, DC: OIDA; Block, F. L., and Keller, M. R. (2011). *State of Innovation: The U.S. Government's Role in Technology Development*. New York, NY: Paradigm Publishers.

a Strategic Computing Initiative to advance computing, which included the optronics programs. In 1987, the Reagan Administration proposed a new high-performance computing initiative, including networking, which evolved into the High Performance Computing and Communications Initiative (HPCCI). DARPA volunteered to take the lead in advancing the technology of networking. DARPA expanded its support of the development of experimental networks and the underlying technologies, including optoelectronics.

Another stream of influence was the evolution of optical communications in the telecommunications and computing industries. Optical fiber-based communications had rapidly been expanding in telecommunications, but using only one frequency of light at a time. Since the mid-1970s, researchers considered the possibility of sending multiple streams of light down the same fiber using different wavelengths to increase the data flow through the fiber, known as wavelength division multiplexing (WDM). Early work was done by both the telecommunications industry, led by AT&T and Bellcore, and the computing industry, led by IBM. In the early 1980s, AT&T used an early version of WDM in a pilot system. However, WDM at this time was limited by two problems. First, to transmit signals over long distances, there was a need to amplify the signals along the way, and this required converting optical signals back into electronic signals, then amplifying them, and subsequently reconverting them back to optical signals. This process was very expensive. A second key challenge was converting data streams into and out of wavelength-divided light signals, through multiplexing and de-multiplexing.

In the late 1980s, there were possible solutions to both of these problems. The development of the erbium-doped fiber amplifier provided the means to amplify light signals without having to convert them to electronics. Advances were also made, primarily by IBM, in multiplexing and de-multiplexing the light signals.

While in the late 1980s there was industry interest and capability in these technologies, both the telecommunications companies and computer companies were under stress. Previously, IBM and AT&T had monopoly or near-monopoly positions that allowed them to generously fund R&D. However, the breakup of AT&T and IBM's weakening competitive position led to reduced R&D funding, and neither saw optical communications as a lucrative market.

Another important part of the context was the decline of U.S. competitiveness in information technologies in the 1980s. This led to Congressional concern and interest in expanding investment in key technologies and in supporting industry. In 1990, Congress gave DARPA extra money in the fiscal year 1991 budget to fund a series of industry-university-government R&D consortia. Congress had earlier provided funds for the SEMATECH consortium, in which the Defense Department and the semiconductor industry shared the cost of a project to improve semiconductor-manufacturing technology. Senators such as Jeff Bingaman (NM) were impressed with the early results of SEMATECH, and decided to extend the model to other areas of technology. Congress did not earmark the new money for any particular technologies or projects, but instead left the decision on what projects to fund to DARPA.

The combination of these influences created a situation in which the DARPA program managers believed it was timely to pursue a program to take advantage of the recent advances in component WDM technologies (light amplifiers and multiplexing), in order to make major progress in digital communications systems that would have both defense and commercial benefits.

The program managers put forth a proposal to spend $20 million of this extra FY (Fiscal Year) 1991 money for optoelectronic consortia. DARPA's leaders agreed, and later they added approximately $10 million of regular FY 1992 agency funds to this effort. Three consortia received this initial funding, with a focus on developing experimental WDM systems. These three DARPA-supported projects helped revolutionize optical communications. They included:

- The Optical Network Technology Consortium (ONTC). Bellcore (later Telcordia) led ONTC. Other participants included Nortel, Rockwell, the Hughes Research Laboratory, United Technologies, Lawrence Livermore National Laboratory, Columbia University, and Pacific Bell. ONTC is generally credited with designing the standard systems architecture for long-distance, telephone based WDM fiber networks. Several of the key technical participants in ONTC went on to play roles in a subsequent DARPA project, MONET, discussed below.

- All Optical Network Consortium (AONC). MIT's Lincoln Laboratory led this research program, with participation by Bell Labs, MIT, and Digital Equipment Corporation. AONC drew heavily on earlier AT&T research and on government-funded R&D investments at Lincoln Lab and MIT. Its goal was to create a high-speed fiber-optics architecture that was entirely optical, with no electronic regenerators needed to amplify weak optical signals, and was well suited to handling computer data rather than phone calls.
- IBM. IBM won the third contract and focused on developing its ideas for key components for WDM, particularly multiplexer/de-multiplexer ("mux/demux") devices that take multiple data streams, mix them into the WDM light streams, and separate them out again at the end of the fiber line. IBM built one of the first practical WDM networks.

These three consortia focused on the development of both (1) key WDM devices, such as mux/demux devices, and (2) systems architectures that would enable an entire WDM network to operate. DARPA's office for electronic devices, MTO and its computing office (now the Information Innovation Office) cooperated in funding and managing this program.

In 1991, during the time of these three initial consortia projects, DARPA also began funding university centers in optoelectronics. These generated graduate students trained in the new technology and continued to advance the technology. They focused on improving devices for fiber-optic networks and other applications.

The 1991 Gulf War reinforced Pentagon and DARPA interest in developing new data communications technology, and this interest, combined with the technical successes of the three consortia projects, led to a new DARPA initiative—the Broad Band Information Technology Program (BIT), also known as Global Grid. An important Global Grid project was MONET—the Multi-Wavelength Optical Network Project, which ran from FY 1994 through FY 1999. Led by Bell Labs, AT&T Labs (which formed after most of the original Bell Labs went to Lucent), and Bellcore, MONET extended the work of the earlier ONTC project.

MONET brought together the key people from telecommunications companies, equipment producers, and government users, and developed a realistic and feasible WDM architecture. It also promoted technical standards and created a community of WDM experts.

Background and Role of the Program Manager

In 1989, Andrew Yang, became the program manager. He was hired to replace John Neff who was PM from July 1983 through September 1988, and who had come to DARPA from the Air Force Office of Scientific Research (AFOSR). Yang came from the Hanscom Air Force Base in Massachusetts, which was the Air Force's center for developing and acquiring command and control, communications, computer, and intelligence systems, and is also the location of MIT's Lincoln Laboratory. He was recruited to DARPA by Sven Rooslid, another Hanscom alumni at DARPA. Yang was considering retirement when the opportunity for the DARPA job came up. He left the Air Force and joined DARPA.

Yang changed the name of the program from optronics to optoelectronics, but did not make other major changes initially, and continued to support the development of new optoelectronic devices. When the consortia money became available from Congress, Yang put together a proposal for this, and was successful.

Yang stressed that it is better for a PM not to push his/her own idea, but rather to find the best ideas and push those,[23] arguing that this will result in better ideas and more support for these ideas. He further stressed the importance of being flexible and pursuing more than one path towards the goal. Developing the right program is largely a matter of timing (technology, needs, and funding all coming together). One needs to be able to adapt if new opportunities come up (or new sources of funding appear).

He noted that there are a lot of personal connections between PMs, researchers, and future PMs. Technical communities recognize that it is good to get their people into DARPA to help keep the funding flowing to their community.

23 Interview of Dr. Andrew Yang, by authors March, 2016.

Yang was followed, in 1993, Anis Husain, as well as Robert Leheny, and Brian Hendrickson. Their programs continued optoelectronic consortia (e.g., MONET project) and invested in optical signal processing technologies and integration of optoelectronics and electronics on chips.

Other Key Roles in Program Formation

Congress played a key role in providing funding for the consortia, as well as support for working on projects that have commercial as well as defense benefits.

Industry played a significant role in shaping the program. There was industry interest in establishing optoelectronics consortia by 1989. In 1991, the industry formed the Optoelectronics Industry Development Association to provide an organized voice for industry. DARPA provided funding for OIDA to create technology roadmaps, which in turn provided information to DARPA about important technology needs. Industry played a central role in establishing the consortia that were the center of the program.

There was little direct influence on the optoelectronic program from the military services or headquarters, but there was substantial interaction with the defense research laboratories, and especially the Air Force laboratories, due to the close connection between the DARPA program managers and Hanscom Air Force Base.

Key Insights

In the 1990s, there was strong emphasis on industrial competitiveness through consortia, and DARPA was given funds to support them. DARPA efforts included a focus on community building and standard setting, in addition to making technology breakthroughs. DARPA funded the optoelectronic industry's technology roadmaps and formed research consortia that developed real world WDM architecture.

DARPA supported optoelectronics in some form from at least 1985 to 2005 in a series of projects that built upon, at least in part, previous projects. In this regard, the important role of the program manager is not necessarily coming up with a completely original idea, but rather in understanding what the right program is to advance the field at a particular time. It is important to sense when component advances make advances in systems technology possible. In this case, it was also important that the advances to expand digital communications capacity occurred just as the Internet was expanding, creating demand for increased bandwidth.

At the time of the program, DARPA management was not highly metric-driven. Broad Agency Announcements (which formally announced funding opportunities to the public) were relatively new—established around 1990.

The case also illustrates some of the networks from which DARPA program managers are drawn. In optoelectronics, several of the PMs came from or had particularly strong links with the Air Force technical community.

High Definition Systems[24]

Overview

The DARPA High Definition Systems program was started in 1989 as the High Definition TV program. It was renamed as the High Definition Systems Program in 1990 and continued until 1993. It was started in DARPA's Defense Manufacturing Office. After this office was discontinued in 1991, the program became part of the Electronic Systems Technology Office. The program supported work on a number of display-related technologies, including materials and manufacturing techniques. One novel technology supported by the program, digital mirror projection technology, became a commercial success in electronic projectors, and led to an Emmy Award in 1998 and an Oscar Technical Achievement Award in 2015.[25]

---

24 Sources for this section include: Interview by the authors of Marko Slusarczuk (DARPA PM—High Definition Systems Program—Defense Manufacturing Office, 1989–1993); Sternberg, E. (1992). *Photonic Technology and Industrial Policy: U.S. Responses to Technological Change.* New York, NY: State University of New York Press, 207–18.

25 Their OSCAR citation read as follows: "To Harold Milligan, Steven Krycho and Reiner Doetzkies for the implementation engineering in the development of the Texas Instruments DLP Cinema digital projection technology. Texas Instruments' color-accurate, high-resolution, high-quality digital projection system has replaced most film-based projection systems in the theatrical environment", http://www.oscars.org/news/21-scientific-and-technical-achievements-be-honored-academy-awardsr.

Context and History of the Program Development

Some key aspects of the context for the HDS (High Definition Systems) programs were that:

1) In the 1980s, the U.S. competitive position in many technology industries, including electronics, appeared to be declining, primarily with respect to Japan. This was a matter of national concern, but also political debate. Democrats, who controlled the Congress, generally advocated a more aggressive government role to help technology industries through R&D, while Republicans, who controlled the White House, were opposed to industrial policies that would support specific commercial industries. SEMATECH (also funded through DARPA), a consortium to help the semiconductor industry and its suppliers, was formed in 1987 with support from both political parties in Congress and the White House.
2) Throughout the mid-1980s, there had been substantial discussion that high definition television would be the next driver of consumer electronics and information technology. Both Japanese and European TV manufacturers were discussing analog standards for the HDTV. U.S. manufacturers had already largely withdrawn from the television market, but some saw HDTV as a way back in. Displays were recognized to be important for a variety of defense applications, but the display industry was also seen as important to maintaining U.S. capabilities in electronics.
3) The U.S. Department of Commerce had considered a program to support HDTV, but this was rejected by the Bush administration as industrial policy. DARPA did not feel limited by this restriction because it could justify support of the technology due to its importance to defense.

DARPA Director Craig Fields initiated the HDTV program. He viewed high resolution displays as critical for defense, but also saw HDTV as important for the U.S. electronics and semiconductor industries. Firms in these industries were viewed as being important to maintain the defense industrial base to produce the technologies the DOD needs.

The one remaining U.S. television maker, Zenith, had contacted DARPA with a proposal for a research project. Craig Fields and others explored this, talked with other companies, and held a workshop on photonics. They started a $30 million program and released a Broad Agency Announcement (BAA) in 1989. It attracted substantial interest, with eighty-seven proposals being submitted.

Background and Role of the Program Manager

Marko Slusarczuk was hired as program manager in 1989 to manage the HDTV program. He came to DARPA from the Institute for Defense Analyses (IDA, a Federally Funded Research and Development Center that serves the Office of the Secretary of Defense, and works closely with DARPA) which he had joined in 1984 as a research staff member after earning an ScD in Materials Science from MIT and a law degree from Boston College Law School and having practiced law.

He was urged to apply to be a PM at DARPA by Ruth Davis, an IDA board member, who recommended him to DARPA Director Fields. Fields had just begun the HDTV program and had an interim PM, but was looking for someone to take it over fulltime. Slusarczuk knew of the program based on a *Washington Post* article and specifically asked that he be its PM, and Dr. Fields hired him for the position.

Slusarczuk was not initially a display technologist, but he had a substantial background in the underlying microelectronics and materials technologies. He stated that his main source of ideas for development came from his interactions with individual companies and academic researchers. Moreover, he had also earned a law degree and understood issues regarding business development. His experience at IDA gave him a perspective regarding defense interests in microelectronics generally. This background helped Slusarczuk see the need to support not just the end-product display technologies, but also the underlying component and materials technologies, which included the highly specialized glass substrates for displays produced by Corning Glass and color filters produced by Brewer Science.

Once hired as PM, Slusarczuk had a high level of autonomy to reshape the program. DARPA had brought in subject matter experts as reviewers from the three military services to assess the proposals that responded to the BAA. These reviewers selected three technologies as inputs that they recommended DARPA pursue: Liquid Crystal Displays (LCDs)—primarily of interest to the Air Force for aircraft cockpits and for use in large screen command centers; Electro-luminescent (EL) displays—primarily of interest to the Army for ground vehicles; and Plasma displays—primarily of interest to the Navy for large ship displays. The reviewers specifically rejected several other more novel display technologies. Slusarczuk reviewed all the submissions to the BAA and the reviews and determined that some of the technologies the reviewers had rejected should be supported. In particular, a proposal that had been rejected was the Texas Instrument (TI) Digital Mirror project, which Slusarczuk decided merited more attention. He consulted with another DARPA PM, William Bandy of the Microelectronics Office, who agreed with him that the digital mirror technology, while risky, had great potential. Slusarczuk funded the TI project as well as the three projects the military services recommended. The funding was not sufficient to fully fund the TI project. Nonetheless, he encouraged TI to proceed and to take on the risk, stating that "DARPA will take on all risks of failure",[26] and thus essentially asserting that TI would be shown as successful. He was able to provide additional funds to TI the following year, and the project was indeed a success. TI further developed the digital mirror technology, which became a commercial success. 80 percent of movie houses and 50 percent of all electronic projectors use the TI technology.

The program was originally focused on High Definition Television (HDTV), which Slusarczuk viewed as too narrow and too commercially oriented (given the political dispute over the appropriateness of DARPA helping commercial industries). He reoriented the program to High Definition Systems (HDS). Fields was removed from the DARPA Director position in 1990 in part due to his disagreement with the Bush White House on DARPA's role in supporting dual-use technologies.

Slusarczuk saw his approach as consisting of (1) providing an overall vision; (2) identifying and filling holes; (3) providing connectivity across the technology area. From his perspective, his role was to seek out potential in what was unproven. Slusarczuk said he saw himself "as the conductor of an orchestra". He was "totally unconstrained" with

26 Interview of Marko Slusarczuk by the authors, March 2016.

no reviews, no specific milestones. He had to demonstrate progress, but was not held to concrete milestones. This flexibility allowed him to adjust program direction as the technologies evolved. He could make decisions without consulting management at each step. He feels that this was the general approach at DARPA at the time.

Regarding how he developed the program, he said he had complete authority within the budget to layout and pursue his research agenda. He mentioned that he even briefed Secretary of Defense Dick Cheney and Chairman of the Joint Chiefs of Staff General Colin Powell without having to review these briefs with anyone at DARPA. He said that today at DARPA that this would be very unlikely.

Management was very hands-off. Slusarczuk stated that he never had to seek approval for any decisions once he became the PM. He informed and consulted with management, but the decision ultimately rested with him. He worked under Michael Kelly, the Director of the Manufacturing Technology Office (MTO) at DARPA.

Slusarczuk said another thrust he took on his own was funding companies to work on underlying manufacturing technologies needed for making advanced displays. This included companies such as Applied Materials, which made production equipment for depositing and etching the amorphous silicon for LCDs, Standish Industries for assembling the glass panels into displays and filling them with the liquid crystal material, and MRS which made lithography equipment for imaging the electronics onto the glass substrates. He supported work on the phosphors needed for plasma displays (Phosphor Center of Excellence at GA Tech, plus individual research efforts to develop blue phosphor). He also conceived an industry consortium (USDC) for providing inputs from display makers on the equipment and materials infrastructure needed.

He encouraged or required participants in his program to work together in a variety of ways. He required university programs that received more than $250,000 from his program to send their principle investigator to a private company working on the DARPA display program to learn what problems commercial firms had in display technologies. He also used annual "information exchanges" in which all participants in his program were required to attend in order to "share and collaborate". During these sessions he said he would hold special meetings with specific participants to encourage linkages between firms based on connections that they might not themselves see. "I could do this because I had knowledge across the program that they didn't".[27]

Other Key Roles in Program Formation

The DARPA Director, Craig Fields, played a key role in establishing the program. There was also strong influence from Congress and the White House. Political sensitivities encouraged the shift from HDTV to HDS.

Congress strongly supported the program, while the White House was initially opposed. With the change from President Bush to Clinton, the White House also strongly supported the program, and enabled additional funding.

Industry also played a role, with early support for a program coming from industry. The idea for the digital mirrors technology, which became one of the most important parts of the program (and perhaps produced the most notable result), came from an industrial proposal in response to the Broad Agency Announcement.

The DOD services had input to the program through their review of proposals in response to the BAA. They each tended to want to continue to support technologies in which they already had some involvement (plasma, LCD, electro luminescent).

Key Insights

In this case, the PM was not the source of the idea for the program, but had a major influence in shaping the direction of the program. The program idea came from the DARPA director, based on his view of what was important to both industry and the defense establishment in the long run. The PMs role was as the conductor of an orchestra and driver of the program; he identified gaps that needed to be filled for the program to succeed. This case also illustrates that sometimes DARPA's originality is not in the idea for the program, but in its ability to support creative ideas within the program (in this case the digital mirror technology).

---

27 Interview of Marko Slusarczuk by the authors, March 2016.

The PM at that time (and in that office) had a high level of autonomy in this case, and was not required to meet rigid metrics. The case also illustrates the importance of Congressional and Administration politics in some areas of DARPA technology at some times.

Spintronics (Quantum Computing)[28]

Overview

Development of magnetics-based and quantum microelectronics at DARPA was initiated and sustained by program manager Stuart Wolf in DARPA's Defense Sciences Office (DSO) from 1993 through 2005. The Spintronics program developed non-volatile magnetic memory (MRAM) devices and led to SPiNS, a project which sought to develop spin-based integrated circuits (ICs). During this period Wolf started a dozen related programs in the field of magnetics and electron spin for microelectronics. Thus, Wolf exemplifies the role of PM as a program initiator—in fact, he was what might be termed a serial instigator of programs, as he sought to develop and build on the initial ideas into increasingly diverse and complex technology developments.

Context and History of the Program Development

Stuart Wolf became a project DARPA manager in 1993 while he was still at the Naval Research Laboratory (NRL), where he was the Branch Head in Materials research. In that capacity he had provided technical consultation to DARPA's Defense Science Office, specifically to program manager Frank Patten on high temperature superconductivity. High temperature superconducting materials had been discovered in 1987 and DARPA wanted a program in this area. Patten asked Wolf to help put together a program. Thus, Wolf was a government scientific expert who advised DARPA on creating this new program.

In 1993, Wolf informed Patten that he was to take a sabbatical from NRL and was considering going to the National Science Foundation for the year. Patten suggested that Wolf instead come to DARPA as a

28 This section is primarily based on Stuart Wolf, interview with Richard Van Atta, March 2016.

PM—but to do this he would have to come for a minimum of two years. To accommodate Wolf, DARPA agreed for Wolf to be "part-time" an NRL while serving as a DARPA PM.

Wolf had specific ideas on developing his own program at DARPA based on developments in magnetic materials and devices. His branch at NRL had explored various aspects of magnetic materials, including work on how to make magnetic thin films. This research had contributed to the development of Giant Magnetoresistance (GMR) in France and Germany. Wolf's idea was to explore possible applications of GMR structures.

Spintronics

Wolf began what he termed a "super-seedling" with $5 million of funds from the Technology Reinvestment Project (TRP).[29] The participants in this program included IBM, Motorola, Cornell University, and Non-Volatile Electronics (NVE). Since IBM had already been working on GMR sensors for hard drives, this application was eliminated from this project. The seedling led to two results: (1) magnetic sensors, and (2) non-volatile magnetic memory (MRAM). The latter, MRAM, was developed by IBM, Motorola, and Honeywell. The program was explicitly dual-use, as "DOD uses a lot of magnetic memory", but the then current technology—plated wire magnetic memory—was comparatively bulky. The argument was that use of MRAM technology would allow a memory device with 128 Kb capacity, that cost $250,000 and weighed 40 pounds, to be replaced by an MRAM megabyte chip that would cost on the order of $1000. Wolf renamed this program "Spintronics" for SPIN TRansport electrONICS. In an interview with TPI, Wolf noted a couple of additional features of this program: (1) it lasted 10 years; (2) it was cost-shared with industry on a sliding scale in which for the first year the funding was 80 percent DARPA and 20 percent industry. The funding then shifted progressively more to industry (70–30, 60–40, 50–50) so that by the end industry was paying the bulk of the costs.

29 The Technology Reinvestment Project was an in initiative of President Clinton to use defense funds to support dual-use technologies, with the intent of helping the defense-related industries shift to non-defense markets following the end of the Cold War. See Congressional Budget Office. (1993). "The Technology Reinvestment Project: Integrating Military and Civilian Industries", July, https://www.cbo.gov/sites/default/files/103rd-congress-1993-1994/reports/93doc158.pdf

Beyond Spintronics

In the TPI interview, Wolf said that he drew upon his background as Branch Head in electronic materials at NRL to conceive of additional programs for DARPA. One of these was Frequency Agile Materials for Electronics (FAME), which drew on NRL work on superconductivity used in tunable filters. The military application that was the initial focus of this program involved replacing phased arrays that were then controlled using costly diodes. The advantage of "paraelectric devices" resulting from the FAME Program was that they varied continuously and were much cheaper. Initially they were manufactured using ceramic materials processing, but later were made with sputtered thin film processes. Devices based on this technology now are used in cellphones.

Wolf said that the process he went through was very straightforward. He would propose an idea to the Office Director of the Defense Science Office, who was very supportive of his ideas, and then he (Wolf) would "pitch the idea" to the DARPA Director. He said his ideas were generated from his role as a Branch Head at NRL and his own technical reading about advanced electronic materials. For example, his reading of research papers on the prospects for magnetic semiconductors—including one from Japan on a GaMnAs magnetic semiconductor that could be tuned using an electric field—led him to believe that this would create a new opportunity for spin-based ICs, which he pursued in his spintronics program. One program that evolved from this was DARPA's SPINS program (SPin IN Semiconductors).

Wolf also funded a consortium to explore whether it was possible to create gate-defined quantum dots as Qbits. They produced a single electron quantum dot as a Qbit using GaAs. Later this was done with silicon. Wolf decided that, while this was one way to produce a Qbit, there were other approaches that were being developed. He conceived of a project called QuIST—Quantum Information Science and Technology—one of which's goals was to identify the best way to produce Qbits. This project has led to on-going research. Furthermore, Wolf has stimulated other programs for other DARPA PMs. One example is a program on "metamaterials", which his colleague from NRL, Valerie Browning, started when she came to DARPA as a PM. One outcome of this program is negative refractive index materials, which are used in specialized lenses and antennae.

Background and Role of the Program Manager

Wolf is an interesting example of DARPA's varying approach to Program Management. He was recruited by a current PM based on his having supported that PM as a technical advisor for several years. Wolf is an expert on electronic materials, magnetism and related superconductivity—an expertise deriving from his having been a scientist and manager at the Naval Research Laboratory. This background was the basis of his knowledge and connections that permitted him to conceive so many DARPA projects. Thus, he brought to the PM position long-standing expertise in the new field of quantum electronics.

He was brought into DARPA's DSO as an employee of the NRL and stayed at DARPA from 1993 to 2005, being renewed year to year by the DARPA Director. By 2003 the DARPA Director, Tony Tether, decided that, since Wolf was essentially fulltime at DARPA, he should sever his ties to NRL. Tether made special arrangements for this to occur. Wolf retired from NRL and joined the faculty at the University of Virginia, but with the agreement that he would stay "on loan" at DARPA for another two years before going to the university.

Wolf is unusual at DARPA not only for his long tenure, but also for creating a number of different projects: Magnetic Materials Devices, followed by Spintronics, FAME, QuIST, and SPinS. He also started programs in Hard Magnetic Materials called AMPS (Advanced Magnets for Power Systems), SuperHyPE for Superconducting Hybrid Power Electronics, ATM for Advanced Thermoelectric Materials, MO-SAIC for Molecular Observation and Imaging using Cantilevers, FASTCARS for Femtosecond Adaptive Spectroscopic Techniques for Coherent Anti-Stokes Raman Spectroscopy, FLAME, for Femtosecond Lasers for Materials Exploitation, and finally FHOENICS for Femtosecond High Output ENergy Integrated Coherent optical Systems. He was also instrumental in initiating CNID, the Center for Nanoscale Innovation for Defense, which included UCLA, UCSB, UC Riverside, and AMRI, the Advanced Materials Research Institute at the University of New Orleans.

Additionally, Wolf's twelve-year tenure at DARPA exemplifies the fact that DARPA exercises considerable flexibility in its program management—in this case renewing him as a PM for three times the normal four-year assignment.

Other Key Roles in Program Formation

The Spintronics case is one in which the PM played the dominant role in program formation. Wolf drew upon ideas from the scientific literature and through interaction with his colleagues, and program ideas were supported by the DSO office director and approved by the DARPA director.

Key Insights

Dr. Wolf constitutes a clear example of a PM being the initiator of DARPA programs. He had technical expertise in a new field of science and technology and through his NRL management perspective was highly connected to leading research and researchers. He took the lab and university-based research and through industry pushed it into initial implementation.

DARPA provided a venue for Wolf to conceive and grow several programs that took an incipient field from the conceptual research stage to development of practical devices. While this drew heavily on his NRL experience, DARPA provided a means for him to organize ambitious implementation programs involving numerous participants, which was beyond what he could do at NRL.

Wolf's twelve-year tenure at DARPA demonstrates that it is an organization that is flexible even within its own "rules"—such as a PM only being hired for four years.

Personalized Assistant that Learns (PAL)[30]

Overview

The Personalized Assistant that Learns (PAL) program was an artificial intelligence (AI) program run through the Information Processing

30 This section is primarily based on interview of Ray Perrault (co-PI of CALO project), by David Cheney, March 2016.

Techniques Office (IPTO) from 2002 to 2009. It consisted of two projects, the CALO[31] project (managed by SRI International), and the RADAR[32] project (managed initially by Carnegie Mellon University). The PAL program (and specifically the CALO project) is best known for leading to the Siri application on the Apple iPhone, but was also transitioned to the military's Command Post of the Future (CPOF) system.

Context and History of the Program Development

DARPA had funded artificial intelligence since the 1960s, with several cycles of optimism and expansion followed by disappointments and contracting funding. Artificial intelligence had been making progress in several different domains, such as speech recognition, cognition, and machine learning, but there had not been a project that integrated advances across all of these domains and shown what AI could do.

The initial impetus for an initiative came from DARPA director Tony Tether, who wanted to do something in cognitive computing systems—systems that can reason, learn from experience, take advice, explain themselves, and respond intelligently to situations never encountered before.[33] He hired Ron Brachman as the IPTO office director for this purpose. Brachman was leader in the AI community. He had worked at BBN and AT&T Bell labs, and then AT&T technologies. He was highly respected in the community and was very strong in knowledge representation, and he had put together a very strong team at AT&T. Changes at AT&T (cuts in their research programs) had put him on the job market, and Tether was able to attract him to DARPA.

Brachman worked with the community to develop the program. He talked to a lot of people in the community and structured the intellectual area. During these discussions, the concept emerged of doing a large project to bring together the various pieces of AI—speech, learning, cognition, etc., all integrated by a prime contractor. The focus was on developing a virtual personal assistant that could help search for and retrieve information, schedule meetings, make appointments, and so on. The idea for a large project integrated by a prime contractor was

31 "Cognitive Agent (or Assistant) that Learns and Organizes".

32 "Reflective Agents with Distributed Adaptive Reasoning".

33 DARPA. (2003). "DARPA Awards Contract for Pioneering R&D in Cognitive Systems", *DARPA News Release*, 16 July, http://www.adam.cheyer.com/pal.pdf

supported by Tether. This was very different from the typical DARPA program.

They held a workshop with members of the technical community. Since Brachman knew the community, he had a good idea of the capabilities of different groups. It was clear early in the process who Brachman wanted in the program, and Tether took Brachman's word for who should participate. A BAA was released, and the original SRI-led proposal for CALO, based on the discussions at the workshop, was focused on a broad integrative system that combined different elements of AI: vision, natural language, planning, learning, etc. The twenty-page proposal was given to Tether, who rejected it and demanded that it be refocused on learning. So, the focus shifted away from the integration of every part of AI, and towards learning in every part of AI. It was clear that there would be a project, and that it would use the same project team as in the original proposal. However, the focus of the work needed to change in order to focus the program—and specifically the metrics and tests to demonstrate progress—around machine learning. The first year's test would be of the system components, but each component had to focus on learning within that component—e.g., learning in natural language; learning in speech recognition, etc. It was a legitimate and interesting approach, but it was not the only possible approach.

The focus on learning did make it clear and specific. They graphed what the system performance was with learning, versus what it would have been without learning, and it helped to sell the program.

A team led by Carnegie Mellon University won a second smaller project known as RADAR that focused more narrowly on helping managers to cope with tasks such as organizing their email, and planning meetings.

The CALO project had four phases with an evaluation at the end of each. The last phase was focused on technology transition, and so the final phase evaluation was based on how the results were transitioned to different applications. CALO was transitioned into part of the Command Post of the Future, for which General Dynamics was the prime contractor. They also developed a version called "CALO Express" that was created for use by DARPA PMs. It was built and demonstrated, but it never got through DARPA's certification process to be put into their IT system. SRI also used the CALO technologies to develop Siri for mobile phones and spun this off (with venture capital funding) as a new company that was later acquired by Apple.

Background and Role of the Program Manager

The PM, David Gunning, was hired after the BAA was out and proposals had been submitted. He did not have a role in the conception of the program. Gunning had previously been a PM at DARPA, and was the PM for the Command Post of the Future project, which was highly successful. He was brought back to manage PAL. He contributed to the project, but he was not hired for his program ideas—he was hired to manage the program that Tether and Brachman had conceived. He managed it throughout the duration, from 2003–2008. It is not uncommon for DARPA office directors to seek and hire PMs who are able to further develop and execute the Office Director's ideas. PMs are often brought in to manage programs that already exist, and then are expected to develop their own new ideas.

Other Key Roles in Program Formation

The DARPA Director and IPTO Office Director played the key roles in forming and shaping the PAL program project. The Office Director came from the community, and the community shaped the program through workshops. There was less direct influence on the program from DOD.

Because this program was much larger than most DARPA programs, it was visible to Congress and received substantial Congressional oversight due to its size. It was threatened with cancellation by the Congress. However, Brachman and others were able to defend the program so that it continued to receive funding.

Zach Lemnios, Brachman's deputy at IPTO, was also influential in forming the project. He came on board in April 2002. He was very good at managing the bureaucracy and ended up as Assistant Secretary of Defense for Research and Engineering, after going to Lincoln Labs. He subsequently became Vice President, Research Strategy and Worldwide Operations at IBM.

Key Insights

1) PAL was driven from the top down, but was also built on the AI community's perception of what was needed, in an area DARPA had long supported.
2) Some DARPA programs are initiated by DARPA Directors and Office Directors rather than PMs.
3) DARPA sometimes uses a prime contractor, which fills some of the functions of the PM, to integrate different research teams towards a common goal.
4) Some DARPA programs are large enough that they receive Congressional scrutiny.
5) In AI, DARPA support has not been continuous but has come in waves. DARPA provided support for five years or so and then stopped, and then later started another program. The technical community can be significantly disrupted when DARPA stops its funding.
6) A challenge for DARPA is when to decide that it has done enough in an area. This can be when progress is slow, or when commercial entities are getting ready to take over. DARPA used to provide more continuous support for fields.

Topological Data Analysis[34]

Overview

Topological Data Analysis (TDA) was a Defense Science Office program from approximately 2004 to 2008. The program developed data analysis techniques for massive data sets. The program spawned TDA research groups at universities and led to the formation of the Ayasdi software firm in 2008, founded by the DARPA-funded principal investigator (Carlsson) and his graduate or post-doc students. Ayasdi (www.ayasdi.com) which is now a 100+ person, venture capital-funded firm that is conducting data analysis for a large number of clients.

34 This section is primarily based on Mervis, J. (2016). "What Makes DARPA Tick?", *Science* 351/6273: 549–53; and Cochran, D. (2016). Personal Communication with David Cheney, April.

Context and History of the Program Development

Gunnar Carlsson, a Stanford mathematics professor, had developed an interest in the possibility of using topological methods for data analysis. He had been receiving NSF support for "pure math" (math developed under its own logic, without thought of applications) aspects of algebraic topology. Benjamin Mann, a program director at NSF, who had known Carlsson in graduate school (under the same advisor, James Milgram at Stanford), was aware of Carlsson's work and interests. Mann arranged for Carlsson to give a lecture at NSF on possible data analysis applications of topology, and arranged for Douglas Cochran, the DARPA program manager of math programs to attend. Mann and Cochran were co-program managers of a joint NSF-DARPA program. Cochran liked Carlsson's ideas and procured "seedling" funding to get him started as a DARPA investigator. The topic was timely because of the explosion in massive data sets (big data) and the need for new techniques to make sense out of the data, which has applications in intelligence and other areas. Cochran also used the possibility of launching a larger DARPA program in topological data analysis as bait to attract Mann to DARPA. He advised Mann on developing a program that would work in DARPA, so that when Mann met Tony Tether, he already had a fairly well-developed proposal for TDA. The seedling produced impressive results, identifying patterns in a data set that had not been identified through existing methods of analysis. These results allowed Mann to get Tether's approval for the full multi-year program.

Background and Role of the Program Manager

Cochran, the first program manager involved with TDA, is a mathematician who has been on the faculty at Arizona State University since 1989 and who served as a PM at DARPA from 2000 to 2005. He received his PhD in Applied Mathematics from Harvard.

Mann, who became the main program manager responsible for TDA, received his PhD in math from Stanford, had held several tenured academic positions and then became a program officer in NSF's mathematics division. While there, he got to know Cochran when they jointly ran an NSF-DARPA program on "Computational and Algorithmic Representations of Geometric Objects".

Mann arrived in DARPA in June 2004 and stayed until 2010. He started up several other programs, including one to establish fundamental mathematical principles in biology. Upon leaving DARPA, he became a vice president in Ayasdi, the new TDA company.

At the time, new ideas came into DSO in DARPA in a bottom-up fashion. A PM (or potential PM) learned of a compelling technology "push" or DOD need "pull" and developed a program concept around it, promoting it to the DARPA Director. New PMs were expected to come to their job interview with fairly well-developed ideas for new programs. In general, candidates were coached by current PMs and the DSO director before meeting with the DARPA Director, who personally made all PM hiring decisions (often after sending candidates away with additional "homework" questions to answer and then meeting with them again). Current PMs often identified and recruited candidates to become future PMs.

Other Key Roles in Program Formation

The key ideas that enabled the program came from the technical community, especially Carlsson. The DARPA director played a role in approving the program. The recognition that data analysis is applicable to anti-terrorism efforts, a key priority in the post-2001 environment, was an important factor in getting approval for the program.

Key Insights

This illustrates a mode of interaction in the more basic science parts of DARPA. PM-driven projects are an important mode in DSO. PMs who are part of a technical community come to DARPA for the opportunity to do bigger and more aggressive things than they can with NSF or NIH funding. The PMs often have a good idea of the opportunities and the performers.

This case also illustrates that there is some important interaction between NSF and DARPA, including both joint programs and movement of people between the agencies. Some university programs are supported by both agencies. Such programs have been occurring for a long time (several decades).

Furthermore, this case illustrates the close and mutually reinforcing network among researchers and PMs, especially in DSO. PMs come from the research community and often fund researchers who they know and may recruit others from the community to be the next PM. PMs may return to the DARPA-supported research community after serving as a PM.

Revolutionizing Prosthetics[35]

Overview

The Revolutionizing Prosthetics program is intended to produce better prosthetic arms, using advances in robotics and brain-machine interfaces. It began in 2006 in the Defense Sciences Office (DSO), and was transferred to the Biological Technologies Office (BTO) when that office was created in 2014. The program continues today. The program has produced a new prosthetic arm that has been approved by the U.S. Food and Drug Administration (which regulates medical devices). It has demonstrated robotic arms that are both brain-controlled and provide tactile feedback to the brain.

Context and History of the Program Development

The U.S. wars in Iraq and Afghanistan have led to many soldiers (as well as civilians in the local population) losing limbs. This was due in part to improved trauma medical care (and advances in body armor) that allowed many soldiers to survive injuries that would have been fatal in previous wars. Much progress had been made in developing workable prosthetic legs, but developing effective prosthetic arms had been much more challenging due to the many directions of movement and sensitivity of control required of arms and hands.

35 Sources for this section include: Belfiore, M. (2009). *Department of Mad Scientists*. New York, NY: Harper Collins; Burck, J. M., Bigelow, J. D., and Harshbarger, S. D. (2011). "Revolutionizing Prosthetics: Systems Engineering Challenges and Opportunities", *Johns Hopkins APL Technical Digest* 30/3: 186–97; Miranda, R. A., et al. (2015). "DARPA-Funded Efforts in the Development of Novel Brain-Computer Interface Technologies", *Journal of Neuroscience Methods* 244: 52–67.

The private market for prosthetic arms was not large or lucrative enough to drive innovation in prosthetics, and so there was a clear need for DOD investment to give injured soldiers a better life.

Previous DARPA programs, such as the Brain Machine Interface (BMI) program and the Human Assisted Neural Device (HAND) program, had developed techniques to enable direct brain control of computers and use of motor neural signals and sensory feedback for control of appendages or robotic devices. These showed that direct brain control of prostheses could be possible, and that the potential existed for much more sophisticated prosthetic arms.

While driven by a clear military need, Geoffrey Ling, the PM and a physician, is generally credited with creating the Revolutionizing Prosthetics program, motivated by his experiences serving as a military doctor in Iraq and Afghanistan. The DARPA Director, Tony Tether, was also strongly encouraging a program in this area. The goal of the Revolutionizing Prosthetics program was originally to create a neurally-controlled device, packed into the size and weight of a native human arm, that could do most or all of the things expected of a human arm. The program was started in 2006 and led to two main projects. The largest was led by the Applied Physics Laboratory (APL) of Johns Hopkins University. APL led a consortium of more than 30 research institutions and private companies in a project focused on prosthetics controlled by neural impulses, either through noninvasive surface electrodes, more-invasive wireless intramuscular implants, or peripheral nerve or cortical implants.

A second project was started that did not require the same degree of neural integration (requiring no surgery), and with prosthetics that would be controlled by muscular contractions. This project went to DEKA Research and Development (the firm headed by inventor Dean Kamen, who is known for creating the Segway transportation device).

Background and Role of the Program Manager

Ling, the founding PM for the program, was an army colonel and intensive care doctor. He had a PhD in pharmacology from Cornell University and an MD from Georgetown University. He joined the Army and was assigned to the Uniform Services University of the Health

Sciences, the military medical school, where he treated patients, taught medical students, and ran a research lab. He did a neurology residency at Walter Reed Medical Center, and trained in neurocritical care at Johns Hopkins University, with a specialty in caring for traumatic brain injury. He was encouraged to consider DARPA in 2002 from a navy commander and intensive care unit doctor, and he was then recruited by the Director of the DSO, Michael Goldblatt. Ling went on a tour of duty in Afghanistan in 2003, where he saw many civilians and military personnel with limb injuries, and this motivated him to join DARPA to try to develop better technologies for limb injuries. He joined DARPA in 2004 but then was deployed to Baghdad in 2005. When he returned, he started the Revolutionizing Prosthetics program, which requested proposals in 2005 and began in 2006.

After establishing and running the Revolutionizing Prosthetics program and several other programs, he became Deputy Director of the Defense Sciences Office, and then became an Assistant Director in the White House Office of Science and Technology Policy. He later returned to DARPA as the first director of the Biological Technologies Offices (BTO).

Other Key Roles in Program Formation

The program development was clearly influenced by military needs that came out of Iraq and Afghanistan and was also built on prior DARPA projects. The DARPA director, Anthony Tether, was actively involved in hiring Ling, in the decision to fund APL as the prime contractor, and in expanding the program to include the second path that became the DEKA project.

Key Insights

This program is larger (over $100 million) and longer (thus far, ten years) than the typical DARPA project. Like the PAL program, it uses a prime contractor (in this case APL) to integrate a large project consortium. While this model is not the typical DARPA program, it also is not unique (both PAL and Revolutionizing Prosthetics were started when Tony Tether was the director and reflect his influence).

This is an example of a program that was motivated by a clear military need but also shaped by a passionate PM.

Findings, Conclusions, and Key Observations

Process of Program Development

It is clear from these cases that there is no single DARPA program development process. Ideas for DARPA programs come from many places (the technical community, military, advisors, and companies). Programs develop in many different ways, and the process differs by program, over time, and with different DARPA directors. The approach also varies according to:

1) The maturity of the technology (whether a completely new area or one that DARPA has supported before)
2) Whether the technology being developed is at a component or system level; and
3) The political environment at the time (whether non-defense applications are a factor in whether to support the technology).

When discussing DARPA, it is important to be clear about which part of DARPA one is discussing, and which time period. What some may think of as the "standard DARPA model"—with programs initiated and driven by the program manager—better represents the more upstream, science and technology driven parts of DARPA (DSO, BTO, MTO) than the defense systems-oriented TTO and STO. The latter often support larger projects and are more likely to be driven from the top down, as can be seen in the Assault Breaker and Stealth cases.

While there is no single program development process, there are several typical patterns of program development. These can be characterized as follows.

Top down assignment from DOD, OSTP, White House, DARPA director, or others. This was especially common in DARPA's early days, when the focus was on satellites, missile defense, and test ban monitoring, and is more common at the systems level (in the TTO and STO offices), when the goal is to develop a new military system that meets an important defense need. In several cases, the drive for new systems came from the Office of the Secretary of Defense, usually the Director of Defense Research and Engineering, rather than from the military services, which tended to be more resistant to new technology. In some cases, DARPA funded a variety of conceptual studies to generate ideas for programs to meet a military need. In other cases, the DARPA director played an important role in supporting an area of technology that he or she thinks will be useful to the services in the long run (whether the services want it or not).

Programs based on the PM's Idea. There are cases where the PM comes up with the idea, wins support for it, and develops a program. In many cases it is not the PM's idea alone, but rather the PM has successfully drawn ideas from the technical community and used those ideas to form a program. PM-initiated programs appear to be more common in the Defense Science Office and in the more basic technology offices. Many such programs start with a seedling to test their viability. In some cases, a PM may be hired for a specific program. In other cases, they may be hired to implement an existing program (often when the current PM is leaving) and then are expected to develop their own program over the next year or two.

Long-standing thrust areas. Some new programs do not appear to be radically new. In some cases, DARPA has supported a community (e.g., mathematics, optoelectronics, and artificial intelligence) for some time. New programs may be similar to old programs that didn't succeed previously, but for which technology advances have made success more likely. Some programs represent the logical (if aggressive) next steps in a field, and there may be consensus in the community about what the next priorities are. In such areas there is a tension between a desire for continuity and the need for originality. The community wants some continuity of support, while DARPA sees its role as disruptive change that may require the disruption of existing communities. DARPA management feels a need to make sure DARPA's work does not become incremental and inappropriate for DARPA.

Many programs represent a combination of these patterns. For example, DARPA may work on a general problem due to top down interest, such as the need to be able to detect improvised explosive devices better, but the program manager may get ideas from the community through workshops and in responses to a request for information (RFI) or BAA to get the more specific ideas that result in a program.

It is important to note that DARPA programs are often (but not always) preceded by smaller studies. These may be conceptual studies of systems or technical studies to test the viability of a key technology (seedlings).

Roles of Program Managers

There is a variety of kinds of PM, and PMs view their roles somewhat differently. Some PMs are visionaries. Some are idea generators. Some are champions/drivers of other's ideas. Some are facilitator/enablers of communities. Some are hired to manage complex programs that have been conceived by someone else, such as an Office Director or the DARPA Director. Each kind can be successful.

PMs generally have a common role in assembling a program, serving as its champion (advocating the program and overcoming whatever obstacles are in its ways), and managing the program, but the PM may or may not be the source of the idea for the program. Some PMs inherited programs or were hired to manage programs. Some PMs see their role as finding the best ideas from the community and supporting them, rather than originating an idea. Some PMs see their role as conducting an orchestra of contractors.

Many DARPA managers are recruited and recycled through a small community. Many PMs come from and return to organizations such as IDA, SRI, Lincoln Labs, other defense labs, as well as universities that receive DARPA funding. We found that some PMs stay more than the standard three to five years, and some have served multiple assignments at DARPA over decades.

PMs have a variety of backgrounds. They are more likely to have an academic/research background in the upstream offices (DSO, BTO, MTO, I2O) and are more likely to have defense or industrial background in the systems offices (TTO, STO).

The autonomy of PMs has varied substantially over time. At some times, the PMs have had a great deal of autonomy in shaping their programs and have had very little oversight. At other times, especially during the directorship of Anthony Tether, the director was actively involved in shaping many programs.

Additional Observations

The cases we examined illustrate that the key element for the success of a DARPA program is not always the originality of the program. In some cases, a program topic may not be surprising, but the program may generate creative proposals for projects. This was the case for the digital mirror projection system in the high definition systems case, as well as in the prototype UAV. In other cases, the DARPA program may be distinctive not for its program idea but because of its unconventional approach—such as the use of a large-scale industrial consortium or using a prime contractor to integrate university, company, and laboratory research. DARPA's impact may also come from the focus with which an idea (which may have already been supported in a small way by other agencies) is executed. DARPA may achieve greater effects by pursuing an idea with greater funding, more urgency, and more aggressive and specific focus.

In many cases, timing is a key element of success. Part of the art of having a successful program is the ability to sense when science and technology advances at the material and component level have advanced enough to enable advances at the systems level, or to detect when an area of technology is at a state such that a concentrated effort in a specific area can enable a major advance. DARPA does not always get this timing right—there have been cases when DARPA has discovered that the alignment of the necessary factors was not in place and after a time, even as short as a year, either cancelled or redefined a program. Often, however, DARPA learns from these cases and establishes another program when further progress has been made.

While DARPA seeks to be largely independent of politics and acts independently of other agencies, some cases did exhibit the importance of Congressional and White House interaction, especially regarding dual-use technology in the late 1980s and early 1990s. The cases also illustrate how DARPA collaborates with other agencies, such as the National Science Foundation and the Defense Nuclear Agency, on various programs and studies.

One of DARPA's strengths is its flexibility and lack of bureaucracy. On the other hand, this leads to a lack of consistency in processes and to a weak institutional memory. Over the period of the cases in this study, DARPA has evolved towards greater systematization. The institutionalized use of the Heilmeier questions, the use of BAAs and Requests for Information (RFIs) to formally solicit input from a broad range of potential participants, the requirement for customer involvement in programs, and the increased emphasis on achieving specific milestones and metrics all reflect some organizational learning and institutionalization.

Concluding Thoughts

This study suggests that the approaches used to initiate DARPA programs have varied over time and in different parts of the agency. A question for organizations that are interested in adapting some aspects of the DARPA model, is "which DARPA does one want to copy?" There are several candidates. One option would be the "dual-use" DARPA that supported key technologies such as semiconductors and optoelectronics and industrial consortia in the late 1980s and early 1990s. This might be appropriate for organizations that are seeking to strengthen the key industries in their domain.

Another option is to follow the model that is most prevalent in the Defense Sciences Office, which emphasizes developing breakthrough new technologies, based on opportunities created by advances in fundamental science and technology. This model may be most appealing to organizations whose purpose is to create more radical innovation.

The third option would be to follow the model of the Tactical Technology Office and Strategic Technology Office, which emphasizes the development of systems in response to well-articulated needs. This model may be most appropriate for an organization whose mission is to meet a well-defined social need, whether defense or health care.

It may also be useful to consider the evolution of DARPA over time and its interaction with its environment. As the cases have shown, DARPA interacts with a diverse community of researchers and technologists in universities, research laboratories, defense contractors, think tanks and the military, and these can be a source of ideas for programs. This community has co-evolved with DARPA, and is better developed now than in DARPA's early years. Organizations that seek to emulate DARPA, may wish to consider both how to develop this community, as well as how to operate before such a community is well-developed. It may be that DARPA's early days, rather than its current state, provide a more useful model for new organizations.

All of these considerations suggest that, rather than copying a single model of DARPA's processes, it may be wise to emulate DARPA's flexibility and adaptiveness, giving freedom to the Director, and subsequently to the Office Directors, to choose the modalities for initiating programs that appear to be the best for the particular circumstances. In reflection, this is the approach that has worked at DARPA and it is hard to argue against its success.

References

Belfiore, M. (2009). *Department of Mad Scientists*. New York, NY: Harper Collins.

Block, F. L., and Keller, M. R. (2011). *State of Innovation: The U.S. Government's Role in Technology Development*. New York, NY: Paradigm Publishers.

Bonvillian, W. B., and Van Atta, R. (2011). "ARPA-E and DARPA: Applying the DARPA Model to Energy Innovation", *The Journal of Technology Transfer*, 36: 469–513, https://doi.org/10.1007/s10961-011-9223-x (Chapter 13 in this volume).

Burck, J. M., Bigelow, J. D., and Harshbarger, S. D. (2011). "Revolutionizing Prosthetics: Systems Engineering Challenges and Opportunities", *Johns Hopkins APL Technical Digest* 30/3: 186–97.

Cochran, D. (2016). Personal Communication with David Cheney, April.

Congressional Budget Office. (1993). "The Technology Reinvestment Project: Integrating Military and Civilian Industries", July, https://www.cbo.gov/sites/default/files/103rd-congress-1993-1994/reports/93doc158.pdf

DARPA. (2003). "DARPA Awards Contract for Pioneering R&D in Cognitive Systems", *DARPA News Release*, 16 July, http://www.adam.cheyer.com/pal.pdf

Drezner, G., et al. (1999). *Innovative Management in the DARPA HAE UAV Program*, MR-1054-DARPA. Santa Monica, CA: RAND Corporation.

Fuchs, E. R. H. (2009). "The Road to a New Energy System: Cloning DARPA Successfully", *Issues in Science and Technology* 26/1, http://issues.org/26-1/fuchs/

Hirschberg, M. J. (2010). "To Boldly Go Where No Unmanned Aircraft Has Gone Before: A Half-Century of DARPA's Contributions to Unmanned Aircraft", *American Institute of Aeronautics and Astronautics* (January): 11–13.

Mervis, J. (2016). "What Makes DARPA Tick?", *Science* 351/6273: 549–53.

Miranda, R. A., et al. (2015). "DARPA-Funded Efforts in the Development of Novel Brain-Computer Interface Technologies", *Journal of Neuroscience Methods* 244: 52–67.

Optoelectronics Industry Development Association. (2001). *Creating Bandwidth for the Internet Age*. Washington, DC: OIDA.

Sternberg, E. (1992). *Photonic Technology and Industrial Policy: U.S. Responses to Technological Change*. New York, NY: State University of New York Press.

Van Atta, R., Deitchman, S., and Reed, R. (1991). *DARPA Technical Accomplishments. Volume II*. Alexandria, VA: Institute for Defense Analyses, https://apps.dtic.mil/dtic/tr/fulltext/u2/a241725.pdf

Van Atta, R., Lippitz, M., et al. (2003). *Transformation and Transition, DARPA's Role in Fostering a Revolution in Military Affairs. Volume 1*. Alexandria, VA: Institute for Defense Analyses, https://doi.org/10.21236/ada422835, https://fas.org/irp/agency/dod/idarma.pdf

9. Some Questions about the DARPA Model

Patrick Windham

Often observers of DARPA ask basic questions about how the agency operates and the role it plays within the U.S. Department of Defense. This chapter provides brief answers to some of these questions.[1]

Is decision-making at DARPA "top-down" or "bottom-up"? DARPA is a mix of the two, but mostly "bottom-up". The agency director and deputy director do identify broad technical areas that they and others in the Defense Department think are important, but program managers, in consultation with the broader technical community, propose and then run specific R&D programs. In the "systems offices" at DARPA, office directors and the agency director talk with DOD officials and identify what they believe are significant long-term technological challenges and opportunities for U.S. national security. But again, the program managers propose and then run the actual R&D programs.

How can DARPA respond to Defense Department needs but still have great autonomy? DARPA asks both senior defense officials and the broad technical community what challenges and opportunities they see in the decades ahead. However, DARPA's job is to think about

1 The questions listed here about apparent paradoxes in the DARPA model were first raised in the fall of 2013 by Hiroyuki Hatada, then Chief Representative of the Washington, DC, office of Japan's New Energy and Industrial Technology Development Organization (NEDO). The editors are grateful to him for raising these questions and helping us to frame this discussion.

 https://doi.org/10.11647/OBP.0184.09

and create long-term technologies. It is not responsible for developing, maintaining, and improving current military systems; other parts of DOD perform those duties. In this way, DARPA has the freedom and funding to identify and create new, long-term technologies.

However, in time of war, senior DOD officials may ask the agency to help solve some difficult and immediate technical problems. For example, during the wars in Afghanistan and Iraq DARPA worked with other DOD agencies on the problem of detecting roadside bombs ("improvised explosive devices") and also helped to improve communications in those war zones.

How can DARPA make long-term progress with new technologies when the agency's programs are only three to five years long? William B. Bonvillian and Richard Van Atta point out that DARPA has "multi-generational programs": if the results of an initial program are promising, then there can be follow-on work. But if the initial program fails or points in a different direction, then the program is terminated or redirected. The use of three- to five-year projects allows great flexibility.[2]

Why does DARPA sometimes fund several different research projects within a single program? While some programs will fund a single large R&D project, such as the development of a prototype military system, other programs fund multiple research projects performed by different research teams. There are at least two reasons for multiple awards within a single program.

First, when trying to develop a new basic technology the agency often funds multiple teams with different technical approaches, to see which approaches are most promising. This is a "portfolio policy," in which the agency funds multiple ideas and then learns which work and which do not. Moreover, funding different teams with different ideas also allows the research teams to learn from each other, further advancing the overall technology. For this reason, a program manager may organize periodic meetings of a program's various R&D performers and ask these researchers to share information and learn from each other.

Second, in some cases the development of a new technology or capability requires several complementary parts. For example, the

2 Bonvillian, W. B., and Van Atta, R. (2011). "ARPA-E and DARPA: Applying the DARPA Model to Energy Innovation", *The Journal of Technology Transfer* 36: 469–513, at 473–74, https://doi.org/10.1007/s10961-011-9223-x (Chapter 13 in this volume).

desired technology might need several hardware components plus associated software. In these cases, a program might fund several R&D teams, with each of them responsible for an important part of the overall effort. If the technology proves promising, then that program or a follow-on program might fund work on additional steps that go beyond the initial R&D work, such as the integration of components or applications or the demonstration of the new technology's applications.

How do DARPA programs maintain continuity and success when program managers change every few years? New program managers have responsibility for existing programs and then make their own judgments about whether and how to continue them.

How can an agency build political support when it will not generate significant new technologies until many years from now? In the U.S., this can be a problem for several reasons: political leaders often want relatively quick results, applicants who do not get grants can complain to Congress, and other agencies or parts of your own department may see your agency as a rival. DARPA succeeds because it has an important defense role, it has a record of successes, and it does not threaten the budgets of other R&D agencies.

The new Advanced Research Projects Agency—Energy (ARPA-E) has thus far built important political support. It has done so by investing in a range of areas that people care about, by having credible processes, by soliciting views from everyone, by being transparent, by helping even losing applicants with valuable advice, and by working hard to convince other parts of the Department of Energy (DOE) that it is a good partner, not a rival.

How can a DARPA-type agency or program avoid rigid internal bureaucratic processes? Van Atta and others emphasize the importance of a very "lean" management structure. At DARPA a program manager needs approval from only two levels to get a new program: his/her office director (and deputy) and the agency director (and deputy). In addition, DARPA does not have a separate evaluation or audit office; evaluation is a constant process of judging which programs and R&D projects within those programs are succeeding or not. Program managers are not required to spend a great deal of time reporting to an audit unit.

Related, how can an agency demonstrate accountability (and create the political credibility that it needs to survive politically) but still

be relatively free of outside bureaucratic processes such as committees, layers of approval, audits, and so forth? Senior government groups oversee (supervise) all U.S. government agencies, including DARPA. For DARPA, these groups include senior DOD officials, the DOD Office of the Inspector General, the President's Office of Management and Budget (OMB), DOD and Congressional audit agencies, and of course the U.S. Congress.

However, DARPA has successfully argued that it does good work, that it follows all government rules, that it is has good internal evaluation processes, and that it needs autonomy and freedom from bureaucratic processes in order to do its job well. These arguments have largely succeeded, and neither senior DOD officials nor Congressional committees try to manage the details of the agency's work. In the long run, DARPA's successes and lack of scandals help it convince Congress and senior administration officials that it is doing a good job and does not need intensive bureaucratic supervision.

Specifically, how can an agency have credible rigorous evaluation of projects without highly bureaucratic and time-consuming reviews? This is a very important question, because evaluation is important not only to the effectiveness of the agency but also to its political credibility, since any agency that does not carry out proper evaluation and maintain high quality will eventually lose political support.

DARPA's process for evaluating programs and R&D projects within those programs differs from other U.S. Government science and technology agencies. Some other agencies use formal evaluation groups, which examine projects and provide useful information on the quality of research and how to improve operations. The former Advanced Technology Program/Technology Innovation Program at the U.S. Department of Commerce had a highly respected evaluation unit. Other agencies, such as the National Science Foundation (NSF) and the National Institutes of Health (NIH), maintain quality through both rigorous competition among applicants and the use of peer review as part of their overall merit review processes. At these agencies, the review processes include examinations of whether those applying for new grants have done good work in the past and therefore are likely to do good work in the future.

DARPA is different. It expects a great deal of its R&D performers, but it also expects that these high-risk R&D projects will not always work as originally planned. Things will "go wrong". Some DARPA directors therefore will *not* judge projects in terms of the original and sometimes unrealistic milestones.[3] Other directors have required that program managers get formal approval to change milestones and metrics. In both cases, however, the problems that arise provide important information that contributes to learning and adaptation. Surprise and change are normal. DARPA program managers therefore help R&D performers learn from problems and adjust research projects.

In this world, evaluation is a constant process, done by program managers and their office directors. Based on this ongoing process of learning, DARPA program managers try to help their R&D performers and discuss changes in projects. However—and this is very important—if a specific R&D project, or even an entire program of projects, fails to produce results, then DARPA will stop this work and move money into other, more promising areas. In addition, every year the agency formally reviews all of its programs. In addition to working with R&D performers, DARPA officials routinely talk with senior Defense Department civilian officials, with leaders of the military services, and sometimes also with leaders of DOD laboratories and research agencies to get their views on the usefulness and quality of the agency's programs. Here, too, the agency engages in continuous process of communication and evaluation.

This process of "continuous evaluation" works for several reasons: DARPA program managers are technical experts who can both help R&D performers and judge whether the performers are making acceptable technical progress or not; office directors and the agency's directors and deputy directors are themselves technical experts who can judge results and are willing to terminate unproductive programs; the agency emphasizes the importance of learning; and both senior Defense Department officials and members of the U.S. Congress see that DARPA does high-quality work.

3 Dugan, R. E., and Gabriel, K. J. (2013). "'Special Forces' Innovation: How DARPA Attacks Problems", *Harvard Business Review* 91/10: 74–84.

How does DARPA survey and analyze needs, technological trends, and future developments? Does DARPA use think tanks or consultants? DARPA has two major processes for gathering information.

First, program managers talk extensively with scientists and engineers in their fields, understanding technology challenges and opportunities. For example, program managers will talk with university scientists, corporate researchers, and experts in government laboratories to understand technology trends and possible future developments.

Second, program managers and DARPA leaders talk extensively with military officers and leading experts to understand what long-term needs the Defense Department might have and what types of technical solutions might help. These conversations take several forms: informal conversations with military officers assigned to DARPA, frequent conversations between DARPA leaders and the top civilian officials in the Defense Department, meetings every three months or so between DARPA leaders and senior military officials, "study groups" that meet regularly over several months to discuss a topic, interactions with the Defense Science Board (DSB) and other high-level advisory groups, and, in some cases, formal studies conducted by outside analysts and think tanks, such as the Institute for Defense Analyses (IDA).

One important issue for DARPA is whether or not other DOD agencies or think tanks have already done a good job of analyzing needs, trends, and opportunities in particular areas of technology or national security. If other agencies conduct useful analyses of, for example, space technologies and needs, then DARPA can use that information. But in some areas no one else has considered which types of new technologies might solve long-term challenges. In these cases, DARPA needs to organize its own meetings with military officers and others and conduct its own analyses.

How does DARPA recruit program managers? And how specific or broad is the subject area that DARPA presents when recruiting program managers? Because program managers usually serve for only a few years, DARPA's office directors and the agency director and deputy director spend much of their time recruiting new program managers. In some cases, departing program managers will recommend people to replace them. In other cases, office directors and the agency director and deputy director will ask colleagues in the technical community for recommendations. DARPA officials will look for candidates who are technically strong, have a good vision of where technology might go in the future, and have strong leadership skills.

These informal recruitment processes work well because the office directors and agency heads are themselves technically-trained individuals who know the R&D community well and can effectively judge the technical qualifications of potential program managers. They also understand what leadership skills are needed.

Recruiting prospective employees can sometimes be difficult, for both professional and personal reasons. University professors usually are not required to give up their current jobs, since they can take "leaves of absence", but they may worry about leaving graduate students or interrupting their own research. Company people face other concerns. DARPA usually requires that company employees leave their jobs before being eligible to join the agency, which is difficult even if they know that they will probably get a good job when they return to the corporate world. In addition, government salaries in the United States are much lower than corporate salaries.

In turn, both university and corporate people may also have personal concerns about joining DARPA. People with school-age children and whose spouses have careers may be reluctant to move to the Washington, DC, area for several years. Some people decide to come to DARPA while their families stay at home, leading to weekly commutes, but not every prospective employee wants to go through that constant travel. On the other hand, older people who are semi-retired and whose children are grown may find it easier to accept a DARPA position. Van Atta provides additional important insights into why individuals may or may not accept a position at DARPA.[4]

A new program manager will work in a specific subject area. He or she will propose and then run new programs in that area, and sometimes may also run existing programs created by earlier program managers. The programs can be quite complex, often involving work that brings together researchers from multiple disciplines. For example, a program that seeks ways to improve how an injured person can use her

4 Van Atta, R. (2013). *Innovation and the DARPA Model in a World of Globalized Technology*. Presentation at the National Institute of Science and Technology Policy and the Center for Research and Development Strategy, Tokyo, July.

brain to control artificial arms will involve physicians, neuroscientists, robotics experts, and others. Meanwhile, an effective program manager will need to understand enough about all of these disciplines to design a sensible research program and identify and select competent researchers. In this way, the work of a DARPA program manager can be both specific (focused on specific questions or challenges) *and* broad (that is, multi-disciplinary).

When DARPA recruits a university professor, does DARPA allow that professor to continue his/her university research and teaching? Usually professors will temporarily stop their academic research and teaching while serving as DARPA program managers. They need to focus on their DARPA responsibilities—not on their previous activities. Of course, this situation can cause difficulties. A professor may have on-going research projects and a number of graduate students working on their PhD projects. Usually, professors coming to DARPA will ask other professors to handle these responsibilities. But in some cases, they may continue working with existing graduate students, advising them from Washington and also reading drafts of their PhD theses. If a DARPA program manager is also a medical doctor, that person sometimes will be allowed to continue working part-time in a hospital or academic medical center.

How does DARPA decide about R&D themes, and what do program managers decide about R&D themes? Four points are important.

First, as mentioned earlier in this book, DARPA has two general activities: (1) maintaining strong leadership in basic technologies and (2) creating and demonstrating new equipment or processes that could help the Defense Department in the future. So, DARPA's technology offices pay attention to promising new technologies and often also pay particular attention to the long-term challenges facing the military services.

Second, within these overall subject areas DARPA's main criterion for selecting R&D themes and programs at any given time is to ask whether the agency can make a significant difference. That is, both program managers and senior agency managers look for game-changing technologies that can contribute to U.S. national security. Selecting themes and programs in any given year is a matter, therefore, of looking at both technological opportunities, and the specific long-term challenges facing the Defense Department at that time. Sometimes, senior Defense

Department officials or even the President will direct DARPA to work on specific topics. For example, when ARPA first began work in 1958 it focused on three key presidential priorities: space, missile defense, and nuclear-test detection.[5] Usually, however, DARPA officials will talk with senior Defense Department officials about long-term challenges, talk with the technical community about new technical opportunities, and then decide itself which projects offer the most potential.

Third, DARPA's priorities do change over time. For example, computer networking was a major priority from the 1960s through the 1980s, and this pioneering work led to the Internet. Subsequently, other agencies and the commercial sector took the lead in building the Internet, and DARPA switched to other opportunities and challenges. It still does work in computing and communications, but now it concentrates on new problems and opportunities, such as cybersecurity and big data. Another example is that, for many years, DARPA did little work in biology; DARPA was a physics and engineering agency. But the combination of bioterror threats, severe brain and other injuries to U.S. soldiers, and exciting new scientific and technical opportunities, has led DARPA to make biology, medicine, and synthetic biology major priorities.

Fourth, the fact that DARPA has no internal laboratories and instead funds temporary three- to five-year programs gives the agency the flexibility it needs to change themes and programs as new challenges and opportunities arise.

References

Barber Associates, R. (1975). *The Advanced Research Projects Agency, 1958–74.* Report prepared for the Advanced Projects Research Agency. Springfield, VA: Defense Technical Information Center.

Bonvillian, W. B. (2013). *Evolution of U.S. Government Innovation Organization: From the Pipeline Model, to the Connected Model, to the Problem of Political Design*. Presentation at the National Graduate Institute for Policy Studies (GRIPS) GRIPS Innovation, Science, and Technology Seminar, Tokyo, April.

5 Van Atta, R. (2008). "Fifty Years of Innovation and Discovery", in *DARPA, 50 Years of Bridging the Gap,* ed. C. Oldham, A. E. Lopez, R. Carpenter, I. Kalhikina, and M. J. Tully. Arlington, VA: DARPA. 20–29, https://issuu.com/faircountmedia/docs/darpa50 (Chapter 2 in this volume).

Bonvillian, W. B. (2009). "The Connected Science Model for Innovation—The DARPA Model", in *21st Century Innovation Systems for the U.S. and Japan*, ed. S. Nagaoka, M. Kondo, K. Flamm, and C. Wessner. Washington, DC: National Academies Press. 206–37, https://doi.org/10.17226/12194, http://books.nap.edu/openbook.php?record_id=12194&page=206 (Chapter 4 in this volume).

Bonvillian, W. B., and Van Atta, R. (2012). *ARPA-E and DARPA: Applying the DARPA Model to Energy Innovation*. Presentation at the Information Technology and Innovation Foundation, Washington, DC, February, https://www.itif.org/files/2012-darpa-arpae-bonvillian-vanatta.pdf

Bonvillian, W. B., and Van Atta, R. (2011). "ARPA-E and DARPA: Applying the DARPA Model to Energy Innovation", *The Journal of Technology Transfer*, 36: 469–513, https://doi.org/10.1007/s10961-011-9223-x (Chapter 13 in this volume).

Chesbrough, H. (2003). *Open Innovation: The New Imperative for Creating and Profiting from Technology*. Boston, MA: Harvard Business School Press.

Christensen, C. M. (1997). *The Innovator's Dilemma: When New Technologies Cause Great Firms to Fail*. Boston, MA: Harvard Business School Press.

DARPA. (2005). *DARPA—Bridging the Gap, Powered by Ideas*. Arlington, VA: Defense Advanced Research Projects Agency, http://www.dtic.mil/cgi-bin/GetTRDoc?Location=U2&doc=GetTRDoc.pdf&AD=ADA433949

Dugan, R. E., and Gabriel, K. J. (2013). "'Special Forces' Innovation: How DARPA Attacks Problems", *Harvard Business Review* 91/10: 74–84.

Heilmeier, G. (1992). "Some Reflections on Innovation and Invention", Founders Award Lecture, National Academy of Engineering, Washington, DC.

National Research Council. (2013). *21st Century Manufacturing: The Role of the Manufacturing Extension Partnership Program*. Washington, DC: The National Academies Press, https://doi.org/10.17226/18448, https://www.nap.edu/catalog/18448/21st-century-manufacturing-the-role-of-the-manufacturing-extension-partnership

National Research Council. (2012). *Rising to the Challenge: U.S. Innovation Policy in the Global Economy*. Washington, DC: The National Academies Press, https://doi.org/10.17226/13386, https://www.nap.edu/catalog/13386/rising-to-the-challenge-us-innovation-policy-for-the-global

Office of the Under Secretary of Defense for Acquisition, Technology, and Logistics. (2001). *"Other Transactions" (OT) Guide for Prototype Projects*. Washington, DC: Department of Defense, www.acq.osd.mil/dpap/docs/otguide.doc

Shinohara, K. (2014), "High-Risk & High-Impact Program in Japan: ImPACT", in *Weekly Wire News from East Asia and Pacific*, National Science Foundation Tokyo Regional Office, July 4, 2014.

Singer, P. L. (2014). *Federally Supported Innovations: 22 Examples of Major Technology Advances That Stem from Federal Research Support*. Washington, DC: Information Technology and Innovation Foundation, http://www2.itif.org/2014-federally-supported-innovations.pdf

Van Atta, R. (2013). *Innovation and the DARPA Model in a World of Globalized Technology*. Presentation at the National Institute of Science and Technology Policy and the Center for Research and Development Strategy, Tokyo, July

Van Atta, R. (2008). "Fifty Years of Innovation and Discovery", in *DARPA, 50 Years of Bridging the Gap*, ed. C. Oldham, A. E. Lopez, R. Carpenter, I. Kalhikina, and M. J. Tully. Arlington, VA: DARPA. 20–29, https://issuu.com/faircountmedia/docs/darpa50 (Chapter 2 in this volume).

PART II

THE ROLE OF DARPA PROGRAM MANAGERS

10. DARPA—Enabling Technical Innovation[1]

Jinendra Ranka

The Role of DARPA

DARPA is a unique institution that is consistently evolving. Every program manager you ask will give you a different view of DARPA. Everyone has a different opinion, whether it be a DARPA program manager, a performer on a DARPA project, a small business, the academic community, other government organizations, or the public. This is important to understand and is a result of how DARPA is structured, and how DARPA works with so many different technical communities on ground breaking high-risk projects that can have enormous potential. There are plenty of failures to criticize, but the successes have changed the world. DARPA is a very individualistic agency and I simply am presenting my view as a former DARPA program manager.

My experience at DARPA was very different from that given in the next chapter, and illustrates the diversity of the agency. I was a program manager in the Strategic Technology Office at DARPA from 2008 to 2013. I had spent the prior few years at MIT Lincoln Laboratory working as a scientist who enjoyed research, but had minimal management experience. Though I had a background in academic, commercial,

1 This chapter is based on a presentation that Dr. Ranka made at the Workshop on "How to Support Disruptive Change: Lessons from the DARPA Model", National Graduate Institute for Policy Studies, Tokyo, 25 February 2014.

 https://doi.org/10.11647/OBP.0184.10

and government research, I had never directly worked on a DARPA program. I knew little about the agency outside of the ARPANET.

DARPA has a relatively flat organizational structure. All program managers have a limited time at DARPA and are driven to accomplish as much as possible in that short time. The need to replace approximately twenty percent of the program managers each year requires the agency to be aggressive in hiring while being careful not to sacrifice technical excellence. Soon after a highly intensive vetting process with a DARPA office, I met with the agency director and had the most unique interview of my career. Three weeks later, I was a program manager at DARPA working on the ideas I wanted to pursue.

DARPA's history is important to understand as it has shaped the agency culture. The government created ARPA four months after the 1957 launch of Sputnik. The key people involved were President Dwight Eisenhower—a former general who distrusted the military industrial complex and who wanted a new agency that could coordinate closely with the technical community and could develop technology rapidly—Neil McElroy—the new Secretary of Defense, a former president of Procter & Gamble, who had absolutely no defense or military experience, was not a scientist or engineer, but knew what it took to run an organization and be effective—and James Killian—a scientist and the president of MIT, who was instrumental in the creation and design of ARPA. It was intentionally created to work with the research community for the Department of Defense, but was created outside of the military services. The agency's original charter was quite simple: "ARPA will do what the Secretary of Defense wants it to do".

DARPA and Innovation?

DARPA's mission is to develop breakthrough technologies for national security. In a similar fashion, ARPA-E was recently formed to advance U.S. energy research and IARPA for the intelligence community. DARPA is part of the Department of Defense and works closely with the different military services, but does not directly serve any of them. DARPA projects focus on the long term, and the agency is willing to take risks the services may not be willing to consider.

Program managers focus on high-risk/high-payoff projects that typically run for four to six years each, with well-defined metrics to measure success and ensure the truly hard problems are being addressed. DARPA looks at a broad range of national security problems and then invests to develop prototype technologies that solve those problems. Many of these technologies have broad uses, civilian as well as military, but, ultimately, all our work is anchored to our defense mission. Research for the sake of advancing scientific understanding is important, but it is not DARPA's mission. DARPA is there to create and prevent strategic surprise for national security.

DARPA focuses on adapting and executing faster than traditional government institutions are structured to do. DARPA must understand how to innovate and evolve rapidly, to address current problems as well as potential future technology gaps. Over the past two decades, advanced technology has shifted focus away from government and military dominance and towards the commercial sector. As such, the threats we face are rapidly evolving. While traditional military threats continue, we now also need to address cyber warfare, communication, encryption, social media, manufacturing, and much more. The traditional defense agencies were not designed to quickly address such disparate and complex technical areas as they rapidly evolve. DARPA has been addressing these problems for years and continues to make further investments.

DARPA is also structured to remove barriers to innovation. True innovation and innovative technologies do not appear based on a prescribed schedule, and can be hampered by aversion to risk, bureaucracy, funding limitations, lack of focus, and poor coordination. In a risk-averse culture, funding is often directed toward incremental technical improvements rather than riskier efforts which may provide dramatic new advancements. Coordination is important as the projects connect the research community to real users of the technology, with real problems and constraints. Similarly, limited user insight can be another barrier. You may have an idea, but you are not sure how to properly transition it to the user community or marketplace. It doesn't matter if you are in government, commercial, or academic research: these are challenges that we all face in technology development. To address this, DARPA programs are designed to be aggressive and focused. The agency provides the resources needed, attracts the best technical people to develop and run well-defined programs, and provides the oversight and coordination to ensure the best chances of success. DARPA doesn't simply fund people to work on hard problems, but funds new attacks to those problems through R&D projects that may be high-risk, but have the potential of achieving high payoff, high-impact results.

After a program is approved by the DARPA director, the program manager is given a significant amount of control. They effectively become the CEO, COO, CTO, and CFO for the program. The PM is provided with a long-term budget, with enough flexibility and finances to actually accomplish the proposed program. But this flexibility is coupled with accountability. The DARPA director and the office director act as the board of directors, and review program progress based on the vision, metrics, and deliverables that were originally proposed. Typically, the agency director and office directors will have had experience as a DARPA program manager. They know what it takes to run a DARPA program and will not hesitate to terminate or require course corrections for a program that does not meet performance metrics, or provide additional resources to programs that are successful at the "DARPA Hard" challenge.[2]

Developing and Running DARPA Programs

How does one actually develop a program idea? You first need to understand the problem you would like to solve and the current solutions. What are the limits to the current approach and what has been tried in the past? Is there a simple path forward, either technical or non-technical? A high-tech solution is not always the best answer. You also have to look and see what is possible. What are the fundamental scientific limits for the problem you are trying to address? What are the potential manufacturing limitations from now to the foreseeable future? What may can be done beyond that? From the answers to these questions, you develop a vision of what is possible, and you define goals, metrics, and a plan on how to actually achieve that vision. In a sense, you are not trying to predict the future, you are the one driving it.

2 The concept of a "DARPA Hard" problem is also discussed in this volume's Chapter 5.

As an example of looking at the future, imagine it is 1990. You look at the growth of computer processing power and communication network speed. The growth in the number of transistors in a commercially produced integrated circuit has been following Moore's Law, doubling approximately every two years. Now ask what is possible twenty years from now, in 2010, if computer processing power continues to increase at the same rate? In 1990, student computer use at universities was not widespread, with most students sharing limited computing resources. Yet, if computing processing power and network bandwidth followed the historical trend, as they were far from the fundamental limits of circuit size or capacity, you could easily predict that within the next decade desktop and laptops would be widespread across campuses. By 2010, hand held computing devices would dominate the marketplace and be capable of streaming high-definition content in real time. In fact, DARPA contributed to many of the technologies in your smartphone—not just the processor, but also displays, voice recognition, and inexpensive GPS receivers. DARPA tries to envision what future technologies and applications are possible, and then sets out to create that future with a specific goal in mind. The mindset is not the six to twelve-month product development cycle that the commercial world is driven to.

Now, the agency understands that many programs will not succeed. Typically, five or ten out of one hundred programs meet their goals and are transitioned to the user community. However, this does not mean the other programs have failed, at least from a DARPA perspective. It just means that the original vision or the transition is not fulfilled. A revised program may succeed, and valuable lessons may be learned from a technical dead-end. A failure, in DARPA's view, happens when a program does not succeed because of lack of due diligence, because a program manager did not understand the problem correctly, did not clearly define the program, did not develop effective goals and metrics, or did not properly understand the risks involved, and did not look at ways to mitigate those risks. A failure occurs because you did not do your job as a technical expert and as the DARPA program manager, not because the problem was too hard to solve at present.

This definition allows the agency to work on very hard and high-impact projects. As a program manager, you are not worried about failure because the task is too difficult. The only thing you need to ensure is that you do the job you joined DARPA to do.

Hard and high-impact projects are always going to be risky. You cannot be fearful and avoid risk that is inherent to a program. You just need to understand what the risks are and find ways to address them. When new risks are identified, do not just push them aside. You identify them and you aggressively attack those risks to make sure the program succeeds.

As a program manager, and as a technologist, you also need to find a certain balance. Fundamentally, I am an optimist. I know technology has enormous potential. But at DARPA it is important to understand the need for a sense of pessimism. At DARPA, so many people come in and present ideas they believe to be new and novel, and in the end, most ideas have resurfaced time and again, and, from your experience, you know that these ideas have fundamental flaws. You realize that good new ideas are rare, and good new DARPA breakthrough ideas even rarer. Despite this, it is important to be optimistic about technology development, and to learn to thoroughly question everything. That is one of the key aspects of the DARPA culture: if you do not look closely and question, you will not understand the nature of the problem, and its possible solutions. You will not understand what difference a new approach might make and you will not understand which ideas are promising and which are not.

A program manager must be respectful of people's ideas. One thing that I learned at DARPA—maybe one of the most important—is that when someone comes to you looking for funding for a new idea, they are actually exposing vulnerability to you. Any good idea, anything that challenges long held beliefs or practice, will have a number of issues. It is always easier to focus on those weak points rather than to try and fully understand the potential and possibilities of a new concept. In that situation, you have the option as a DARPA program manager to focus on all of the potential faults, or you can take a balanced approach and look at the possibilities as well as the questions that need answers. Observe the substance of the idea, the supporting science, and the implications. If you always focus on the faults, or mock an idea, people will be hesitant to approach you with new concepts. Though very few ideas will be of interest, and most will either be poorly thought out or presented like a TED talk, you have to be very respectful for every idea brought to you. Understand that your job is not only to pursue your own ideas, but to foster and select good ideas that are presented to you.

In turn, the ideas you pursue must focus on producing tangible results. At DARPA you need to be program- and project-oriented, rather than an investigative researcher. DARPA develops prototypes to show what the future can be. If your end result is simply a paper or presentation, you have not proved what is possible. If you demonstrate a robot climbing a wall to the world, on the other hand, then people truly will believe it is possible.

New ideas must have a valid approach. You must demonstrate that the physics and science are valid and at least have gone through the first-order calculations. At DARPA, a program manager will often fund "seedling" efforts prior to a program. Seedlings are quick efforts that provide evidence for a program that moves an idea from disbelief to doubt. There will always be missing pieces that you know must be worked on. This is acceptable, as long as you have a possible approach and develop metrics to measure your progress. You need to have thought in detail of at least one possible technical approach, a straw man solution. From this, you can estimate cost and schedule. Also, if you are going to ask people to propose solutions to your program, you need to be reasonably confident there is one possible approach. However, you should never limit a program to only that one approach. A diversity of approaches is important to any successful program.

Key to making this process work is by clearly defining goals and metrics. Metrics specified for the early phases of a program help ensure you are making progress and tackling the "DARPA Hard" challenges by focusing on the technical problems. Concise final goals and metrics in R&D provide a clear definition of what are you trying to accomplish to the outside world. This is immensely important to DARPA program manager, as the metrics define what she or he is investing in, the capability, what are the hard challenges, how is success measured, and what is the impact. For potential proposers, it provides clear guidance on what their solution must be capable of achieving.

Performance metrics help identify the key technical challenges and capabilities, independent of possible solution. This allows a program manager to gauge how different solution in the program are progressing and how well they are overcoming the key challenges. Metrics may change as a program progresses and the technical challenges evolve, and as you learn more about the problem and have a better understanding of the missing pieces to the solution. There are always missing elements to a program, where even with your straw man design and scientific vetting, you are not sure they are technically possible. These DARPA Hard challenges are valuable as they provide new capabilities, assuming you succeed.

How each DARPA program is managed can vary greatly between different program managers. What the program managers have in common is the freedom and flexibility to be successful in a high-risk effort, and a fixed timeline to succeed.

Important Questions to Ask

All DARPA programs have to answer a basic set of questions, known as "The Heilmeier Catechism" (or "Heilmeier questions"), named after former DARPA director George Heilmeier. They are listed in Box 11-1. They are fundamental questions that any technology development effort should be asked. If there are parts that do not have an answer, the program is not yet ready to start.

> George H. Heilmeier (DARPA director 1975-1977) developed a set of questions known as the "Heilmeier Catechism" to help Agency officials think through and evaluate proposed research programs:
>
> - What are you trying to do? Articulate your objectives using absolutely no jargon.
> - How is it done today, and what are the limits of current practice?
> - What is new in your approach and why do you think it will be successful?
> - Who cares? If you succeed, what difference will it make?
> - What are the risks?
> - How much will it cost?
> - How long will it take?
> - What are the mid-term and final "exams" to check for success?

Box 11-1. "The Heilmeier Catechism".
Source: https://www.darpa.mil/work-with-us/heilmeier-catechism

As a program manager, you need to have something new in your approach and you have to know what difference that is going to make. Is it a 2x improvement with 10x the cost or is it a 10x improvement at half the cost? It must make a difference that the end user will care about. It must have a significant impact. You have to be able to estimate how much it will cost and how long it will take. You need specific goals, metrics, and milestones in order to clearly define the program. Answering these questions does not imply a program has a high chance of success. In all honestly, I don't know of any way to determine if a new program is going to be successful, but I think there are ways to determine if a program has significant flaws and has not been properly thought out. High-risk projects need a clear vision.

Timelines

When DARPA hires a program manager, that decision does not mean that the agency has approved that person's proposed programs. Hiring you means they like the ideas that you are presenting and understand that you are a technical expert in that field. It could take anywhere from a few weeks to a few months or even years to actually get approval for your program.

Program managers come to DARPA because they have an ambitious idea to pursue and their interest in supporting national security. They typically do not come to run other people's programs, though managing existing programs is part of the job. The true excitement in DARPA is in seeing your own idea from start to completion.

All DARPA program managers are term limited. A PMs initial employment contract with DARPA is for two years. At any time if you do not believe that the agency is adequately supporting you, then you can leave. If the agency does not think you are doing a good job, then they will not renew your contract. The typical tenure at DARPA ranges from three to five years. Having a limited tenure for program managers is important as it fosters a sense of urgency for a PM in pursuing their program vision. For the agency, this helps ensure that creativity and productivity remain fluid.

After a program is approved by the Director, the program manager writes a Broad Agency Announcement (or BAA) that describes the program, the metrics, how companies and universities can propose to the program, what is required in a proposal, how proposals will be evaluated, timelines, and the general governing rules. It usually takes about one month to complete and get approval to release the BAA to the public. Proposals to the program BAA are typically due between forty-five and sixty days after the BAA is released. The program manager will often hold a workshop to review the program vision and metrics and provide a forum for others to discuss and form teams to respond to the BAA.

The amount of technical information required for a proposal is substantial. A proposal needs to describe how a performer plans on meeting the program metrics, with a detailed technical analysis, schedule, and cost estimate. It takes several people about a month of time to put together a good proposal—a substantial resource commitment. Companies that have an idea or solution to propose are willing to invest the resources because they believe in DARPA, and they know that there will be a fair and extensive evaluation process that focuses on the technical merits of the proposed solution. Companies know that if their proposal is selected, DARPA has the resources and funding to see the program to completion. This reputation is important, as it encourages a wide range of companies to propose to a BAA. They also trust DARPA to properly protect the intellectual property and ideas disclosed in a proposal.

It will typically take two months for the proposals to be evaluated by the PM and a team of government experts, and final selection of the proposals to be funded. Another three months is required to put a company or university under contract for the program. In total, it takes about six months from approval by the Director to when the selected performers start their technical work on a complex, multimillion-dollar R&D effort. That is incredibly efficient for any government agency, and this speed is one of the beauties of DARPA and its structure. If a PM is only going to be at DARPA for four years, a slow bureaucratic government process would be a problem. The DARPA structure is there to ensure that technical excellence and speed are not orthogonal.

Additional Thoughts on Why DARPA is Needed

I wish to add a few more points about why DARPA is needed and why it is valuable. If you look at many of the modern technical innovations that drive the world economy, they are based on fundamental scientific developments that arise from periods of intense investment. Truly new fundamental technologies and innovative ideas that can change society are rare. They also tend to be highly reliant on government support, especially from long-term basic and applied research. The U.S. Government spends over $3 billion each year to ensure that DARPA continues to push the limits of science and technology, plus many billions more at other federal R&D agencies.

DARPA is one part of the government S&T funding structure. DARPA's role is to show the world what is possible, building prototypes that demonstrate new capabilities. In the process, DARPA advances science to overcome technical roadblocks. It is a place of ideas. It is not afraid of risks, as risks are inherent in any innovative idea. At the same time, it is not an academic institution, and it is important to stay within the realm of reality. Rigor must be maintained, with well-defined goals and milestones. It is essential that it is understood, from both the management and technical perspective, that these are hard programs.

DARPA has followed this approach for over sixty years, and has earned the respect from the technical community based on what has been accomplished. This reputation provides very important political capital that no agency can ever afford to lose. DARPA is one grand, continually evolving experiment, which observes what works and what does not work, and which continues to persevere, making changes where necessary. As long as DARPA maintains that culture, DARPA maintains that political capital.

DARPA does not work alone. It relies on the technical performers whose proposed program solutions DARPA funds, and extensive collaboration with the military user community. Once a new technology is developed, it has to be transitioned to the military and commercial realm. In the end, it is the need and market that drives the transition, with the Department of Defense as the targeted customer.

The DOD is an early adopter for expensive, high performance system and often continues the support until the commercial space becomes sufficiently mature.

DARPA continues to push future innovations. In 2004, DARPA held a grand challenge to look at autonomous vehicles. In the first challenge, the best team only completed 12 km of the 240 km route. One year later, at the second challenge, five teams crossed the finish line. Ten years later, we have the initial glimpses of commercial autonomous vehicles, based largely on the work of those teams. Ten years from now, autonomous vehicles will be common in the commercial world. DARPA is continuing with robotics today, with the DARPA Robotics Challenge (see Chapter 11, below). What will all this work lead to in 2023? DARPA looks at the future, see what's possible, and then tries to drive technology and create that future.

A Flexible and Supportive Agency

The DARPA Director and Office Directors are responsible for developing the agency strategy and technical thrusts. They need a continual influx of program managers to come in with ideas to build programs in those areas. If an advanced technology agency wants to be successful, it not only needs to hire technical experts as program managers, but also provide a way for these people to succeed. DARPA has found an effective way to do both of these things. It enables its program managers to succeed.

When you come in as a DARPA program manager, there is no rulebook to guide you, but rather you learn what you need to do with the help of the DARPA support staff. This is one of the reasons that the agency is such an individualistic organization. A DARPA program manager can often spend up to 25 percent of their time on the road. A typical tour at DARPA can be exhausting, but the agency makes sure you have the support you need.

A friend once told me that if you have something that you wanted to get done, needed to get done, then DARPA was the place to go. There are not many places that would give you a greater opportunity to change the world while working with the best and brightest. The DARPA mix of innovation, speed, and human experience is singularly unique, and when it is time to depart, you will leave with pride in what you, your colleagues, and the agency have accomplished.

11. Program Management at DARPA

A Personal Perspective[1]

Larry Jackel

In this chapter, I provide a perspective on my experiences at DARPA as a program manager. In Chapter 10 above, Jinendra Ranka recounts his experiences as a DARPA program manager. These two perspectives expound highly different experiences. While there are certainly common themes, there is also a large degree of variance.

How does one create programs at DARPA? When I started at DARPA, I had a general charter to work in the area of applying machine learning to robotics, which had not been done to a significant extent up to that time. Before I joined DARPA, I knew machine learning groups at Bell Laboratories, although I had no direct experience in robotics. At DARPA, I was first assigned to take over two programs that were concluding. One program dealt with autonomous navigation, and one program concerned vehicle mobility. For the first six months, I was busy familiarizing myself with these fields: I watched the field tests, paid attention, and asked lots of questions.

During these first six months at the agency, I was able to identify factors that limited the performance of robots, and I then proposed three

1 This chapter is based on a presentation that Dr. Jackel made at the Workshop on "How to Support Disruptive Change: Lessons from the DARPA Model", National Graduate Institute for Policy Studies, Tokyo, 25 February 2014.

 https://doi.org/10.11647/OBP.0184.11

new programs to overcome limitations largely by applying machine learning to robotics.

Getting New Programs Approved

To get the programs approved, I first worked with the directors of two offices. One of my proposed programs received approval from the director of the Tactical Technologies Office (TTO), which develops prototype systems. I also worked in a second office, the Information Processing Techniques Office (IPTO, now called the Information Innovation Office, I2O), and my office director there helped me with the two other programs that I managed. Once we had developed the briefs, it was the role of the agency director at that time, Tony Tether, to make the final decisions. This was fairly typical, with the office directors coaching program managers and helping them prepare before they seek the agency director's approval.

Soliciting and Reviewing Proposals

After the concept for the program is approved and established, DARPA puts out a document called a Broad Agency Announcement (BAA)—a request for proposals. We would typically get between five and ten times as many submissions as we could fund. I wish to point out that this is not necessarily a positive, because putting together a proposal requires a tremendous amount of work on the part of the proposers. If we only fund a tiny fraction, then many researchers waste a large amount of their time preparing proposals. A better policy has developed in later years in which people submit short papers, known as whitepapers. On the basis of these short papers, program managers encourage those researchers who look likely to be funded to submit larger proposals, and those who look less likely to receive funding are not encouraged to submit full proposals. This saves time both for the researchers and also for the program managers.

In my own case, I had perhaps a hundred proposals to read. This was time-consuming, and, at times, painful, in that it was frustrating to observe the amount of effort people had put into proposals that would not be funded.

In order to pick the proposals that would be funded, I, as a program manager, appointed a team of government employees to read and review the proposals. This is not a peer review process akin to that conducted by the National Science Foundation, in which university professors review the proposals. Instead, a handful of government employees, led by the program manager, conduct the review. Our review was a rank-ordered list of the various proposals. Then we took that list to the DARPA director's office, and he would say: "Okay, I have so much money, I can fund up to this level."

Managing Programs

Next, I want to discuss managing programs.

On the one hand, program managers need humility. When I came to DARPA, I received some advice from my predecessor, who had managed the programs that I inherited. He informed me: "When you're a DARPA PM, you'll be treated like a king by those who depend upon you for funding. Do not act like a king. Stay humble. Your job is to serve the taxpayer". When you become a program manager, you suddenly attract many new friends, and people never contradict you. This can lead to a false impression of your own intellect and ability. It is therefore paramount that you understand that you are not as smart as the people surrounding you pretend that you are. It also means that you have to treat the people who you will fund with respect and must not treat them badly.

On the other hand, unlike NSF, the programs at DARPA are actively managed by the program manager. First, we set clear and realistic goals and schedules. There are milestones that R&D performers have to meet for the program to continue. In addition, we provide technical support and guidance when possible. In the robotics area, for example, I was not an expert in robotics *per se*, but I knew a great deal about machine learning. I would help the performers in their research by giving them suggestions on how to improve the behavior of the robots by incorporating learning. I was also actively involved in testing and evaluation. I greatly enjoyed the fact that, most of the time, I never wore a suit and a tie. Usually, I would be out in the mountains or the desert with hiking boots and blue jeans testing the robots.

For example, one time we brought some of the principal investigators to Fort Carson in Colorado, which is right at the foothills of the Rocky Mountains (where we conducted some of the robot testing). I am an experimental physicist, and this meant that when it came to the testing, I used the methodology that I had learned in physics while testing robots. I was actively involved in planning the actual test. It gave me the opportunity to exercise my pleasure in being a scientist.

It is very rare that programs proceed as the program manager expects: changes must be made to programs to ensure that progress is made to larger goals. In this way, program managers need to learn and adapt, and help their performers learn and adapt. They must, therefore, be both humble *and* active.

A typical program manager will have the necessary technical expertise and research experience. Very often, they will also have managerial experience. It is essential that they have a good understanding of relevant technology, along with the ability to lead a research community.

Independence, Responsibility, and Accomplishments

DARPA trusts its program managers and gives them great independence. In my own case, I was required to report on each program to the upper management about once a year. During the intervals between these reports, I largely had full autonomy in running the programs.

DARPA program managers also have considerable resources. For example, consider Dr. Gill Pratt, who later became the program manager for robotics, the leader of the 2012–2015 DARPA Robotics Challenge, and the person I helped in my subsequent role as a consultant to DARPA. Dr. Pratt had a total budget of about $50 million per year over six years—roughly $290 million in total. With that money, he ran programs in robotics, neuromorphic computing and computer vision. This money was adequate funding to make significant advances in the targeted technology.

One example of where the agency and its R&D performers made progress was the DARPA Robotics Challenge that I mentioned earlier, which was budgeted at $80 million over several years. The goal for that program was to develop robots capable of assisting humans and responding to natural and manmade disasters. Much of the inspiration came from Dr. Pratt's experience trying to help at the Fukushima nuclear plant, after the 2011 accident there. The Robotics Challenge led to some impressive improvements in robots and became an example of how a DARPA program, with good leadership and adequate funding, can make real progress.

PART III

APPLYING THE DARPA MODEL IN OTHER SITUATIONS

12. Lessons from DARPA for Innovating in Defense Legacy Sectors[1]

William B. Bonvillian[2]

As World War II grew and the U.S. production machine began to ship war supplies to Britain in every available ship, an enduring transfer was occurring in the opposite direction. The critical moment was in August 1940: British science leader Henry Tizard landed in Halifax and took a train to Washington, leading a small scientific team on a multi-month mission. In a suitcase they carried perhaps the most critical technology of the war: an early prototype of the microwave radar.

However, it was not the technology alone that was so important, but rather, the innovation organization model. The American team, led by industrial organizer and technologist Alfred Loomis and reporting to Vannevar Bush, Franklin D. Roosevelt's science czar, immediately realized the importance of the small radar device, and they also learned about and replicated parts of the system that led the British to operational radar. The essentials were replicated at the Rad Lab at MIT, where microwave radar advances exploded into a galaxy of electronic applications, then transferred to Los Alamos. As explored below, the

1 This paper originally appeared in modified form in 2015 in *The American Interest* 11/1, as "All that DARPA Can Be", https://www.the-american-interest.com/2015/08/01/all-that-darpa-can-be/

2 William B. Bonvillian is indebted to his Georgetown colleague Prof. Charles Weiss for numerous insights behind this article.

 https://doi.org/10.11647/OBP.0184.12

organizational lessons included: form critical innovation institutions, organize them on an "island/bridge" model, create a thinking community, and link technologists to operators.

Thirteen years after the end of the war, these innovation organization lessons were translated directly into the Defense Advanced Research Projects Agency (DARPA), perhaps the most successful federal R&D agency ever. We review a series of questions: what are the foundations of the DARPA model? What is the context of contending innovation models it operates in? What do the four innovation organizational lessons cited above look like up close? DARPA is famous for sponsoring much of the R&D that led to the information technology revolution, innovating in a "frontier" technology sector. However, it has also brought innovations to a "legacy" sector, the conservative military bureaucracy. This kind of innovation is much more difficult because launching it is contested. Moreover, it is rare—legacy sectors rarely undertake disruptive innovations. How did DARPA do this? DARPA's efforts in this legacy territory are much less understood, but because legacy sectors constitute most of the U.S. economy, may provide wider lessons about the landscape of innovation organization.

The Underlying Innovation Models

Like all R&D agencies, DARPA has an organizational genealogy. Initially, then, we turn to the fundamentals—four models for how innovation is organized in the U.S. to put the DARPA model into the larger context.

The most familiar U.S. innovation model evolved in the immediate postwar; it is the so-called *pipeline* or linear model, developed by Vannevar Bush.[3] It holds that basic research operating at the frontiers of knowledge and supplied by government research investment leads to applied research and development. This, in turn, leads to invention, to prototyping, and, finally, to innovation and corresponding broad commercialization or deployment.

While subsequent literature showed that this process wasn't really linear—technology influenced science as well as the other way

3 Bush, V. (1945). *Science: The Endless Frontier*. Washington, DC: Government Printing Office, https://www.nsf.gov/od/lpa/nsf50/vbush1945.htm

around[4]—"pipeline" is still the term generally associated with this technology supply approach.

The World War II-era success of atomic energy, radar, and other technologies, derived from advances in fundamental scientific knowledge,[5] inspired the model; it led to a host of technology advances.[6] It is a "technology push" model, with the government supporting initial research with only a limited role in pressing these advances toward the marketplace. Therefore, it is inherently a disconnected model, with researchers separated from industry implementers.

The second of these models is the so-called induced innovation concept explored by economist Vernon Ruttan[7] in which technology and technological innovation respond to changes in the market, generally to market niche opportunities and price signals. It is typically industry led. New products in this model often generate from modifications of existing technologies to meet new market needs—incremental advances—rather than emerging from basic research. This model involves "technology pull"—the marketplace pulls technology innovations from firms toward implementation in the market.

The third model, which is a variation of the first, can be called the "extended pipeline", a new term. This model enabled many of DARPA's greatest successes. It describes the role of the U.S. Defense Department (DOD), which could not live with the inherent inefficiency of the pipeline model, where the innovation institutions are disconnected. In this model, DOD not only funds the early stages of research, but also sponsors the follow-on stages. To obtain the technologies it requires to meet national security needs, DOD often will fund the research, the development, the prototype, product design, the demonstration, the

---

4 Stokes, D. E. (1997). *Pasteur's Quadrant, Basic Science and Technological Innovation*. Washington, DC: Brookings Institution Press, 1–25, 45–89.

5 Buderi, R. (1997). *The Invention that Changed the World*. Sloan Technology Series. New York, NY: Simon & Schuster.

6 National Research Council. (1999). *Funding a Revolution: Government Support for Computing Research*. Computer Science and Telecommunications Board. History, Commission on Physical Sciences Mathematics and Applications. Washington, DC: National Academy Press, 85–157 (chapters 4–5), https://doi.org/10.17226/6323, https://www.nap.edu/catalog/6323/funding-a-revolution-government-support-for-computing-research,; Waldrop, M. M. (2001). *The Dream Machine: J. C. R. Licklider and the Revolution that Made Computing Personal*. New York, NY: Viking Press.

7 Ruttan, V. (2001). *Technology Growth and Development: An Induced Innovation Perspective*. New York, NY: Oxford University Press

testbed, all the way to funding implementation and serving as the initial market. Important parts of the information technology revolution—the Internet for example—were developed in this way, but this development was not unique.

Ruttan has noted how DOD also led aviation, electronics, space, nuclear power and computing using this model.[8] These constitute most of the major technology innovation waves of the twentieth century. This model links the initial research stage with a governmental role in the follow-on technology development stages, connecting the institutional actors that dominate each. Agriculture and space advances also employ the extended pipeline, and other R&D agencies are starting to emulate this more connected system.[9] Unlike the pipeline model, it operates at all stages of innovation, not simply the early stages.

The fourth model of innovation dynamics, "manufacturing-led" innovation, describes innovations in production technologies, processes and products that emerge from expertise informed by experience in manufacturing. This is augmented by applied research and development that is integrated with the production process. It is typically industry-led, but with strong governmental industrial support. While countries like German, Japan, Taiwan, Korea and now China have organized their economies around "manufacturing-led" innovation systems, the U.S. in the postwar period did not. It is a major gap in the U.S. innovation system. This system gap is now starting to affect the ability of DARPA and other R&D agencies to translate their technologies into actual innovation.

When the U.S. was constructing its innovation system in the postwar period, it paid little attention to manufacturing-led innovation. This had been the U.S.'s innovation strength since the nineteenth century; it had created the mass production system that had played a central role in winning World War II. Production was not the problem, since the U.S. dominated it. Instead, the U.S. focused on its research system, the front end of innovation, which had emerged at scale during the war, but needed to be retained and augmented. This was the system

8 Ruttan, V. W. (2006). *Is War Necessary for Economic Growth? Military Procurement and Technology Development*. New York, NY: Oxford University Press.

9 Bonvillian, W. B. (2013). "The New Model Innovation Agencies: An Overview", *Science and Public Policy* 41/4: 425–37, https://doi.org/10.1093/scipol/sct059, https://academic.oup.com/spp/article-abstract/41/4/425/1607552?redirectedFrom=fulltext

Vannevar Bush, as President Roosevelt's science advisor, focused on. Others countries, such as Germany and Japan, emerging from wartime chaos, had to concentrate on rebuilding their industrial bases, and thus developed and extended their manufacturing-led innovation systems. As their economies emerged, Taiwan, Korea and China needed to build their industrial bases, and also followed the manufacturing-led innovation path.

Innovation Organization

These first four models exist and can be seen at work at varying degrees of efficiency in the U.S. economy. The fifth model, which can be termed *innovation organization*, is more a conceptual framework that includes the other three and builds on them. It is not a subject in the innovation literature.[10] However, innovation requires not only technology supply and a corresponding market demand for that technology, but also organizational elements that are properly aligned to link the two. There must be concrete institutions for innovation, and organizational mechanisms connecting these institutions, to facilitate the evolution of new technologies in response to the forces of technology push and market pull. This fifth element is essential in our innovation framework: the idea that innovation requires organizations anchored in both the public, academic and private sectors, to form the new technology and to launch it, if innovation theory is to be practical, creating both ideas and means to actually implement them. The focus in the science policy literature is on idea creation; detailed evaluation of implementation is largely ignored.

In other words, while the first four innovation models—pipeline, induced, extended pipeline, and manufacturing-led—are descriptive of existing ways of organizing innovation in the U.S., they are limited in their reach. The fourth provides the organizing methodology that

10 Bonvillian, W. B. (2009). "The Connected Science Model for Innovation—The DARPA Model", in *21st Century Innovation Systems for the U.S. and Japan*, ed. S. Nagaoka, M. Kondo, K. Flamm, and C. Wessner. Washington, DC: National Academies Press. 206–37, https://doi.org/10.17226/12194, http://books.nap.edu/openbook.php?record_id=12194&page=206 (Chapter 4 in this volume); Weiss, C. and Bonvillian, W. B. (2009). *Structuring an Energy Technology Revolution*. Cambridge, MA: The MIT Press, 26–28; Nelson, R. R. (1993). *National Systems of Innovation*. New York, NY: Oxford University Press, 3–21, 505–23.

encompasses the first three and reaches beyond them to the innovation implementation system. It includes the full innovation ecosystem—from research to deployment, but also the forces of culture, political and economic systems, technological routines, and social structures for innovation. This also means the mechanisms and *change agents* needed to surmount the obstacles in that ecosystem to enable innovation. These forces are especially profound in complex, established "legacy" economic sectors—like energy, transport, health care delivery, manufacturing, higher education, agriculture—and also in defense.[11]

These tend to lock in established technologies and resist technology advances that are different from and disrupt their existing economic and technological model. They use political, economic and social systems in their defense against disruptive innovation. By recognizing that there are institutions and mechanisms operating within an innovation system, legacy or otherwise, the innovation organization model enables a richer evaluation of innovation and of potential policies to improve the overall system. The innovation organization model, then, moves beyond the institutional "linkage" idea of the extended pipeline model to embrace a series of elements to provide a bigger picture of innovation: connecting public and private sectors, from research through implementation; merging pipeline and induced innovation, radical and incremental; overcoming structural barriers to innovation particularly relevant to legacy sectors; and consciously embracing change agents.

These five models fit into an historical context. The manufacturing-led model was embodied in the mass production system that the U.S. was the first nation to fully develop, and is also embodied in Japan's quality production system. The pipeline model was inspired by the dramatic advances seen in World War II deriving from basic science, such as nuclear energy from particle physics and electronics from radar advances, in the 1940s-50s. The induced technology model has long dominated industry's role in innovation, with advances derived largely from incremental gains in existing technology, such as, in the 1960s and 1970s, from

11 Bonvillian, W. B., and Weiss, C. (2009). "Taking Covered Wagons East, A New Innovation Theory for Energy and Other Established Sectors", *Innovations* 4/4: 289–94, http://www.mitpressjournals.org/userimages/ContentEditor/1259694503297/Bonvillianinov.pdf; Bonvillian, W. B., and Weiss, C. (2011). "Complex Established 'Legacy' Systems: The Technology Revolutions that Do Not Happen", *Innovations* 6/2: 157–87, https://doi.org/10.1162/inov_a_00075, https://www.mitpressjournals.org/doi/pdf/10.1162/INOV_a_00075

automobiles, consumer electronics and jet aviation. Throughout that era, the kind of innovation described by the extended pipeline model was humming along, bringing out a personal computing and internet revolution in the 1990's after decades of government R&D inputs. While the induced model best fits incremental innovation, the pipeline and extended pipeline models best fit breakthrough or radical innovation. These breakthrough innovations supply the ingredients for waves of innovation that create "frontier" economic sectors that periodically form new parts of the economy. Underlying these developments in technology advance is the innovation organizational model described here and its additional series of elements, vital for understanding our innovation system yet largely unexplored. These innovation organization elements in the model are important in particular for any analysis of the entry of technology innovation into legacy sectors.

Beyond Pipeline

The dominant literature on technological innovation has remained focused on the strengths and weaknesses of the pipeline model, because of the perception that the frontier economy is key to growth. The innovation waves in information technology and biotechnology, for aspects of which the pipeline model provides a description, command most of the analytical focus to date. This pipeline literature has not confronted the problems involved in bringing innovation into established legacy economic sectors. It pays too little attention to how the overall economic and policy environment affects technological innovation in complex networks of both related and unrelated technologies. While the extended pipeline is not a term in the innovation literature, there is some work describing that model,[12] although it is still focused on the frontier economy. The induced technology model often pays too little attention to the governmental role.[13] The literature on induced

12 See, for example, Bonvillian, W. B. (2006). "Power Play, The DARPA Model and U.S. Energy Policy", *The American Interest* 2/2: 39–48, at 40–47, https://www.the-american-interest.com/2006/11/01/power-play/; Alic, J., et al. (1992). *Beyond Spinoff: Military and Commercial Technologies in a Changing World*. Cambridge, MA: Harvard Business School Press.

13 Although Vernon Ruttan was a leading theorist of the induced model, in his last book he turned to an exploration of what we call here the extended pipeline model (Ruttan. (2006). *Is War Necessary for Economic Growth?*).

technology has rested primarily on market pull theory, and on the role of firms in filling technology needs based on changing market signals, ignoring governmental R&D and policy interventions. The two pipeline and the induced models have been viewed as separate and distinct paths; to date none has focused on what is described here as the fourth direction, innovation organization. If we are to adequately describe the framework required for innovation in the range of technologies to be introduced into complex and legacy sectors, the organization model suggests we must combine and integrate the other three models. The systemic barriers to legacy sector innovation also arguably require change agents—institutional and individual actors prepared to push innovations through the sector barriers at each innovation stage.

To summarize, we have described a series of models of innovation. We have noted how they apply to both the frontier as well as the legacy sectors that, combined, make up most of the economy. We have developed a broad new model that encompasses and adds new considerations to the other models to meet the challenge of optimizing the organization of innovation. We have a new framework, then, in which to understand the functioning of innovation systems and the actor institutions that perform within them, including DARPA.

While we placed DARPA in the discussion above within the sweep of the extended pipeline model, it also has developed features that have enabled it to innovate in the legacy defense sector. This means that it represents, as well, key features of what we term the innovation organization model. It is this new way of analyzing DARPA's role that is the primary focus of this article.

First Things First—The Front End of the Innovation System

There is an obvious rule functioning here: no innovations, no innovation system. Innovation requires not only an understanding of the overall system for its development, as set out above, but the first problem concerns the earlier stages of the innovation system where the innovations originate. Later come the problems of overcoming the structural barriers to innovation and creating the linkages between the innovation actors at the subsequent stages of the innovation process, including the role of change agents, where ideas move to implementation. First, however, we must tackle the problem of how to bring about innovation, whether into legacy or frontier sectors. To put the horse before the cart, we must begin with the "front end" of the innovation system, the research, development, prototyping and early demonstration stages.

This means we must move beyond the long-standing focus of pipeline theorists on the valley-of-death stage between research and late-stage development[14] because innovation requires what we can term *"connected science and technology"*—linkages between innovation stages and actors—an integrated consideration of the entire innovation process, including research, development, and deployment or implementation, in the design of any program to stimulate innovation in any complex, established technology sector. As noted, this requires drawing on the two pipeline models, the manufacturing-led model, and the induced innovation model. In addition, we see deep system issues of organization for innovation, because new organizational routines are required across both the public and private sectors to facilitate integrated policies that will support innovation.

These considerations lead to a new approach to innovation policy, aimed at what Avery Sen and others call *transformative innovation*.[15] This transformational task of innovation for both frontier and legacy sectors is usually particularly dependent on the strength of the front end of an innovation system. While, by definition, this will be the case for frontier sectors—which initially require new innovations—it will not always be the case in legacy sectors, where both breakthrough and incremental advances may be needed. For example, in the health legacy sector, incremental advances in electronic medical records could lead to dramatic improvements in the health care legacy sector, although breakthrough medical devices and nanoscale drug delivery are also required. Or, in

14 Branscomb, L., and Auerswald, P. (2002). *Between Invention and Innovation, An Analysis of Funding for Early-State Technology Development*. NIST GCR 02–841. Washington, DC: National Institute of Standards and Technology, 2, https://link.springer.com/article/10.1007%2Fs10961-011-9223-x

15 Bonvillian, W. B., and Van Atta, R. (2011). "ARPA-E and DARPA: Applying the DARPA Model to Energy Innovation", *The Journal of Technology Transfer* 36: 469–513, at 470, https://link.springer.com/article/10.1007%2Fs10961-011-9223-x; https://doi.org/10.1007/s10961-011-9223-x; Sen, A. (2014). "Transformative Innovation: What 'Totally Radical' and 'Island-Bridge' Mean for NOAA Research", PhD thesis, George Washington University, Washington, 18–56.

the energy legacy sector, "smart" devices are evolving incrementally for the electric power grid, even if technology breakthroughs in power electronics are needed as well. Other legacy sectors, such as defense or advanced manufacturing, require more breakthroughs. In this way, the legacy sector transformational task will be both breakthrough and incremental, pipeline and induced. Regardless, we need to focus in depth on understanding and strengthening the front-end system; otherwise, creation of frontier sectors and transformation of most legacy sectors will be largely curtailed.

Strengthening the Front End

Strengthening the "front end" of the innovation system requires an innovation capability analysis of the research development, prototyping and early demonstration elements, and of the institutions that support them. Is the system capable of generating the innovations required to bring change to complex and legacy sectors? A series of evaluations is needed, and may require implementing system improvements. Since the front end of innovation is typically driven, initially, by the pipeline or extended pipeline models, we must consider these and their application to the optimal innovation organization approach required in taking this first step.

A series of factors for consideration in this step are reviewed below, and the application of each to DARPA is discussed.

1) *Form critical innovation institutions.* If R&D is not being conducted at an adequate scale by talented researcher teams, innovations will not emerge. However, talent alone is not enough—talent must be operating within institutional mechanisms capable of moving technology advances from idea to innovation. *Critical innovation institutions* represent the space where research and talent combine, where the meeting between science and technology is best organized. Arguably, there are critical science and technology institutions that can introduce not simply inventions and applications, but significant elements of entire innovation systems.[16]

16 Bonvillian. (2009). "The Connected Science Model".

This is where DARPA takes center stage, with its history of attracting outstanding research talent, and of spurring remarkable technology advance.[17] In promoting innovations, it has long played within both frontier sectors, through its role in the information technology (IT) wave, and the defense legacy sector, through its role in such defense advances as precision strike, and unmanned aerial vehicles (UAVs). As the most successful U.S. R&D agency operating in the innovation space, and because it represents more of a "connected science and technology" approach than other agencies, our initial focus is on lessons that can be learned from the characteristics of the DARPA model.

Formed in 1958 by President Eisenhower to provide more unified defense R&D in light of the separate, stove-piped military services' space programs that had helped lead to America's Sputnik failure, DARPA became a unique entity, aimed at both avoiding and creating "technological surprise".[18] In many ways, DARPA directly inherited the "*connected science and technology*" (linking science research to implementation stages) and "*challenge*" (pursuing major mission technology challenges) organization models of the Rad Lab and Los Alamos projects stood up by Vannevar Bush, Alfred Loomis and J. Robert Oppenheimer in World War II. Building on the Rad Lab example, it built a deeply collaborative, flat, close-knit, talented, participatory, flexible system, oriented to breakthrough radical innovation. Its challenge model for R&D, moved from fundamental, back and forth with applied, creating connected science and technology, linking research, development, and prototyping, with access to initial production. In other words, it followed an innovation path not simply a discovery or invention path.

However, innovation requires not only a process of creating connected science and challenges at the *institutional level*, it also must operate at the *personal level*. People are innovators, not simply the overall institutions where talent and R&D come together. Warren Bennis and Patricia Biederman have argued that innovation, because it is more

17 Van Atta, R. (2008). "Fifty Years of Innovation and Discovery", in *DARPA, 50 Years of Bridging the Gap*, ed. C. Oldham, A. E. Lopez, R. Carpenter, I. Kalhikina, and M. J. Tully. Arlington, VA: DARPA. 20–29, https://issuu.com/faircountmedia/docs/darpa50 (Chapter 2 in this volume).

18 Discussion drawn from Bonvillian. (2009). "The Connected Science Model", 207, 209, 215.

complex than the earlier stages of discovery and invention, requires *"great groups"*, not simply individuals.[19] However, unlike other federal R&D agencies, DARPA has attempted to operate at *both* the institutional and personal levels. DARPA became a bridge organization connecting these two institutional and personal organizational elements.[20]

At the heart of the DARPA ruleset is what Tamera Carleton has termed a *"technology visioning"*[21] process, which appears to be particularly key. It uses a *"right-left" research model*—its program managers contemplate the technology breakthroughs they are seeking to emerge from the right end of the innovation pipeline, and then go back to the left side of the pipeline to look for proposals for the breakthrough research that will get them there. As noted, it uses a *challenge-based* research model—seeking research advances that will meet significant technology challenges. It looks for *revolutionary breakthroughs* that could be transformative of a technology sector. All of these elements go into a process where agency program managers develop a vision of a technology advance that could be transformative, then work back to understand the sequence of R&D advances required to get there. If these appear in range of accomplishment, the agency has processes that allow very rapid project approvals by the agency directors. This technology visioning process is very different from the way industry undertakes step-by-step down-selection of technology options known as the "stage-gate"[22] process, where budget and market gain are factors used to weed out which incremental advances to pursue. The visioning process is also very different from how other federal R&D organizations work; these place more emphasis on research for the sake of research. In the context of attempting to bring innovation into legacy sectors, the visioning process may be particularly apt.

---

19 Bennis, W., and Biederman, P. W. (1997). *Organizing Genius: The Secrets of Creative Collaboration*. New York, NY: Basic Books.

20 Bonvillian and Van Atta. (2011). "ARPA-E and DARPA", 483–84. See also, on the origins of ARPA-E, Weiss and Bonvillian. (2009). *Structuring an Energy Technology Revolution*, 161–65, 185–86, 206, 260n9, 262nn17–19.

21 Carleton, T. L. (2010). "The Value of Vision in Radical Technological Innovation", PhD Thesis, Stanford University, Palo Alto, http://purl.stanford.edu/mk388mb2729; Bonvillian and Van Atta. (2011). "ARPA-E and DARPA", 485 (italics added).

22 See, for example, Cooper, R.G., Edgett, S. J., and Kleinschmidt, E. J. (2002). "Optimizing the Stage-Gate Process", *Research Technology Management* 45/5, 43–49, https://doi.org/10.1080/08956308.2002.11671532

Other DARPA characteristics enhance its ability to operate at both the institutional and personal innovation organization levels. The following list is largely drawn from DARPA's own descriptions of its organizing elements:[23]

- Small and flexible—DARPA consists of only 100–150 professionals; one can refer to DARPA as "100 geniuses connected by a travel agent".
- Flat—DARPA is a flat, non-hierarchical organization, with empowered program managers.
- Entrepreneurial—DARPA's emphasis falls on selecting highly talented, entrepreneurial program managers, willing to press their projects toward implementation, often with both academic and industry experience. They serve for limited (three- to five-year) duration, which sets the timeframe for DARPA projects.
- No laboratories—DARPA's research is performed entirely by outside performers, with no internal research laboratory.
- Focus on impact not risk—DARPA's projects are selected and evaluated on what impact they could make on achieving a demanding capability or challenge.
- Seed and Scale—DARPA provides initial short-term funding for seed efforts that can scale to significant funding for promising concepts, but with clear willingness to terminate non-performing projects.
- Autonomy and freedom from bureaucratic impediments—DARPA operates outside the civil-service hiring process and standard government contracting rules, which gives it unusual access to talent, plus speed and flexibility in contracting for R&D efforts.

23 This list is drawn from DARPA. (2008). *DARPA—Bridging the Gap, Powered by Ideas*. Arlington, VA: Defense Advanced Research Projects Agency, http://www.dtic.mil/cgi-bin/GetTRDoc?Location=U2&doc=GetTRDoc.pdf&AD=ADA433949; DARPA. (2003). *DARPA Over the Years*. Arlington, VA: Defense Advanced Research Projects Agency. For a more detailed evaluation of DARPA's ruleset, see, Bonvillian and Van Atta. (2011). "ARPA-E and DARPA".

- Hybrid model—DARPA often puts small, innovative firms and university researchers together on the same project so that firms have access to breakthrough science and researchers see pathways to implementation.
- Teams and networks—at its best, DARPA creates and sustains highly talented teams of researchers, highly collaborative and networked to be "great groups", around the challenge model.
- Acceptance of failure—DARPA pursues a high-risk model for breakthrough opportunities and is very tolerant of failure if the payoff from potential success is great enough.
- Orientation to revolutionary breakthroughs in a connected approach—DARPA is focused not on incremental innovation, but on breakthrough/radical innovation. It emphasizes high-risk investment, moves from fundamental technological advances to prototyping, and then attempts to hand off the production stage to the armed services or the commercial sector.

The above rules are part of the established DARPA culture as a critical innovation institution. But there are other important foundational and underlying features that DARPA has adopted, not as well understood, but more central to building a strong, up front-end innovation system that it exemplifies. These provide broad, overall front-end organization lessons.

2) *Use the island/bridge model.* Bennis and Biederman[24] have argued that innovation requires locating the innovation entity on an "island" and protecting it from "the suits"—the bureaucratic pressures in larger firms or agencies that too frequently repress and unglue the innovation process. Nonetheless, they note that there must also be a "bridge" —the innovation group must also be strongly connected to supportive top decision-makers who can press the innovation forward, providing the needed resources. Sen has argued this is a foundational innovation model.[25]

---

24 Bennis and Biederman. (1997). *Organizing Genius,* 206. See also, Sen. (2014). "Transformative Innovation", which expands and builds on the Bennis-Biederman concept.

25 Sen. (2014). "Transformative Innovation".

The island/bridge model has been, from the beginning, a key to DARPA's success. Indeed, other innovative organizations use it as well. Lockheed's Skunk Works,[26] Xerox's PARC (Palo Alto Research Center)[27] and IBM's PC project[28] have exemplified island/bridge at the industry level, severing innovation teams from interference from the business/ bureaucratic side. As noted in point (4), below, some of the ideas for this approach came from the British in the 1940s. While the Skunk Works and IBM PC groups also had strong bridges back to "mainland" decisionmakers, PARC did not, and exemplifies the need for the bridge. DARPA exemplifies the island/bridge model at the federal R&D agency level.[29] It has initiated innovation in frontier sectors, particularly IT, as noted, where it operated largely outside the Pentagon's legacy systems, working with and helping to build emerging technology private sector firms. It has also worked within the defense legacy system. It has operated as an island there but also used strong links with the Secretary of Defense and other senior defense leaders as the bridge; these Defense decisionmakers helped bridge technology advances from DARPA researchers to the implementing military services.

There are alternative models to the island/bridge model. The "open innovation"[30] approach is well-known, where firms drop reliance on in-house R&D labs and reach out to groups at other, often smaller, firms (through acquisitions, technology licensing or partnerships) or at universities (linking to public sector funded researchers at these institutions and licensing their work or creating collaborations). This is primarily, however, a tool for more mature firms facing global competition and less able to afford in-house R&D, or their rivals attempting to out-compete them. Robert W. Rycroft and Don Kash pose a similar model, and broaden it, arguing that innovation requires "collaborative networks" at a series of levels that must reach outside the organization for a kind of heightened R&D situational awareness, and

26 Rich, B, and Janos, L. (1994). *Skunk Works: A Personal Memoir of My Years of Lockheed*. Boston: Little, Brown & Company.

27 Hiltzik, M. (1999). *Dealers of Lightning: Xerox PARC and the Dawn of the Computer Age*. New York, NY: Harper Business. 153.

28 Chposky, J., and Leonsis, T. (1986). *Blue Magic: The People, Power and Politics Behind the IBM Personal Computer*. New York, NY: Facts on File.

29 Bonvillian and Van Atta. (2011). "ARPA-E and DARPA", 486.

30 Chesborough, H. W. (2003). "The Era of Open Innovation", *MIT Sloan Review* 44/3, http://sloanreview.mit.edu/article/the-era-of-open-innovation

can be less face-to-face and more virtual.[31] Neither approach obviates the need for an originating innovation "great group" applying an island/bridge approach.

3) *Build a thinking community.* A prerequisite for the ongoing success of the island/bridge model is building a community of thought. In science, it is well understood that each contributor stands on the shoulders of others, building new concepts on the foundations of prior concepts. Ernest Walton and John Cockcroft, for example, working at Cambridge's Cavendish Laboratory, built an early particle accelerator using a strong electrical field. They became the first people to split the atom, changing the atomic nucleus of one element (lithium) into another (helium) in 1932.[32] They built on the active work of a host of other contemporary physicists, from the Cavendish's director Ernest Rutherford, to Ernest Lawrence, Merle Tuve, Peter Kapitza, James Chadwick, George Gamow and Niels Bohr, to name only a few. The group at the Cavendish was a remarkable "great group" itself, but it was also part of a powerful *thinking community* that was constantly contributing ideas to each other. This community was exemplified by the forty physicists who attended the 1933 Solvay Conference, half of whom won the Nobel Prize (including Cavendish attendees Rutherford, Walton, Cockcroft and Chadwick).

Building a sizable "thinking community" has also been key to DARPA's success, as a source of contributing ideas but also for talent and political support.[33] Composed of multiple generations of DARPA program managers and researchers working in a field supported by DARPA, at its best this community becomes a group of change agents and advocates. J. C. R. Licklider, a tech visionary of the first magnitude, in his two stints at DARPA brought in a succession of office directors and program managers and built supporting university research teams that initiated a series of multi-generational technology breakthroughs

31 Rycroft, R. W., and Kash, D. E. (1999). "Innovation Policy for Complex Technologies", *Issues in Science and Technology*, https://issues.org/rycroft/

32 Cathcart, B. (2004). *The Fly in the Cathedral.* New York, NY: Farrar, Straus & Giroux.

33 Bonvillian and Van Atta. (2011). "ARPA-E and DARPA", 476–77, 492.

that, over time, led to personal computing and the Internet.[34] Building a thinking community around a problem takes time to evolve, but reaches a density and mass where ideas start to accelerate. For example, in the field of nanotechnology, physicist Richard Feynman arguably initiated the community with a 1959 noted talk entitled "There's Plenty of Room at the Bottom", urging work at the smallest scale where quantum properties operate. In 1981 researcher Eric Drexler published the first journal article on the subject, and by 2000 over 1800 articles using the term nanotechnology had accumulated, showing a thinking community had formed and was starting to accelerate advances.[35]

4) *Link Technologists to Operators.* Another key organizational feature of successful innovation organizations involves connecting the technologists to the operators. This approach perhaps is best exemplified by the relationship between British scientists and the military on the eve of, and during, World War II. In the early 1930s the assumption of all, from the Prime Minister down, was that "the bomber will always get through"—there was no adequate defense to bomber aircraft, which could devastate both military and civilian targets virtually at will.[36] With Hitler building 4000 aircraft in 1935, and with England only a few miles across the Channel from the European mainland, the ramifications of this assumption in the 1930s' appeasement policy were profound.

However, a small group began to investigate whether air defenses could be created. At the behest of the Royal Air Force's (RAF) scientific Tizard Committee, a scientist team, under Robert Watson-Watt (scientist supervisor of a small defense lab) began investigating radio beam technology that became radar. However, the technology alone did not create an air defense against the bomber; extended trial and error testing with RAF pilot teams led by physicist Henry Tizard, Rector of Imperial College, developed the operational routines that enabled the British to maximize the utility of radar technology for air defense and win the

34 Waldrop. (2001). *The Dream Machine,* chapters 2, 5–7, and 466–71.

35 Milunovich, S., and Roy, J. M. A. (2001). "The Next Small Thing—An Introduction to Nanotechnology", Merrill Lynch Industry Comment, 4 September, p. 2, https://www.slideshare.net/tseitlin/intro-to-nanotechnology-merrill-lynch

36 Clark, R. W. (1962). *The Rise of the Boffins.* London: Phoenix House, 23–31.

Battle of Britain.[37] In this way, it was the constant testing and evaluation with air force operators—fighter interceptor pilots and what became ground control groups—that linked the technologists to the operators, using new but demonstrated technology-based operating systems. Tizard, a World War I pilot as well as leading scientist, famously spoke the pilots' language from shared experience, and the experimental regimens he helped devise and the RAF implemented between 1935 and 1938, coupled with continuing incremental improvements in the technology to meet evolving operator needs, changed the course of the war.[38]

Along with Tizard, three members of his RAF committee, A. V. Hill, A. P. Rowe and Patrick Blackett, developed a doctrine for linking scientists and technologists with operators. This became known as Operations Research.[39] This approach used statistical analysis of operations, applying a range of variable technology and operational approaches to find optimal solutions to operational challenges. Operations Research had World War I precedents in optimizing anti-aircraft artillery developed by Hill[40] and was written up by Blackett in 1941 as a chapter in a short edited book entitled *Science in War*, advocating its widespread use by the military.[41] Blackett, as director of Naval Operational Research, subsequently applied the techniques he helped develop to the war against U-Boats, which were threatening to cut off Britain's wartime food and supplies. Research by his team (known as "Blackett's Circus") resulted in dramatic improvements to optimal convoy size and air-sea convoy protection, with a corresponding dramatic reduction in incidences of U-boat ships sinking.[42]

The British approach to applying science in World War II was to isolate and protect its scientists from military hierarchies—the island/bridge approach—but also to integrate them with the military operators when the outcomes of their research appeared promising. Inventing and

37 *Ibid.*, 33–54.

38 Clark, R. W. (1965). *Tizard*. Cambridge, MA: The MIT Press, 23–48, 105–92.

39 The term "Operational Research" was coined by A.V. Rowe in 1937, while working as assistant director at the RAF radar research and testing center at Bawdsey; "Operations Research" is the American term. Budiansky, S. (2013). *Blackett's War*. New York, NY: Alfred A. Knopf, 87.

40 Clark. (1962). *The Rise of the Boffins*, 8–9.

41 Budiansky. (2013). *Blackett's War*, 117–18.

42 Budiansky. (2013). *Blackett's War*, 113–66, 221–49.

using Operations Research analysis, it found that the scientists must be informed, involved in, and linked to the decision making not just on technology but also on related strategy and tactics. The British model for using scientists, then, was to keep them out of uniform working in separate research centers (from the RAF's radar operational experiments at Biggin Hill and Bawdsey, to the codebreaking at Bletchley Park) as islands, but with strong ties to the mainland—the service operators.

Tizard, leading the 1940 Tizard Mission that brought vital British microwave radar advances to the Americans before they entered the war, spent two months in discussions with American scientists and military that year, including extensive exchanges with science leaders Vannevar Bush and Alfred Loomis.[43] Tizard and his team apparently explained to Bush and Loomis the science organizational model he and other British science leaders had developed.[44] Bush and Loomis ended up creating largely the same island/bridge model in the U.S. with links to operators, implementing it in such famous projects as the Rad Lab for microwave radar advances at MIT[45] and atomic weapons development at Los Alamos.[46] These projects in turn became central to the subsequent organization of post-war U.S. science.

DARPA, in its work on major defense technology advances, also exemplifies an effort to link technologists with operators, to transform operations. Its work on personal computing and the Internet, which shattered the arm's length relationships in mainframe computing between technologists and operator/users, exhibits the same drive to produce technologies that connect with operators. DARPA's Tactical

43 Clark. (1962). *Tizard*, 248–72.

44 MIT's history of the Rad Lab states, that "Running conferences [with Tizard Mission members] continued till October 13 [1940], and by that time practically everybody was agreed that what the program needed was a central laboratory built on the British lines: staffed by academic physicists, committed to fundamental research but committed even more than that to doing anything and everything needed to make microwaves [radar] work." MIT Radiation Laboratory. (1946). *Five Years at the Radiation Laboratory*. Cambridge, MA: The MIT Press, 12, https://archive.org/details/fiveyearsatradia00mass. See also, Clark. (1962). *Tizard*, 265, 267 (Tizard meetings with V. Bush), 268–69 (Mission meetings with Loomis).

45 Conant, J. (2002). *Tuxedo Park: A Wall Street Tycoon and the Secret Palace of Science that Changed the Course of World War II*. New York, NY: Simon & Shuster, 178–289.

46 Bird, K., and Sherwin, M. J. (2005). *American Prometheus, The Triumph and Tragedy of J. Robert Oppenheimer*. New York, NY: Alfred A. Knopf, 205–28, 255–59, 268–85, 293–97.

Technologies Office (TTO) is specifically designed to bring technologies into military tactical systems, using rapid prototyping to transition to air, ground and naval operators.

To summarize the first step of building front-end innovation capabilities, one of the important lessons from DARPA's ability to bring innovation into a defense sector with deep legacy characteristics has been the importance of *critical innovation institutions*. To perform at a critical level, these institutions should attempt to embody a series of characteristics. They should undertake both *"connected science and technology"*—linking science research to implementation stages—and *"challenge"* approaches—pursuing major mission technology challenges. As discussed, and as DARPA exemplifies, innovation requires not only a process of creating connected science and technology and related challenges at the *institutional level*, it also must operate at the *personal level*. The critical stage of innovation is face-to-face not institutional, so while institutions where talent and R&D come together are required, personal dynamics, usually embodied in *"great groups"*, are a necessity. The DARPA *"right-left" research model* can be important to reaching the innovation stage, where program managers contemplate the technology breakthroughs they seek to emerge from the right end of the innovation pipeline, then go back to the left side of the pipeline to look for proposals for the breakthrough research that will get them there. This process tends to lead to *revolutionary breakthroughs* that could be transformative of a technology sector. A technology *"visioning"* process at the outset of the effort appears to be particularly key. The approach results in *high-risk but high-reward* projects.

The island/bridge organizational approach for innovation institutions also appears to be important. The innovation team should be put on a protected island apart from bureaucratic influences so it can focus on the innovation process. The strength of the innovation process will also depend on building on forming a solid *thinking community* as a source for ideas and support. Because innovation must span numerous steps from research through initial production, means for *linking technologists to operators* appear to be critical. Again, DARPA, more than any other U.S. R&D agency, exemplifies these approaches.

These rules apply to the important first step of front-end innovation organization. They take in the key features of the extended pipeline model: strong initial research and linkages between researchers and the institutions that can lead an innovation through the later stages toward implementation. But what about the additional issues presented by the innovation organization framework? These include not only the front-end research and the institutional linkages, but also overcoming the barriers to innovation presented within an innovation ecosystem by legacy sectors and the role in that ecosystem of change agents.

In summary, despite its ruleset and the way it exemplifies optimal front-end innovation, DARPA is part of a defense innovation system; it is an entrepreneurial innovator, but *within* DOD. To foster implementation, it must still rely on the military services, and face the legacy pressures they can embody, for the follow-on stages. How DARPA, and its allies, have undertaken this innovation within a legacy sector provides important lessons for the overall U.S. innovation system.

DARPA Innovation within the Defense Legacy Sector

The defense sector has often led U.S. technological advance. Yet historically, militaries have often been the most conservative of organizations, seeking to refight the last war, suppressing innovation in the name of discipline and reliability, and therefore famously subject to technological surprise—Sputnik (which led to DARPA's creation) is a good example. The U.S. military, like all others, exhibits these legacy sector tendencies. However, in the late 1970s, after almost three decades of Cold War, a remarkable effort began in the Defense Department to introduce transformative technologies. That process contains important lessons for innovation organization within legacy sectors.

When Harold Brown became Defense Secretary and William Perry Undersecretary for Defense Research and Engineering (DR&E) in the Carter Administration in 1977, the nation faced a major Cold War dilemma. Starting under Eisenhower and Kennedy, the U.S. had developed a superiority in nuclear weapons and their missile delivery systems that offset Soviet advantages in conventional forces in Europe. However, by the mid-1970s, that advantage had faded, with the U.S. and Soviets in rough parity in these systems. With its deterrence threat eroding, and the Army's capability in decline as a result of the terrible pressures of the Vietnam War, Perry and Brown were deeply concerned about the possible outcome of a conventional warfare confrontation in Europe. Concern about mutual destruction blunted the ability to use nuclear weapons as a deterrent, and the Soviets had built a three-to-one advantage in force levels, tanks, armored fighting vehicles and artillery in Europe. As Perry later put it, "We thought they had a serious intent to use them, to send a blitzkrieg down the Fulda Gap [the anticipated route of the Soviet ground invasion of Western Europe then thought possible]".[47] This imbalance in conventional forces could have forced the U.S. into a situation where it would have had to employ nuclear weapons, with all of their devastating consequences.

Since equaling Soviet force levels in Europe was not feasible, Perry and Brown developed an "offsets" theory as the basis for a new U.S. defense strategy.[48] They decided to achieve parity and therefore deterrence in conventional battle through systematic technological advance in order to offset the Soviet advantage in force levels. They began a process of translating advances in computing, information technology, and sensors, which had been initiated and long-supported by defense research investments, through DARPA in particular, into three areas of advance: stealth, precision strike, and unmanned aerial vehicles (UAVs). These capabilities later became known as the Revolution in Military Affairs (RMA).[49]

How did this RMA come about? Although this revolution suggests the power of DOD's innovation system, it is also possible, as noted, to characterize much of DOD as a legacy sector. The existing military paradigms within DOD are averse to the risk of innovation. In many cases, this group of RMA capabilities was seen as threatening to vested technologies and capabilities and to the officers and their organizations that had spent their careers developing and using them. In each case, the new technologies faced difficulty in obtaining needed investment and support, just like disruptive technologies in civilian firms that are

47 Perry, W. J. (1997). "Perry on Precision Strike", Air Force Magazine 80/4: 75–76, at 76, http://www.airforcemag.com/MagazineArchive/Documents/1997/April%20 1997/0497perry.pdf

48 *Ibid.*

49 Marshall, A. W. (1993). "Some Thoughts on Military Revolutions—Second Version", DOD Office of Net Assessment, Memorandum for the Record, 23 August, p. 3; Krepinevich, A. F. Jr. (2002). *The Military-Technical Revolution: A Preliminary Assessment*. Washington, DC: CSBA, 3, https://csbaonline.org/uploads/documents/2002.10.02-Military-Technical-Revolution.pdf

organized around older technology—picture clunky electromechanical calculating machines and their support systems at the advent of electronic calculators. Still, in each case, DOD found a way around these legacy challenges, in ways explored below.

DOD does have a series of institutions that can enable a technology to emerge from research into production and procurement. At its best, these can operate as an integrated innovation handoff system. In practice, however, this system can break down, particularly in the links between the military services—the Army, Navy, and Air Force—and the central functions of the Office of the Secretary of Defense. Despite its best efforts to create a connected system that can smoothly incorporate disruptive technology, DOD is, after all, a half-trillion-dollar annual economy dating from 1789, and inevitably has developed significant features of a legacy sector. DOD is dominated by its services, which can have the characteristics of *vested interests defending existing paradigms*, as is typical of all legacy sectors. The services can employ a series of means to assure legacy paradigm dominance, including, to briefly summarize: *budgeting processes* dominated by the services that protect their established technologies, from aircraft carriers to tanks; a *cost structure* that commits DOD long term to these established weapons platforms; service *institutional architectures* that limit cross service collaboration; and established service-led *knowledge/human resources structures* that are heavily hierarchical, service-oriented, and that limit bottom-up ideas.

These, and related characteristics, have led to four major challenges to the defense innovation system: (1) problems in linking innovators (such as DARPA research teams) with service-led implementation; (2) lack of clarity on security threats the nation faces, thereby creating corresponding difficulty in developing department-wide technology strategies (for example, the U.S. currently faces both monolithic and distributed threats); (3) barriers because of defense business practices that curtail innovation, resilience and adaptability (for example, through "Lowest Price, Technically Acceptable" (LPTA) procurement requirements that sacrifice long term value for short term price gains); and (4) too long of an innovation timeline—platform procurements can be twenty-five years or longer, which limits experimentation and the ability to move technological advances into procurement programs. These problems translate into competitive challenges. China, the upcoming peer competitor, currently has some nine jet fighter programs ongoing compared to one in the U.S., and dozens of UAV programs against less than ten in the U.S. Yet in recent decades, the U.S. has been able to overcome comparable problems.

Against this background, we can now explore the legacy sector problems faced by three of the major sets of technologies behind DOD's Revolution in Military Affairs to see how these obstacles to innovation work out in practice within the defense establishment. In each case, DARPA played a critical role, operating, along with key defense leaders, as a change agent, to overcome these structural obstacles.

Stealth Aircraft

Air superiority has been a fundamental doctrine of U.S. defense since World War II.[50] However, Soviet air defense systems by the late Vietnam War were making U.S. aircraft ever more vulnerable. This forced the Air Force to employ vast air armadas of mixed-purpose aircraft, undertaking jamming and electronic counter-measures, chaff dropping, and radar attack, so as to protect a smaller number of attack aircraft that were actually undertaking the strike mission. As early as 1974, discussions began between DOD's office of the Director of Defense Research and Engineering (DDR&E) and DARPA about the need to develop a "Harvey" aircraft (named after the invisible rabbit in the play and film) that would have a greatly reduced radar, infrared, acoustic and visual appearance. The then Director of DDR&E, Malcolm Currie, sent out a memo inviting DOD organizations to develop radical new ideas for such an aircraft. These ideas became known in DARPA, borrowing a term from anti-submarine warfare, as "stealth", and DARPA began to pursue a research agenda around it.

In 1975, a Lockheed engineer, Denys Overholser, located a research paper by the Chief Scientist at the Russian Institute for Radio Engineering on "Method of Edge Waves in the Physical Theory of Diffraction" and realized that from these concepts a computer program

50 This section draws extensively on chapter 1 (on stealth) in Van Atta, et al. (2003). *Transformation and Transition*; and on Rich, B, and Janos, L. (1994). *Skunk Works: A Personal Memoir of My Years of Lockheed.* Boston: Little, Brown & Company, 16–41. The author is indebted to the IDA studies cited for much of the analysis in the three subsections on defense technologies.

could be developed for geometric shapes that would minimize the radar cross section of an aircraft. Lockheed created the program and brought it to DARPA.

DARPA staff understood the importance of the findings, and jumped on them. DARPA Director George Heilmeier, however, insisted that if the concepts were going to become an aircraft, the Air Force would have to take the lead in developing it because developing and buying aircraft was not a DARPA role. Currie supported the stealth approach and used contacts he had built up in the Air Force leadership to try to bring them on board. However, a major Institute for Defense Analyses study found that,

> Air Force support was highly uncertain, as the Air Force saw limited value in a stealthy strike aircraft, given the severe operational limitations that [meant it] would be relatively slow and unmaneuverable, giving it limited air-to-air combat ability, and it would have to fly [only] at night—a far cry from the traditional Air Force strike fighter. There were also competing R&D priorities, most notably the Advanced Combat Fighter program (which eventually became the F-16).[51]

Currie was able to get the Air Force to go along only by securing extra funding for the project, so that stealth development would be in addition to existing Air Force R&D efforts, and, in particular, would not curtail the F-16 program.

William Perry, who succeeded Currie in leading DDR&E, continued to press the stealth program forward because it fit perfectly with his "offsets" strategy. Lockheed's noted "Skunk Works" won the development contract for what became the F-117 strike fighter. Skunk Works used its famous skills in experimentation, flexible problem solving, strong engineering and collaboration to successfully push the F-117 from idea to break-through reality.[52] Northrop, the other defense contractor working in the stealth field, embarked on a follow-on project that became the B-2 stealth bomber. To retain support from a still skeptical Air Force, Defense Secretary Harold Brown made development of stealth aircraft "technology limited" as opposed to "funding limited". In other words, the funding for this secret program was open-ended and was to continue unless a technological barrier emerged.[53] In Desert

51 Van Atta, et al. (2003). *Transformation and Transition. Volume 1*, I–4.
52 Rich and Janos. (1994). *Skunk Works*, 16–41.
53 Van Atta, et al. (2003). *Transformation and Transition. Volume 1*, I–5–6.

Storm, the F-117 enabled the U.S. to obtain air dominance at the outset of the conflict despite being up against the same type of Soviet air defense system that had created such difficulty for U.S.-built aircraft in Vietnam.

Because the services had limited interest in such a radical and different concept that potentially made many of their existing and upcoming aircraft platforms obsolete, stealth overcame the service legacy sector barriers listed above (from powerful vested interests, to cost structure, to institutional architecture to established knowledge/human resource structures) only because of DARPA's highly innovative organizational and technical capabilities, which operated outside the established defense service hierarchies. DARPA, in turn, required support from the highest levels of DOD's civilian leadership, including Secretary Brown and the heads of DDR&E, and from a separate funding stream. Thus, a series of change agents came to bear on the problem, led by DARPA but linked to the DOD senior leadership and to Lockheed, a major defense contractor with its own unique island/bridge innovation organization, its Skunk Works. The Air Force, however, did embrace the technology over time. Interestingly, initial attempts to introduce stealth technology into Navy ship-building—Lockheed's Skunk Works developed the "Sea Shadow"—failed because of Navy opposition for reasons very similar to the Air Force's concerns.[54]

Precision Strike

The mix of defense capabilities known as precision strike developed as part of DOD's focus on the RMA, responding to the confrontation between Cold War forces in Europe. Faced with much larger Soviet forces, William Perry formulated precision-strike objectives as the capability to "see all high value targets on the battlefield at any time; make a direct hit on any target we can see; and destroy any target we can hit".[55] While armies before the RMA had relied on the massed force of as many individual weapons as possible and a few overwhelming nuclear weapons, precision-strike doctrine focused on the ability to both see and select critical high-value targets and to rapidly cripple them in order to break down the enemy's operating capabilities, without major

54 Rich and Janos. (1994). *Skunk Works*, 271–80.

55 Van Atta, et al. (2003). *Transformation and Transition. Volume 1*, IV-35.

casualties on either side and without significant civilian casualties.[56] While the wars Clausewitz wrote about were those between mass armies inflicting mass casualties on a massive scale, the RMA used precision strike to scale this way back.

To achieve precision strike required "joint" efforts between services. Air Force and Navy weapons systems would have to work in intimate coordination with Army systems. This coordination is never easy between rival stovepipes, and weapons procurement itself remains service controlled. Again, DOD's efforts began with DARPA working initially outside the service R&D systems. The "Assault Breaker" R&D program was envisioned to break up any Soviet charge through the Fulda Gap, and was led by a series of related DARPA technological development efforts over many years.[57] Over time, the technologies contemplated in Assault Breaker were modified and evolved into DOD's "1997 Joint Warfighting Science and Technology Plan (S&T) Plan".[58] The precision-strike system came to include JSTARS, a large aircraft packed with powerful radars to "see" much of the battlefield and acquire and track ground targets. These were tied to Army Tactical Missile System (ATACMS) that could hit mobile targets well behind battle lines, as well as to a range of other precision guided missiles and aircraft-launched precision "submunitions" (smaller weapons carried in a missile warhead) and "smart bombs"—all linked to a "Battlefield Control Element" (BETA) to collect and integrate battlefield information.

In summary, the Joint Warfighting S&T Plan entailed a combination of technologies for surveillance, targeting and precision-guided munitions, all resting on earlier DARPA-led advances in information technology. Again, there was service resistance at a number of stages in the implementation process. Leadership from the Office of the Secretary of Defense was required to build and mount the operating systems, and was crucial in pressing for more service "jointness". The retrospective Institute for Defense Analyses study found:

---

56 Department of Defense. (1996). *Joint Warfighting Science and Technology Plan*. Washington, DC: Department of Defense (chapter 4, Part B, Precision Force, 1. Definition), https://apps.dtic.mil/dtic/tr/fulltext/u2/a310991.pdf

57 Van Atta, et al. (2003). *Transformation and Transition. Volume 1*, VI.

58 Department of Defense. (1997). *Joint Warfighting* (chapter 4, Part B).

> Perhaps even more important than the testing and developing of specific technologies [led by DARPA] was the conceptual breakthrough in getting the Services to work together across the barriers of roles and missions to attack the Warsaw Pact tank threat. This cooperative approach was resisted by… the Services, but facilitated by parts of the Army because they understood that the Service needed to work more closely with the Air Force to meet the European threat… The Services had other priorities. The Army continued developing and deploying tanks and helicopters and many in the Service did not want to invest in the new missile technology. So too the Air Force. The larger Service had more important acquisitions: the F-15 and F-16, for example. When competing with Service programs, even good new ideas will not get through the system without a powerful advocate—and for a Joint concept as sweeping as Assault Breaker the advocate had best be the Secretary of Defense.[59]

The combination of an innovative entity, DARPA, and pressure from the Secretary's Office constituted the change agents required to get around the legacy sector problems—from vested interests in the services, to cost structure problems through service commitments, to other programs to problems in creating collective action between services—that afflict the defense establishment.

Unmanned Aerial Vehicles

The idea for unmanned aerial vehicles (UAVs) began, and went through limited development stages in both World Wars, as attack devices, before the advent of guided missiles. While there were early Cold War efforts by the Navy and Air Force, with some remotely piloted vehicles (RPVs) used in Vietnam, the Air Force shut down its UAV efforts in 1976 and shifted focus to cruise missiles. Work on a Navy anti-submarine rotor aircraft ("Dash/Snoopy") was undertaken in the late 1960s and used on ships and by Marines in Vietnam, but subsequently the program was terminated.[60] Despite this early history, today's UAVs are pervasive on the U.S. battlefield, including for counter-terrorist operations. They undertake a wide range of roles: reconnaissance (using cameras, sensors and radar), electronic intelligence gathering, long term surveillance, target designation, communications relays, and now, carrying on-board

59 Van Atta, et al. (2003). *Transformation and Transition. Volume 1*, IV.

60 The developments discussed in this paragraph are detailed in Van Atta, et al. (2003). *Transformation and Transition. Volume 1*, VI–1–11.

weapons, attack on specific targets. The U.S. military is approaching the point where it will have more UAVs than manned aircraft.

Starting in the mid-1970s, DARPA played a key role in developing the enabling technologies that lay behind later UAV success. It funded R&D in sensors, radar, signal location systems, controls, lightweight and low-visibility airframe structures, long endurance propulsion, and new operating concepts. In the 1980s, working with a highly innovative designer, Abraham Karem, and his small company, DARPA also funded a critical UAV technology development program that built and tested the Amber UAV. After initial flight demonstrations, Navy Secretary John Lehman, a UAV advocate, provided support for the program.

However, Amber was terminated in 1990, rejected by the services as not meeting their durability requirements. Nonetheless, the prototypes for Amber pushed the state of the art, developing critical technologies that were fundamental to subsequent development. This was an example of DARPA pushing outside the box of its R&D role and undertaking product development traditionally left to the services. DARPA played a significant role in the development of other UAV prototypes during this period, and the Navy learned lessons from Israeli drones, which were adopted as the "Pioneer UAV" for spotting ship gunfire.[61] However, UAVs were not scaling up. Frustrated with service failures in developing UAV technologies, Congress intervened in 1988 and forced the consolidation of service UAV programs into a joint project office, which led to a third generation of UAV technology.[62]

Following the remarkable performance of RMA technologies in the 1991 Gulf War, the Defense Science Board, the leading DOD technical advisory body, highlighted military problems that could be resolved by improved UAV capabilities. And in the subsequent Clinton administration, the trio of defense and intelligence agency leaders, Secretary of Defense William Perry, Undersecretary of Defense John Deutch, and CIA Director James Woolsey, pushed together for a renewed UAV effort. In cooperation with DARPA, a new "Advanced Concept Technology Demonstration" (ACTD) process was created under Deputy Undersecretary for Advanced Technology (and later

61 Polmar, N. (2013). "The Pioneering Pioneer", *Naval History* 27/5: 14–15.

62 Developments discussed in paragraph detailed in Van Atta, et al. (2003). *Transformation and Transition. Volume 1*, VI–11–26.

DARPA Director from 1995–1998) Larry Lynn, to streamline and accelerate defense technology development and management, but with early cooperation with service users. In effect, Lynn, Perry and Deutch created a new process outside of but involving the services to implement new defense technologies, using UAVs to test the approach.

The result was two deployed UAVs, Predator and Global Hawk, both of which proved highly successful. Predator proved its worth in Bosnia, then in Kosovo, in Iraq no-fly zones, and in Afghanistan, where it was also armed with Hellfire missiles, becoming an attack as well as a surveillance system. Global Hawk was developed by DARPA (using its unique "Other Transaction Authority" to waive traditional acquisition laws and requirements in order to speed development) and initially deployed in Afghanistan as a highly sophisticated reconnaissance tool.[63]

The Institute for Defense Analyses study reached several conclusions about the on-again-off-again UAV experience:

> As occurred with [precision strike and stealth], successful demonstration of the technology for RPV/UAVs did not lead to early acceptance and deployment of the vehicles... There were often differences between the expectations of the DARPA [program manager] and those of the Services on performance (unprepared field verses prepared airstrip) and the level of development (proof of principle verses the need for extensive engineering) needed to transition a program. These differences had an impact on the ability of the system to successfully continue into a deployed system... The systems did not fit within the existing force structure and did not have strong service champions. Without better planning they could not survive the budget battles. The developments often did not fit with existing [Service] operations and doctrine.[64]

When UAV programs started, DARPA's role was to transition the technology to the Services after the proof- of-concept stage, with DARPA doing the R&D and the services and industry doing the engineering and development. Then, with the Amber project, DARPA undertook to actually do the development, but the handoff to the services still proved difficult. After two decades of problems, the technology transition mechanism changed to the "Advanced Concept Technology Demonstration" (ACTD), where a new technical transition entity in

63 Developments discussed in this paragraph detailed in Van Atta, et al. (2003). *Transformation and Transition. Volume 1*, VI–26–38.

64 *Ibid.*, VI-39.

the Secretary's Office, using DARPA's highly-flexible procurement authority and building in service participation, undertook a more extended process. In effect, a new organizational mechanism was created outside the existing system as a change agent that finally succeeded in getting around the legacy sector problems between the services and the repeated efforts at senior levels of DOD to push innovative technologies.

Change Agents

Innovation does not just happen. Even if the elements cited here for a strong innovation system are assembled, someone or some entity must serve as the catalyst for change; those *change agents* can be persons and/ or organizations. Change agents, like innovation itself, must operate at both the institutional and the personal, face-to-face level. As usual in human affairs, there is no substitute for leadership.

If the front end of the innovation system generally is a prerequisite to innovation in legacy sectors, then the concept of change agent, suggested in the above discussion of DOD's technology advances, is a requirement as well. In this way, the innovation system needs strengthening, including through specific approaches cited here such as critical innovation institutions, island/ bridge organization, thinking communities, and linking innovators to operators. None of these steps alone will implement innovation, particularly in thorny legacy sectors, unless there are institutions and accompanying individuals prepared to act as change agents. DOD in the past has been able to initiate change through (1) competition between services (for example, through competing missile programs), (2) struggles between competing groups in a service (such as between "brown shoe" aviators and "black shoe" battleship sailors in the Navy), or (3) through directives from defense civilian leadership. (such as through the DARPA-led advances noted above). In each, change agents were critical.

To return to an example cited above, the Royal Air Force in the 1930s could be viewed as a legacy sector. Like its German counterpart, it was dominated by an emerging air power ethos led by its bomber force, which was not focused on generating defenses against bombers—a task it considered largely hopeless. It took a defense R&D organization, led by defense scientists under Tizard and others, to take on this assumption.

To bring on the transformative technology innovation of radar, they built a strong research group, made links to political authorities prepared to support the effort, and created a working testing process with fighter pilot operators. Allied with civilian and RAF leaders, as change agents they implemented war-changing technologies and practices.

DARPA led similar changes in UAV's, precision strike and Stealth in similar ways. Nonetheless, here too, change agents were critical. William Perry, allied with DARPA in two different tours of duty at DOD, guided a series of major innovation efforts though the Department. Moreover, he helped initiate a change agent system, putting in place the structures and policies that enable the change agents to do their jobs. Other defense sector examples include Malcolm Currie at DR&E who supported GPS, Stealth and smart weapons in the 1970s, early DARPA Director Jack Ruina, who guided its early contributions, and J. C. R. Licklider, the first Information Processing Technologies Office Director at DARPA and the visionary of personal computing and the Internet. President Eisenhower might rate as change agent for putting DARPA in place, and Herbert York, the first DARPA chief scientist (and first Director of DR&E) for helping to envision its initial structure.

Without such change agents, it is hard to see how innovations, particularly in legacy sectors, can emerge out of the innovation pipeline.

Conclusion: Innovation in the Defense Legacy Sector

The stories of the three core breakthrough technologies behind the Revolution in Military Affairs illustrate that the defense sector has many of the attributes of a legacy sector. However, the important point is that DOD found a way to still put these revolutionary technologies into place and bring on significant innovation. Unlike most legacy sectors where breakthrough and disruptive innovations languish, DOD actually implemented them.

DOD turned out to have two major advantages in managing change in its change-resistant, entrenched legacy sector. First, it developed DARPA, a unique innovation entity aimed not only at radical technological advance but also at innovation as a system and trying to solve profound puzzles surrounding implementation.

DARPA operates outside the pressures of the military legacy sector and was created and designed as a result of Sputnik to bring innovative change to a Defense Department affected by legacy problems. In effect, DARPA (and its allies) came to play the role that Hyman Rickover and his group played for atomic submarines and that Bernard Schriever and his group played for ballistic missiles.

It appears vital, then, to bring front-end innovation capabilities to influence legacy sectors. An important lesson from DARPA's ability to bring innovation into a defense sector with deep legacy characteristics has been the importance of *critical innovation institutions*. These institutions should attempt to embody both *"connected science and technology"*—linking scientific research to implementation stages—and *"challenge"* approaches—pursuing major mission technology challenges. As discussed, innovation requires not only a process of creating connected science and technology challenges at the *institutional level*, but it also must operate at the *personal level*.

The critical stage of innovation is face-to-face, not institutional, so, while institutions where talent and R&D come together are required, personal dynamics, usually embodied in *"great groups"* are a necessity. The DARPA *"right-left" research model* can be important in reaching the innovation stage, where program managers contemplate the technological breakthroughs they want to emerge from the right end of the innovation pipeline, then go back to the left side of the pipeline to look for proposals for the breakthrough research that will get them there. This process tends to lead to *revolutionary breakthroughs* that could be transformative of a technology sector. A technology *"visioning"* process at the outset of the effort appears to be a particular key. The approach results in seeking *high-risk but high-reward* projects.

As discussed, the *island/bridge* organizational approach for innovation institutions also appears to be important. The innovation team should be put on a protected island apart from bureaucratic influences that can ruin it, so that it can focus on the innovation process. The strength of the innovation process will also depend on building a solid *thinking community* as a source for ideas and support. Because innovation must span numerous steps—from research through initial production—the means for *linking technologists to operators* appear to be critical. Finally, *change agents* will be required to move the innovation toward implementation.

Second, DARPA alone was not enough. Unlike most legacy sectors, DOD has an official, the Secretary of Defense, who must by law be a civilian, who can exercise authority to force change. If the Secretary sees the need for a technology shift, he or she can muster the power, despite all the legacy sector checks in the system, to direct it. DARPA has been successful when it ties its technological advance to a senior defense leader in the Office of the Secretary who is prepared to override legacy pressures and be a *change agent*. Of course, DOD faced an additional intense pressure for change—meeting national security needs—but these two characteristics, a strong front-end innovation linked to change agents, remain central.

There are important lessons here for other legacy sectors: a "connected" innovation agency, using the extended pipeline model which is outside the legacy system, and then linked to a source of power that can direct change—a change agent—has proved to be a vital combination in the defense sector's ability to innovate. The longstanding perspective on DARPA has been that its successes have been in the "frontier" sector; it is rightly acclaimed for its foundational role in the IT revolution. But there is a less understood perspective on DARPA that constitutes the other side of the coin: it has brought disruptive, radical innovation into a legacy sector.

In this way, DARPA does not only belong in the "extended pipeline" model; it also has developed features that have enabled it to innovate in the legacy defense sector. This means that it also represents key features of what we term the "innovation organization" model. Legacy sectors use political, technological, economic and social system barriers in their defense against disruptive innovation. The innovation organization model recognizes that there are many institutions and mechanisms operating within an innovation system, particularly in legacy sectors; this mandates a richer evaluation of innovation and of potential policies to shift the overall system. DARPA and its senior Department allies have found ways, delineated above, to impose this richer mix of policies. This mix of strong front-end innovation capability and change agents provides basic lessons for innovation in other legacy sectors that go far beyond defense to other key parts of the economy.

References

Alic, J., et al. (1992). *Beyond Spinoff: Military and Commercial Technologies in a Changing World*. Cambridge, MA: Harvard Business School Press.

Bennis, W., and Biederman, P. W. (1997). *Organizing Genius: The Secrets of Creative Collaboration*. New York, NY: Basic Books.

Bird, K., and Sherwin, M. J. (2005). *American Prometheus, The Triumph and Tragedy of J. Robert Oppenheimer*. New York, NY: Alfred A. Knopf.

Bonvillian, W. B. (2015). "All that DARPA Can Be", *The American Interest* 11/1, August, https://www.the-american-interest.com/2015/08/01/all-that-darpa-can-be/

Bonvillian, W. B. (2013). "The New Model Innovation Agencies: An Overview", *Science and Public Policy* 41/4: 425–37, https://doi.org/10.1093/scipol/sct059, https://academic.oup.com/spp/article-abstract/41/4/425/1607552?redirectedFrom=fulltext

Bonvillian, W. B. (2009). "The Connected Science Model for Innovation—The DARPA Model", in *21st Century Innovation Systems for the U.S. and Japan*, ed. S. Nagaoka, M. Kondo, K. Flamm, and C. Wessner. Washington, DC: National Academies Press. 206–37, https://doi.org/10.17226/12194, http://books.nap.edu/openbook.php?record_id=12194&page=206 (Chapter 4 in this volume).

Bonvillian, W. B. (2006). "Power Play, The DARPA Model and U.S. Energy Policy", *The American Interest* 2/2, November/December, 39–48, https://www.the-american-interest.com/2006/11/01/power-play/

Bonvillian, W. B., and Van Atta, R. (2011). "ARPA-E and DARPA: Applying the DARPA Model to Energy Innovation", *The Journal of Technology Transfer*, 36: 469–513, https://doi.org/10.1007/s10961-011-9223-x (Chapter 13 in this volume).

Bonvillian, W. B., and Weiss, C. (2011). "Complex Established 'Legacy' Sectors: The Technology Revolutions that Do Not Happen", *Innovations* 6/2: 157–87, https://doi.org/10.1162/inov_a_00075, https://www.mitpressjournals.org/doi/pdf/10.1162/INOV_a_00075

Bonvillian, W. B., and Weiss, C. (2009). "Taking Covered Wagons East, A New Innovation Theory for Energy and Other Established Sectors", *Innovations* 4/4: 289–94; https://doi.org/10.1162/itgg.2009.4.4.289, http://www.mitpressjournals.org/userimages/ContentEditor/1259694503297/Bonvillianinov.pdf

Branscomb, L., and Auerswald, P. (2002). *Between Invention and Innovation, An Analysis of Funding for Early-State Technology Development*. NIST GCR 02–841. Washington, DC: National Institute of Standards and Technology, https://link.springer.com/article/10.1007%2Fs10961-011-9223-x

Buderi, R. (1997). *The Invention that Changed the World*. Sloan Technology Series. New York, NY: Simon & Schuster.

Budiansky, S. (2013). *Blackett's War*. New York, NY: Alfred A. Knopf.

Bush, V. (1945). *Science: The Endless Frontier*. Washington, DC: Government Printing Office, https://doi.org/10.1002/sce.3730290419, https://www.nsf.gov/od/lpa/nsf50/vbush1945.htm

Carleton, T. L. (2010). "The Value of Vision in Radical Technological Innovation", PhD Thesis, Stanford University, Palo Alto, http://purl.stanford.edu/mk388mb2729

Cathcart, B. (2004). The Fly in the Cathedral. New York, NY: Farrar, Straus & Giroux.

Chesborough, H. W. (2003). "The Era of Open Innovation", *MIT Sloan Review* 44/3, http://sloanreview.mit.edu/article/the-era-of-open-innovation/

Chposky, J., and Leonsis, T. (1986). *Blue Magic: The People, Power and Politics Behind the IBM Personal Computer*. New York, NY: Facts on File, https://doi.org/10.2307/3115979.

Clark, R. W. (1962). *The Rise of the Boffins*. London: Phoenix House.

Clark, R. W. (1965). *Tizard*. Cambridge, MA: The MIT Press.

Conant, J. (2002). *Tuxedo Park: A Wall Street Tycoon and the Secret Palace of Science that Changed the Course of World War II*. New York, NY: Simon & Shuster.

Cooper, R.G., Edgett, S. J., and Kleinschmidt, E. J. (2002). "Optimizing the Stage-Gate Process", *Research Technology Management* 45/5, 43–49, https://doi.org/10.1080/08956308.2002.11671532

DARPA. (2008). *DARPA—Bridging the Gap, Powered by Ideas*. Arlington, VA: Defense Advanced Research Projects Agency, http://www.dtic.mil/cgi-bin/GetTRDoc?Loca-tion=U2&doc=GetTRDoc.pdf&AD=ADA433949

DARPA. (2003). *DARPA Over the Years*. Arlington, VA: Defense Advanced Research Projects Agency.

Department of Defense. (1996). *Joint Warfighting Science and Technology Plan*. Washington, DC: Department of Defense, https://apps.dtic.mil/dtic/tr/fulltext/u2/a310991.pdf

Hiltzik, M. (1999). *Dealers of Lightning: Xerox PARC and the Dawn of the Computer Age*. New York, NY: Harper Business.

Krepinevich, A. F. Jr. (2002). *The Military-Technical Revolution: A Preliminary Assessment*. Washington, DC: CSBA, https://csbaonline.org/uploads/documents/2002.10.02-Military-Technical-Revolution.pdf

Marshall, A. W. (1993). "Some Thoughts on Military Revolutions—Second Version", DOD Office of Net Assessment, Memorandum for the Record, 23 August.

Milunovich, S., and Roy, J. M. A. (2001). "The Next Small Thing—An Introduction to Nanotechnology", Merrill Lynch Industry Comment, 4 September, https://www.slideshare.net/tseitlin/intro-to-nanotechnology-merrill-lynch

MIT Radiation Laboratory. (1946). *Five Years at the Radiation Laboratory*. Cambridge, MA: The MIT Press, https://archive.org/details/fiveyearsatradia00mass

National Research Council. (1999). *Funding a Revolution: Government Support for Computing Research*. Computer Science and Telecommunications Board. History, Commission on Physical Sciences Mathematics and Applications. Washington, DC: National Academy Press, https://doi.org/10.17226/6323, https://www.nap.edu/catalog/6323/funding-a-revolution-government-support-for-computing-research

Nelson, R. R. (1993). *National Systems of Innovation*. New York, NY: Oxford University Press.

Perry, W. J. (1997). "Perry on Precision Strike", *Air Force Magazine* 80/4: 75–76, at 76, http://www.airforcemag.com/MagazineArchive/Documents/1997/April%201997/0497perry.pdf

Rich, B, and Janos, L. (1994). *Skunk Works: A Personal Memoir of My Years of Lockheed*. Boston: Little, Brown & Company.

Polmar, N. (2013). "The Pioneering Pioneer", *Naval History* 27/5: 14–15.

Ruttan, V. W. (2001). *Technology Growth and Development: An Induced Innovation Perspective*. New York, NY: Oxford University Press.

Ruttan, V. W. (2006). *Is War Necessary for Economic Growth? Military Procurement and Technology Development*. New York, NY: Oxford University Press.

Rycroft, R. W., and Kash, D. E. (1999). "Innovation Policy for Complex Technologies", *Issues in Science and Technology*, 16/1, https://issues.org/rycroft/

Sen, A. (2014). "Transformative Innovation: What 'Totally Radical' and 'Island-Bridge' Mean for NOAA Research", PhD thesis, George Washington University, Washington.

Stokes, D. E. (1997). *Pasteur's Quadrant, Basic Science and Technological Innovation*. Washington, DC: Brookings Institution Press.

Van Atta, R., Lippitz, M., et al. (2003). *Transformation and Transition, DARPA's Role in Fostering a Revolution in Military Affairs. Volume 1*. Alexandria, VA: Institute for Defense Analyses, https://doi.org/10.21236/ada422835, https://fas.org/irp/agency/dod/idarma.pdf

Waldrop, M. M. (2001). *The Dream Machine: J. C. R. Licklider and the Revolution that Made Computing Personal*. New York, NY: Viking Press.

Weiss, C. and Bonvillian, W. B. (2009). *Structuring an Energy Technology Revolution*. Cambridge, MA: The MIT Press, https://doi.org/10.7551/mitpress/8161.001.0001

13. ARPA-E and DARPA

Applying the DARPA Model to Energy Innovation[1]

William B. Bonvillian and Richard Van Atta

Overview

The United States faces powerful economic challenges in the interlinked and contradictory nexus of the economy, energy, and environmental issues. In this arena, transformative innovation is understood to be a key public policy response.[2] One element of the response has been the creation of an energy-DARPA (ARPA-E).

DARPA was formed to address the problem of transformative innovation. Instigated in 1958 by the Sputnik shock, the Advanced Research Projects Agency (subsequently renamed the Defense Advanced Research Projects Agency) was created with an explicit mission: to ensure that the U.S. never again faced a national security "technological surprise", like Sputnik, due to failure to pay adequate attention to and

1 This paper originally appeared in 2011 in the *Journal of Technology Transfer* 36, at 469–513.

2 Bonvillian, W. B., and Weiss, C. (2009). "Taking Covered Wagons East, A New Innovation Theory for Energy and Other Established Sectors", *Innovations* 4/4: 289–94, http://www.mitpressjournals.org/userimages/ContentEditor/1259694503297/Bonvillianinov.pdf

 https://doi.org/10.11647/OBP.0184.13

stay focused on break-through technological capabilities.[3] DARPA itself can be categorized as a disruptive innovation, creating an approach to fostering and implementing radically new technology concepts recognized as transformational.[4]

Innovation is recognized as the linchpin for U.S. economic growth,[5] transforming the economy based on new products that provide new economic and social functionality. DARPA was at the center of that innovation process in the second half of the twentieth century, playing a keystone role in the computing and internet innovation waves.[6] However, the process of innovation aimed at such transformation is recognized as highly risky and extremely difficult to implement. In a complex innovation system laced with market failures between the stages of fundamental research and technology transition, governmental support increasingly has been viewed as a necessary element. This risk and difficulty are captured in the term "disruptive innovation"[7]—where the potential novel capabilities offered by new innovations are impeded

---

3 Van Atta, R. (2008). "Fifty Years of Innovation and Discovery", in *DARPA, 50 Years of Bridging the Gap*, ed. C. Oldham, A. E. Lopez, R. Carpenter, I. Kalhikina, and M. J. Tully. Arlington, VA: DARPA. 20–29, https://issuu.com/faircountmedia/docs/darpa50 (Chapter 2 in this volume); Bonvillian, W. B. (2009). "The Connected Science Model for Innovation—The DARPA Model", in *21st Century Innovation Systems for the U.S. and Japan*, ed. S. Nagaoka, M. Kondo, K. Flamm, and C. Wessner. Washington, DC: National Academies Press. 206–37, https://doi.org/10.17226/12194, http://books.nap.edu/openbook.php?record_id=12194&page=206 (Chapter 4 in this volume).

4 Van Atta, R., Lippitz, M., et al. (2003). *Transformation and Transition, DARPA's Role in Fostering a Revolution in Military Affairs. Volume 1.* Alexandria, VA: Institute for Defense Analyses, https://doi.org/10.21236/ada422835, https://fas.org/irp/agency/dod/idarma.pdf

5 See, for example, Solow, R. M. (2000). *Growth Theory, An Exposition.* 2nd ed. Oxford: Oxford University Press, http://nobelprize.org/nobel_prizes/economics/laureates/1987/solow-lecture.html; Romer, P. (1990). "Endogenous Technological Change", *Journal of Political Economy* 98: 72–102, https://doi.org/10.1086/261725; and Jorgenson, D. (2001). "U.S. Economic Growth in the Information Age", *Issues in Science and Technology* 18/2, http://www.issues.org/18.1/jorgenson.html

6 Ruttan, V. W. (2006). *Is War Necessary for Economic Growth? Military Procurement and Technology Development*. New York, NY: Oxford University Press.

7 The term was developed by Christiansen, and reflects Schumpeter's economic concept of capitalism (Christiansen, C. (1997). *The Innovator's Dilemma: When New Technologies Cause Great Firms to Fail.* Boston, MA: Harvard Business School Press, xviii–xxiv). See, also, Schumpeter, on the concept of "creative destruction" in capitalism in which new technologies and processes create or alter firms and markets (Schumpeter, J. (1942). *Capitalism, Socialism, and Democracy*. New York, NY: Harper & Row).

not only because they are new and different, raising unknowns and risks, but because they often entail the potential of disrupting existing markets, products, practices and approaches. They bring with them two "shocks": (1) they are sufficiently different that the existing system of investment and development is risk averse to them; and (2) in many cases they actually are seen as threats to vested products and capabilities, and thus face further difficulty in achieving needed investment and support.[8] The threat and risk of such innovations are significantly expanded when they attempt to enter in complex, established "legacy" sectors such as energy.[9] Overcoming these double impediments requires strategically-focused technology development and management approaches.

ARPA-E, the Advanced Research Projects Agency—Energy, was established in 2007 as part of the America COMPETES Act[10] and initially funded under the 2009 economic stimulus, the American Reinvestment and Recovery Act.[11] ARPA-E was created to foster disruptive innovation in the complex, established "legacy" sector of energy, exactly the model described above. Although threatened by the partisan budget environment, ARPA-E obtained funding for FY 2011 from Congress at close to the same level it was funded in the two previous fiscal years.[12] It has already emerged as a dramatically new model in the energy innovation space, worthy of in-depth examination.

This paper looks first into DARPA as a model, asking a series of questions: What about DARPA has enabled its success? Is DARPA's success transferable to other arenas? In particular, this paper reviews less well-known features of the DARPA model not widely commented on to

---

8 Van Atta, R., Bovey, R., et al. (2003). *Science and Technology in Development Organizations.* Alexandria, VA: Institute for Defense Analyses.

9 Bonvillian and Weiss. (2009) "Taking Covered Wagons East"; Weiss, C., and Bonvillian, W. B. (2011). "Complex, Established 'Legacy' Sectors: The Technology Revolutions that do Not Happen", *Innovations* 6/2: 157–87.

10 ARPA-E was first proposed in National Academies of Sciences. (2007). *Rising above the Gathering Storm: Energizing and Employing America for a Brighter Economic Future.* Washington, DC: The National Academies Press. 152–57, https://doi.org/10.17226/12537, https://www.nap.edu/catalog/11463/rising-above-the-gathering-storm-energizing-and-employing-america-for#toc

11 American Recovery and Reinvestment Act (ARRA). (2009). P.L. No: 111–15 (signed by President 17 February 2009).

12 ARPA-E received $400 million in initial funding from the 2009 stimulus legislation (ARRA 2009) for FY's 2009 and 2010; it did not therefore seek additional funding in FY2010, The Administration's budget sought $550 million for ARPA-E in FY2011; Congress funded it at $180 million for FY 2011 (Continuing Resolution FY2011).

date. Secondly, the paper looks at ARPA-E, raising similar questions: Is ARPA-E designed to effectively emulate the DARPA model? Are there significant differences in the energy arena that inhibit or prohibit the success of this model? Are there new elements in the ARPA-E approach modifying and adapting the DARPA approach to increase ARPA-E's chance of success? Do some of the less well-known features of DARPA, as noted, provide lessons for ARPA-E?[13]

There is an additional question behind this inquiry. What about DARPA and ARPA-E could or should be emulated by other organizations seeking to foster and effect transformative technological change? Both agencies represent a different model for technology advance. While standard model R&D agencies focus on research not technology, rely on a peer review process for selecting awardees, and do not use what could be called a technology visioning step in their process, DARPA, and now ARPA-E, reverse all these rules. They focus early in their processes on developing a vision of new technologies, then on developing a research program to achieve that vision, and on using empowered program managers, not a disparate peer review process, for award selections. DARPA's remarkable string of technology success has demonstrated the power of its model, and early successes at ARPA-E suggests it is dynamic and replicable. Features of this model may be of interest to other parts of the US R&D system. For example, the Department of Education and the National Science Foundation are considering an ARPA-Ed for education research, NIH is considering a translational research program, and the Department of Homeland Security and the intelligence agencies are working on implementing existing authority to implement their own DARPA clones. Thus, a careful review of the DARPA and emerging ARPA-E

13 Other attempts have been made in recent years in the federal government to create DARPA "clones", particularly HS-ARPA for the Department of Homeland Security, I-ARPA for the intelligence community and BARDA within the Department of Health and Human Services for biothreats and health emergencies. While there are questions whether these attempts successfully emulated DARPA, this paper does not specifically explore these organizations. HS-ARPA was never fully implemented; a number of the reasons, and implications for ARPA-E, are discussed in Bonvillian, W. B. (2007). "Will the Search for New Energy Technologies Require a New R&D Mission Agency?—The ARPA-E Debate", *Bridges* 14, 12 July, https://ostaustria.org/bridges-magazine/volume-14-july-12-2007/item/2297-will-the-search-for-new-energy-technologies-require-a-new-r-d-mission-agency-the-arpa-e-debate

rulesets may offer lessons not only to each other, but to innovation ecosystems more broadly.

The paper then looks at a challenge faced by both DARPA and ARPA-E: technology implementation. Both agencies move technologies down the innovation pipeline to the prototype or small-scale demonstration stage. Neither agency has direct authority to enable commercialization of its potentially breakthrough technologies. DARPA often relies on procurement programs by military services to form initial markets; ARPA-E has no counterpart to the services within DOE. How could this implementation hurdle be overcome? The paper concludes by reviewing this question, including a model at DOD for achieving this. Throughout the paper, drawing from detailed evaluations of both agencies, we discuss and make recommendations on the role of government in fostering transformative technology energy and national security, including in the current and future world of globalized businesses, economies and technologies.

I. The DARPA Model

Well-Known Elements in the DARPA Culture

DARPA Deputy Director Ken Gabriel has suggested that several features central to DARPA are best explained by its name.[14] DARPA, (1) is not a broad research organization or lab but a "*projects*" agency pushing particular technology projects; (2) is primarily a "*defense*" agency that should take full advantage of its presence in DOD to move its technologies, and (3) works primarily on the "*advanced*" stage of breakthrough innovation not on incremental or engineering efforts that other parts of DOD focus on. This gives us a broad-brush portrait, but how does it actually operate?

Michael Piore has suggested that DARPA program managers exemplify a form of what are known in organizational literature as "street-level bureaucracies".[15] Like cops on a beat or school teachers,

14 Discussion by Kaigham Gabriel, DARPA Deputy Director, at forum on Leveraging DOD's Energy Innovation Capacity, at the Bipartisan Policy Center, Washington, DC, 25 May 2011.

15 Piore, M. (2008). *Learning on the Fly: Reviving Active Governmental Policy in an Economic Crisis*. Presentation at How Will a New Administration and Congress Support Innovation in an Economic Crisis? Sponsored by the Economic Policy

or welfare caseworkers, the identity of line officers in bureaucracies, even one as creative and flat as DARPA, is best understood by the roles they play and the informal rules adopted by their colleagues in their professional communities. Piore writes,

> … in street-level bureaucracies, a series of tacit rules emerge which the agents apply in making their decisions. These rules grow out of the culture of the organization as it is shaped by the backgrounds… from which the agents are drawn, by the training which they undergo within the organization itself, by the discussion and debate which shapes the interpretative community in which they operate, and ultimately by the way in which their decisions are reviewed by their colleagues informally and their superiors formally.[16]

DARPA is widely understood to embody a series of unique tacit rules implemented at the "street level" by its program managers, that reflect the organizing principles of its culture, and these are not typical of similar such rules at other R&D agencies. The DARPA ruleset includes:[17]

- a flat, non- hierarchical organization, with empowered program managers;
- a challenge-based "right-left" research model;
- an emphasis on selecting highly talented, entrepreneurial program managers (PMs) who serve for limited (three- to five-year) duration;
- research that is performed entirely by outside performers, with no internal research laboratory;
- projects which focus on a "high-risk/high-payoff" motif, selected and evaluated on what impact they could make on achieving a demanding capability or challenge;
- and initial short-term funding for seed efforts that scale to significant funding for promising concepts, but with clear willingness to terminate non-performing projects.

Institute, ITIF, Breakthrough Institute, University of Calif., Washington Center and Ford Foundation, Washington, DC, 1 December, https://www.longviewinstitute.org/blockvideo/view/index.html. See generally, Chapter 3 in this volume.

16 *Ibid.*

17 See, for greater detail on these features, Van Atta. (2008). "Fifty Years of Innovation and Discovery"; and Bonvillian. (2009). "The Connected Science Model for Innovation".

These rules are widely understood and have been previously explored in various studies. However, we believe these are not the only rules that need to be understood about DARPA and its culture.

Other Important Elements in the DARPA Model

The model goes beyond the above well-understood features—historically it has embodied a number of other deep features that should be accounted for. Within the overall context of its organizing framework DARPA's structure and focus has ebbed and flowed. In fact, its ability to flexibly adapt to changing circumstances is one aspect of its success: it has generally avoided becoming entrenched in particular technology pursuits, problem focus areas or organizational approaches and structures. Thus, as a living, institutional organism there have been aspects of its management and implementation that have not necessarily been enduring attributes for all of DARPA's history, but are notable as contributing to its success. A discussion of these additional DARPA features follows below.

Multigenerational Technology Thrusts—DARPA does more than undertake individual projects. It has, in many instances, worked over an extended period to create enduring technology "motifs"—ongoing thrusts that have changed the technology landscape. Some of the notable examples of this are DARPA's work in information technology (IT), stealth, and standoff precision strike. Some of these foci are what might be termed broad technology stewardship over a family of emergent technologies—including new sensing systems, such as infrared sensing, or new electronics devices.[18] In these thrust areas DARPA has been able to undertake *multigenerational* technology thrusts and advances over extended periods to foster multiple generations of technology.

The DARPA IT thrust area is the most notable in the context of ARPA-E's model. The IT thrust began with the now well-known vision of the first director of the Information Processing Technology Office (originally prosaically labeled the Information Processing Techniques Office), J. C. R. Licklider.[19] It has endured over decades as "the ambitious vision

18 Van Atta, R., Deitchman, S., and Reed, S. (1991). *DARPA Technical Accomplishments. Volume III.* Alexandria, VA: Institute for Defense Analyses, III, IV-1-IV-5.

19 Waldrop provides a detailed illumination of Licklider's role in fostering the revolution in information technology which has been duly recognized as one of

of Licklider for revolutionizing information processing and applying it to problems of 'human cognition' [that is now] being progressively realized".[20] Importantly, the information technology thrust was implemented largely outside of the Department of Defense through universities and small startup firms that emerged from this research. As will be discussed below, these startup enterprises were nurtured by the DARPA program through research grants and importantly through early purchase of their products as inputs into other DARPA and DOD programs.

The long-term support of a thrust area is neither a given nor an endowment at DARPA. The thrusts are defined generally as challenges appropriate to potentially needed "breakthrough" capabilities. In some cases, as with high-energy particle beams, a thrust might be pursued for more than a decade, and then be terminated due to its lack of progress or impracticality. However, in the case of particle-beam weapons, the program reemerged during the 1980s in conjunction with the Strategic Defense Initiative.[21] The basis of the thrust lay in its promise as a possibly revolutionary technology with the prospect of a transformational impact and evidence of its progress. The information technology thrust clearly demonstrated both this impact and this progress. The particle beam thrust, on the other hand, was terminated when it was unable to overcome increasing issues of technical complexity relative to performance, especially in conjunction with mounting costs.

It should be noted that persistence in an area is not universal in DARPA's ethos or history. DARPA was founded, as noted, as a projects agency—not a technology thrust agency. However, implicit in its organizational structure of program offices is (at least at specific times) a set of basic technical-implementation themes. The specificity and focus of these themes have changed substantially over DARPA's history. Moreover, there have been program offices with very dispersed and unrelated projects that sometimes raise the issue whether there is

DARPA's most remarkable and compelling accomplishments (Waldrop, M. M. (2001). *The Dream Machine: J. C. R. Licklider and the Revolution that Made Computing Personal.* New York, NY: Viking Press).

20 Van Atta, Deitchman and Reed. (1991). *DARPA Technical Accomplishments, Volume III*, IV-7.

21 Van Atta, Deitchman and Reed. (1991). *DARPA Technical Accomplishments, Volume III*, IV-9-IV-10.

adequate coherence or focus. In this way, it must be understood that only some of DARPA's research, and only some of the time, can be identified as an ongoing thrust. Other parts of the agency even at the same time may be pursuing a very eclectic set of individual projects that at least at that time appear to be disconnected. DARPA, then, remains a predominantly project-oriented office, except when it needs to periodically launch a new technology thrust.

From a lessons-learned perspective, in relationship to ARPA-E, it is worth noting that DARPA itself began with a set of explicit, but very high-level programmatic themes, termed "Presidentials": issues of space, missile defense, and nuclear test detection that met Presidential priorities. Although the first—"space"—was quickly transferred to the newly created National Aeronautics and Space Administration—NASA, the others galvanized a set of research programs aimed at a broad objective.[22] Given their high-level imprimatur, these programs were sustained over many years, as they addressed key, daunting challenges. Notably, it took senior DOD management (John Foster, Director of Defense Research and Engineering) to terminate DARPA's early missile defense work and get it transitioned to the Army. The result was that under his mandate the DEFENDER program for ballistic missile defense was transferred to the Army for implementation—including the transfer of staff from the DARPA program to the Army. This represented an early attempt to confirm DARPA's role as a technology development organization, not as a technology implementer.[23]

A crucial management issue at DARPA, therefore, is how to keep such thrusts, themes, or foci from becoming entrenched resource allocations that weigh down the organization at the expense of innovation. DARPA has not been immune from this phenomenon. Iteratively, it has taken

---

22 Licklider's IPTO Office initiatives in computing gathered momentum when Kennedy and McNamara concluded they had a major "command and control" problem from their experience in the Cuban missile crisis; expanding Licklider's program was the DARPA response (Waldrop. (2001). *The Dream Machine*, 200–03).

23 For DEFENDER transition to the Army, see Van Atta, Deitchman and Reed. (1991). *DARPA Technical Accomplishments Volume II*, I-17, I-26. Interesting as a contrast, DARPA's initial assignment for detecting Soviet nuclear tests, the VELA program, remained within DARPA for many years, and the operation of the Large Aperture Seismic Arrays remained under DARPA for decades, as no appropriate agency could be identified to take on the responsibility (Van Atta, Deitchman and Reed. (1990). *DARPA Technical Accomplishments, Volume I*, XIII-14.

high-level intervention from the agency director and above DARPA to overcome this tendency—and this at times has caused rancor. A more recent instance of this management issue was the controversy over DARPA's substantial reduction in funding for university computer science programs under Director Anthony Tether in the 2003–2008 timeframe. While leaders in the computer science programs at universities protested this as undermining DARPA's noted successes in this field and cutting DARPA off from access to ensuing IT talent generations, Tether argued that the research in question was not high enough on the high-risk/high-payoff metric to be appropriate for DARPA.[24] This is evidence of the strong tensions that can emerge between those who have received funding for a successful thrust over the years and the agency's management, which has to make choices on what to fund. To keep such overarching thrusts dynamic, as opposed to institutionalized, is a major ongoing management challenge for ARPA-E leadership. The ability of researchers and their supporters within universities, industry research labs and in Congress to press these interests should not be underestimated.

Complementary Strategic Technologies—DARPA has repeatedly launched related technologies that complement each other, and which help build support for the commercialization or implementation of each. This concept of complementary technologies also ties to the notion of program thrusts.

One way of thinking about this category is that DARPA is not in the "thing" business; rather, it is in the problem-solving business. While a specific innovation may have a major impact, it is unlikely that one such project by itself will adequately address a major challenge or problem. While DARPA may support an individual invention, it usually does so because that invention may be an element of an overall solution to a challenge. From an historical perspective, DARPA has almost never started out with a coherent program thrust (the exception being its inaugural Presidential issues, which were really more articulated overarching challenges than technical thrusts). The usual history is that a DARPA Program Office will pursue several disparate concepts initially and then, as the concepts shake out and begin to show promise, those that emerge will begin to cohere, and opportunities to integrate and link their developments will be identified. Also, as a program

24 Bonvillian. (2009). "The Connected Science Model", 225–33.

becomes better defined it becomes clearer what is missing in the ability to bring it into fruition, and thus, more targeted technology programs can be formed to seek alternatives to these barriers. This approach was certainly evident in DARPA's information technology programs as well as its microelectronics research. The key, then, to such developments is that they do not start out explicitly as coherent, multi-stage inter-linked development programs. Rather, they become this as merited and determined by the progress of the unfolding research.

One example of this inter-linking or complementarity is DARPA's funding of the development of computer workstations for integrated circuit design. DARPA program managers realized that the complexity of the designs that they were seeking in computer chips was outstripping the CAD tools that then existed and they therefore funded the development of advanced inter-netted design capabilities at various universities. Two results of this project were the Sun Microsystems and the Silicon Graphics workstations, both of which were developed out of Stanford University. As these capabilities became demonstrated, DARPA then urged their commercialization, but also supported their researchers doing novel chip design to acquire these systems. Notably, the resulting more advanced chips became available to developers of more powerful workstations, internet servers and routers and PCs, thus fostering a virtuous cycle of technology development and adoption.[25]

The management lesson here is that such complementarity is not predetermined nor necessarily obvious at the outset. However, if the research is defined too narrowly and without some overarching integrated perspective (such as a thrust) it is less likely that, as the projects emerge, with some succeeding and others not, that the linkages and complementarities amongst them will be identified. To reiterate, the purpose of the research efforts is not "things" *per se*, but to solve overarching and daunting challenges or problems.[26] The evolving efforts to do that then help define the synergies and linkages as the projects evolve, with resulting complementary technologies.

---

25 See Van Atta, Deitchman and Reed. (1991). *DARPA Technical Accomplishments, Volume II* (chapter 13), for discussion of this approach, especially Annex B to this chapter on the SUN Workstation.

26 Van Atta, et al. (2003). *Transformation and Transition. Volume 1*, S-12-S-13.

Confluence with an Advocate Community—DARPA has spawned new economic sectors, enabling new firms that have garnered venture capital (VC) support. Accordingly, DARPA has been able to make its advances reinforce each other—it has been able to play an intermediary role with industry in part by building an advocate community across sectoral lines. How has it accomplished this?

Since DARPA itself does not implement the results of the research it sponsors, its main path to effecting implementation is by fostering and supporting the community of what we can term "change-state advocates" as a convener and instigator. In many fields initiated by DARPA research, the initial extant research capabilities are disparate and dispersed. Since the field is new, it is not well supported within the university science community and since it only has, at best, nascent technology to demonstrate, there is little in the way of investor or industry support. Thus, a key element of DARPA's success in such areas as information technology, sensor systems, advanced materials, and directed energy systems is building the community of change agents—a broad community fostered over time from its program managers, from "graduates" of the DARPA program who go on to roles in academia and industry, and from contractors in universities and industry trained in the DARPA model and technology approaches. Importantly, this creates a close-knit network of individuals who know and trust each other, breaking down information/collaboration barriers. This community confluence, in turn, creates a connection with the private sector and its ability to spur implementation.

A former DARPA office director explains one way this community builds itself:

> Good DARPA PMs create the conditions for their individual contractors to cohere into a technical community. The most visible way is through regular (usually annual) program reviews at which all contractors present their work and where the really good conversations take place in the hallways as participants start seeing how they can connect and further their work. This isn't forced by the PM but s/he plays a vital role in nurturing the process. I got to participate in this in semiconductor process technology and a couple of other areas, but there are dozens of examples of technology communities that started this way. I see ARPA-E doing the same.[27]

27 Prabhakar, A. (2011). Personal Communication, 27 May.

A corollary to this community development, then, is that it fosters DARPA's role as an intermediary between companies to get them to consider working together in value-add ways against their own near-term individual interests. Companies developing new ideas generally value their intellectual property and proprietary position over the value of collaboration. The Federal government R&D agencies, and DARPA in particular, however, tend to have a more detached "50,000 foot" view of the innovation landscape and despite industry's tendencies, DARPA uses this community confluence to incentivize high value collaborations without violating the confidentiality of what they learn from industry behind closed doors.

Connected to Larger Innovation Elements—Going beyond the confluence with its support community, DARPA has been an actor within larger innovation efforts, where it is often instrumental, but seldom a sole actor. This is important to DARPA's effectiveness because it does not have its own research facilities and its program managers do not perform their own research. Thus, DARPA PMs' most important function is to identify and support those who have the potentially disruptive, change-state ideas and will ably perform the research. Thus, the PM is an opportunity creator and idea harvester within an emerging technology field. From this concept-idea scouting perspective DARPA has spawned groups of researchers, and from that, new firms that act to help effectuate the program's overall vision.

The information technology thrust initiated by J. C. R. Licklider is the most notable example of DARPA creating a dynamic iterative innovation eco-system, based on university programs and start up enterprises. The major initial research centers for this evolving and expanding nexus of technologies included MIT (with Project MAC), Carnegie-Mellon, Stanford, Berkeley, UCLA, USC (with ISI), and CalTech. However, early-on entrepreneurial private firms, such as Bolt, Beranek and Newman (BB&N), DARPA's internet contractor, also played key roles.[28]

In the IT sector, DARPA followed a conscious "dual-use" approach, recognizing that IT (unlike, say, stealth) would be relevant to civilian as well as military sectors, and that by spinning technologies into civilian sectors which could focus more capital than defense procurement

28 Van Atta, Deitchman and Reed. (1990). *DARPA Technical Accomplishments, Volume I* (chapter 20).

on development and applications, military IT needs could leverage off civilian development and the resulting wealth of applications. Its connections to larger innovation elements enabled DARPA and DOD many more fronts of technology advance than defense development alone could have evolved. Since U.S. Cold War success arguably derived from its IT advantage and the "Revolution in Military Affairs" it enabled, "dual-use" was a profoundly advantageous leveraging success.

Thus, the emerging technical opportunities from the DARPA IT programs had synergistic effects with the high-tech investment communities in IT springing up principally in the San Francisco and Boston areas around Stanford, Berkeley and MIT. While initially early DARPA director Robert Sproull felt that computer developments should be left to the dominant firm in the market, IBM, Licklider convinced him that IBM was mainly interested in large-scale batch processing applications and not interested in the technology for the new concepts of time-sharing and individualized computing that Licklider was championing.[29] Thus, DARPA fostered research at these key universities, initially at MIT, and this helped such firms as Digital Equipment (DEC) expand footholds in a domain that was dominated by IBM's presence. From DARPA-funded IT research starting in the 1960s can be traced an expanding number of firms and commercial applications from DEC to TELENET (the original internet ISP), Xerox's Ethernet, Apple's desktop computing (following leads from Xerox PARC), CISCO Systems (internet protocol routers), Sun Microsystems, Silicon Graphics, MIPS, Thinking Machines, Mentor Graphics, Vitesse Semiconductor, and TriQuint Semiconductor.[30] From these firms second and third order spin-offs and derivative firms can be identified, such as Juniper systems, UUNET, and eventually even Google and Facebook, that built their businesses on the underlying technologies and capabilities, as well as the investment structure of the earlier firms (for example, Anders Bechtolsheim of Sun Microsystems was one of the ground floor investors in Google).

29 Van Atta, Deitchman and Reed. (1990). *DARPA Technical Accomplishments, Volume I* (chapter 29).

30 See, for example, Fong, G. R. (2001). "ARPA Does Windows; the Defense Underpinning of the PC Revolution", *Business and Politics* 3/3: 213–37, https://doi.org/10.2202/1469-3569.1025 (Chapter 6 in this volume); National Research Council. (1999). *Funding a Revolution, Government Support for Computing Research.* Washington, DC: National Academies Press, https://doi.org/10.17226/6323

Moreover, DARPA programs, such as the VLSI program in advanced microelectronics, had impact by providing underlying technologies that not only spurred new companies, but also raised the competency and capabilities across the entire industry. The support of the VLSI design, production, and higher level computer architecture and design was both infrastructural—such as the MOSIS program and the support of VLSI design courses at universities—and technological, supporting new integrated circuit design concepts of Carver Mead and Lynn Conway, that had cross industry impact on major incumbent firms, such as Intel, as well as startup firms, such as MIPS.[31]

The impact of DARPA pressing innovation onto existing firms, while difficult to assess, is exemplified by the following comment by a noted venture capitalist:

> If DARPA had not been available, university researchers would have had to use 'free' equipment from companies like Digital and IBM to do their research. DARPA funding of research was essential in providing an ability to make independent choices. (Vinod Khosla, 1991).[32]

In addition to connecting with companies, venture capital firms (VCs) played a crucial role in the commercialization of DARPA's information technologies. Many of the most prominent California VCs[33] literally grew up with DARPA, with DARPA-based technologies playing a key role in their success and creating deep synergy between these two innovation elements. There are numerous very specific examples, including Vinod Khosla and Sun Microsystems, FED Corp in displays and Silicon Graphics. VCs also followed DARPA programs as a basis for identifying the "next big thing;" because of the dynamism of the research award process, DARPA awards tended to give their small and startup firms a "halo effect", effectively marking them for follow-up support by VCs. VCs and DARPA became symbiotic in the IT sector and more broadly, operating as mutual enablers.

31 Van Atta, Deitchman and Reed. (1991). *DARPA Technical Accomplishments, Volume II* (chapter 28).

32 Interview with Vinod Khosla (Van Atta, Deitchman and Reed. (1991). *DARPA Technical Accomplishments, Volume II*, 17-B-11).

33 See Gupta, U., ed. (2000). *Done Deals, Venture Capitalists Tell Their Stories*. Cambridge, MA: Harvard Business School Press, 1–11, for a discussion of the technology innovation orientation of west coast VCs.

However, this connection process with industry has not always worked; DARPA has had difficulty in staging its technologies for entry into industries at times. For example, DARPA's "High Definition Display" program focused on creating capabilities in the emerging Flat Panel Display technology was not able to successfully intercept Japan's lead in commercializing this technology. Other examples of unsuccessful endeavors include DARPA's investments in the Thinking Machine development of the Connection Machine parallel processing computer and efforts to commercialize digital gallium arsenide computer chips. These less successful endeavors point to the facts that:

1) DARPA is a proof-of-concept technology agency, focused on high-risk disruptive capabilities—success in such efforts is not guaranteed;
2) There are crucial factors beyond technology development and demonstration that impinge on success.

However, this downward and outward linking into the research community and commercial industry is only one aspect of DARPA's connectivity to larger innovation elements. DARPA, as an agency of the Department of Defense, is part of a broader innovation structure within and for DOD. Crucial here is that DARPA is an independent organization under the Secretary of Defense and is explicitly separate from the military service acquisition system. While the Secretary of Defense and the underlying Office of the Secretary of Defense (OSD) bureaucracy rarely directly involve themselves in DARPA's individual research programs, OSD leadership elements at various times have played a strong role in identifying the mission challenges upon which they want DARPA to focus (see further discussion below on "Ties to Leadership"). Whether the challenge is "get us into space", "offset the Soviet advantage in numbers and mass", or "overcome terrorist abilities to strike within the US", OSD leadership has periodically turned to DARPA with broad but explicit mission charters unique from those of the existing military research structure. In addition, and relevant here, DARPA, working with OSD, has been able to tie its advances to the larger innovation elements in DOD, often implementing its technologies through service procurement programs.

To summarize, DARPA has been an actor and creator within larger innovation systems that include emerging industry sectors, the venture support that backs them, and entities within DOD itself.

Takes on Incumbents—DARPA at times has invaded the territory occupied by powerful companies or bureaucracies. As discussed above, it drove the desktop personal computing and the Internet model against the IBM mainframe model. On the military side of the ledger, cooperating with others in DOD, it drove stealth, unmanned systems, precision strike and night vision capabilities, despite the lack of interest and even express objections of the military services. At times, this has taken special mechanisms beyond or outside of (but in coordination with) DARPA to achieve.

For example, the Advanced Concepts Technology Demonstrator (ACTD) program was created in 1993 by OSD Deputy Undersecretary for Advanced Technology Larry Lynn (later DARPA Director from 1995–1998) to move unmanned high altitude, long endurance unmanned air vehicles (UAVs) into initial use. This was used to get the Predator and the later Global Hawk systems into the hands of combat units when the military services would not further their development after DARPA completed its proof-of-concept developments.

Thus, DARPA, at critical technology junctures, has not been afraid to push-back against powerful incumbents, both leading industry firms and the military, to press disruptive technologies forward.

First Adopter/Initial Market Creation Role—in addition to ties to demonstration capabilities, DARPA also has undertaken a technology insertion or adoption role. In coordination with other parts of DOD, DARPA has been able to create initial or first markets for its new technologies. DARPA and DOD were the first adopters of many of the IT advances DARPA supported—e.g., work stations (Sun Microsystems, Silicon Graphics), and ARPANET as MILNET.

> DARPA then provided a critical assist to the launching of Sun by extending funds to a number of academic institutions to permit them to acquire workstations for their own institutional users and networks. According to Khosla, academic institutions (particularly the University of California at Berkeley, Stanford, and Carnegie-Mellon) accounted

> for roughly 80 percent of the orders received by Sun in its first year of business, thanks to this DARPA funding.[34]

DARPA has relied, as discussed above, on both its confluence with its advocacy community, and its ties to larger innovation system elements, to achieve this.

Ties to Leadership—DARPA has been particularly effective when it is tied to senior leaders who can effectuate its technologies through DOD or elsewhere. Just as Vannevar Bush and Alfred Loomis in World War II were able to press developing defense technologies into implementation using their direct links to President Roosevelt and Secretary of War Stimson, DARPA at critical technical moments has been able to call on senior allies. Several examples follow below:

- Undersecretary for Defense Research and Engineering William Perry: Perry supported stealth and precision strike, as initiated and backed under Secretary of Defense Harold Brown. Initial interest in what became stealth technologies was driven by OSD seeking capability to overcome Soviet air defenses. Precision strike was driven by the Brown/Perry desire to develop technological "offsets" to Soviet advantages in mass force, building on earlier DOD-DARPA efforts driven by the previous director for R&E, Malcolm Currie.
- Director of Defense Research and Engineering John Foster: Foster was a strong supporter for DARPA's night vision program and spearheaded DARPA's initial involvement in UAVs.
- DARPA Director of the Tactical Technologies Office Kent Kresa: Kresa went to industry (Northrop) to head an Advanced Technology Division, where he focused on bringing precision strike and stealth to fruition. Subsequently, Kresa became CEO of Northrop-Grumman.

Because DARPA operates at the front end of the innovation process, it historically has required ties to senior DOD leaders to align with the follow-on back end of the innovation system.

34 Van Atta, Deitchman and Reed. (1991). *DARPA Technical Accomplishments, Volume II*, 17-B-8.

Doesn't Necessarily Launch into a Free Market—DARPA embodies what is termed "connected R&D;"[35] DARPA researchers do not simply throw their technology prototypes over their monastery wall, hoping a company's product development and manufacturing units might find them. That separation of research and production uses a theory of "benign neglect" in the face of markets. Instead, DARPA often uses DOD procurement to further its advances, and it funds, as discussed above, creative companies that can attempt to commercialize its products. Thus, DARPA tries to guide its successful developments into commercialization, and builds portfolios of technologies to build depth for a technology thrust in emerging markets. It is in the business of creating opportunity, in some cases picking technology "winners". In DARPA's exploration of radical innovations, it is generally recognized that its developments are ahead of the market; the research it is fostering does not meet an existing market need, but instead is creating a capability—a new functionality[36]—that may (if successful) create a new market or application.

A military example of this dynamic of technology push rather than demand pull is DARPA's sponsorship of high altitude, long endurance UAVs, which created new capabilities for which, at the time, there were no service "requirements". In fact, when these systems, such as the Predator and, later, Global Hawk, were first demonstrated, the military services actively delayed and discouraged their transition and development. Similarly, DARPA's fostering of internetted personal computing was ahead of any market foreseen by the incumbent computer firms, such as IBM and later, DEC. In developing such unanticipated new capabilities, it is unlikely that current market mechanisms and especially current firms meeting those existing markets will be the primary means to bring the capabilities to fruition. From a DARPA standpoint the question is how and in what manner should it foster the transition—and for how long. As the opportunity creator, DARPA has, in many instances, developed a new capability up to a technology demonstration and then found that the potential recipients—either in the military or commercial environments—are not ready or interested to take them further. Several

35 See discussion of this term in Bonvillian. (2009). "The Connected Science Model", 206–10.

36 See discussion of this functionality concept in Weiss and Bonvillian. (2009). *Structuring an Energy Technology Revolution*, 185–90.

times DARPA has essentially backed off further support and other times it has supported additional research that it sees as overcoming the risks that impede the technology's transitioning. However, there is a well understood rule that DARPA itself is not in the business of transitioning—it is in the business of inventing. In summary, DARPA does not simply throw its technologies out into the world and pray that markets will pick them up, it uses its ties to defense procurement and to emerging companies to try to align its technologies with institutions in the difficult back end implementation stage.

Even when it is focused on "connected R&D" and a specific technology, it should be noted, however, that DARPA is "all about competition" amongst ideas and has multiple mechanisms to identify, assess and evaluate alternatives and options, and will restructure or terminate ideas that are judged to not be panning out—especially those that appear not to be making a big enough difference relative to the status quo. This does raise an important question: who decides? At DARPA, as well as other research organizations, individuals often have trouble "letting go" of their vision, even when it isn't working. DARPA's main decision focus is the program manager, but the Office Directors and the Director play important roles in reviewing progress and assuring that programs are scrutinized. Since a PM typically only has three to five years on station to "make something happen", he or she has an incentive to make choices based on what projects appear to have the best chance of success. Just recently, DARPA's F-6 "Distributed Satellite" program was completely restructured with the initial contractor's program halted, when the new program manager decided that the effort was not likely to achieve the objectives.

Ahead of the Game—there is one additional aspect that should be noted, since DARPA is a government agency. While there is a history of Congress pressing a recalcitrant military to implement innovations, such as aviation and the aircraft carrier, Congress has also, at times, played the opposite role, interfering in technology advances. DARPA, historically, has tried to stay ahead of Congressional interference and micro-management by developing coherent narratives about its technology approaches that show DARPA's projects to be "ahead of the game". It has worked to avoid a situation where Congress has captured the narrative and forced DARPA involvement. As a result, DARPA has been almost "earmark" free.

Problems in Tech Paradise

It should be noted as a caveat that the above discussion is of DARPA on a good day; it has bad days, as well. There have been periodic problems at DARPA. It is in the end a human institution, and these concerns provide further lessons. It also faces challenges from **new realities**. The globalization of U.S. industry has created challenges for an agency charged with technology leadership to avoid technological surprise.[37] DARPA, for example, is only now working on a multi-element strategy to adequately respond to the global erosion of the U.S. manufacturing capabilities and depth which will affect the abilities of the U.S. to field advanced technologies.[38] With the imminent tightening of defense budgets, DARPA will face also increased challenges on how to use military procurement to create initial markets to transition its technologies.

In recent years, as briefly cited above, DARPA defunded a significant part of its university research base for advanced IT research, in turn affecting the strength of the DARPA IT technology community and the flow of outstanding university IT talent into DARPA,[39] although the current leadership is attempting to improve this situation. For example, **strong office directors** with extended experience aiding talented PMs with less DARPA and DOD experience, have often been key to DARPA, providing in-depth technology management and leadership for shorter-term and usually less-experienced PMs. However, it appears that in recent years this strong office director model has been cut back in favor of a strong director; while current DARPA leadership has been working to restore this position's authority, it is not an easy task. The strong director approach also affected the ability of its PMs to act as advocates for their technologies—as vision enablers—a traditional key

37 Bonvillian, W. B. (2009). "The Innovation State", *The American Interest* 4/6: 69–78, at 72–75, https://www.the-american-interest.com/2009/07/01/the-innovation-state/

38 DARPA is working to build a manufacturing technology portfolio; see summary of some of these elements in MIT Washington Office. (2010). "Survey of Federal Manufacturing Efforts", 4–6, https://dc.mit.edu/sites/default/files/pdf/MIT%20Survey%20of%20Federal%20Manufacturing%20Efforts.pdf. See, generally, Tassey, G. (2010). "Rationales and Mechanisms for Revitalizing U.S. Manufacturing R&D Strategies", *Journal of Technology Transfer* 35/3: 283–333, https://www.nist.gov/sites/default/files/documents/2017/05/09/manufacturing_strategy_paper_0.pdf; and Pisano, G., and Shih, W. (2009). "Restoring American Competitiveness", *Harvard Business Review* 87/7–8: 114–25, https://hbr.org/2009/07/restoring-american-competitiveness

39 Bonvillian. (2009). "The Connected Science Model", 225–33.

PM role important to DARPA's capabilities. The strong director with a strategic approach may at times run contrary to DARPA's tradition of "hire smart PMs and empower them".

Another problem has been the **hand-off between program managers**. Demand pull from military procurement has helped keep DARPA work moving somewhat smoothly during transitions from one program manager to another. Even so, shifts in DARPA PM management sometimes become "disruptive" themselves, in an undesirable way. For example, these problems have affected DARPA's Strategic Computing program and, recently, the Ultra High Performance Computing Program, where program continuation has not been smooth.

If military demand pull is weak—as it certainly sometimes has been, when DARPA prioritized technology over services (for example, with the arsenal ship and UAVs)—the DARPA model can suffer from too many stops, starts, and changes in direction. These obstacles are often the result of a combination of true belief and whimsy on the part of DARPA management, rather than careful reassessment of the technological frontiers, and of what direction should be taken. **New leaders** often have an instinct to make big changes even if operations are optimal so that they can "leave their mark". When this occasionally happens at DARPA, there can be a real issue whether the culture can supersede such occasionally disruptive leadership, although its history as a highly flexible, small and intimate program, with fewer bureaucratic controls than more traditional R&D programs, helps it.[40] Thus, what has been an overall DARPA strength—talented directors—can at times be a weakness.

There is another issue area, too, that requires notice: the **transition to implementation**. DARPA has worked hard to make this work, as discussed in many of the points above. But as a radical innovation organization, inevitably it is only secondarily concerned with transitioning the results into implementation. Some have accused DARPA in the past of too often being a "hobby shop" for talented PMs.

Sometimes, this transition capability has been DARPA's greatest strength, and, at other times, its biggest weakness. It has been noted that, "if fielded disruptive capabilities are the objective, it will be insufficient

40 Some observers counter that DARPA's productivity generally has remained high, despite such concerns. See Fuchs, E. R. H. (2011). "DARPA Does Moore's Law: The Case of DARPA and Optoelectronic Interconnects", in *The State of Innovation: The US Government's Role in Technology Development*, ed. F. Bloch and W. Keller. Boulder, CO: Paradigm Publishers. 133–48.

to generate an example... [such as a proof-of-principle prototype]... and then rely upon the traditional DOD/Service acquisition system to recognize its value and implement it".[41] This problem will be further explored in detail in this chapter's Section III, below.

In conclusion, the above discussion of DARPA cited the well-known elements of its innovation culture and focused on a number of less well-understood elements that have been important to its strength and capabilities. Both offer lessons in the energy technology sector to ARPA-E, which will be explored below. In addition, DARPA, like any human-created and run organization, is not perfect and a number of problems it has faced offer lessons in addressing how to best organize and manage ARPA-E.

II. ARPA-E—A New R&D Model for the Department of Energy

Replicating Basic DARPA Elements[42]

ARPA-E was consciously designed by Congress to apply the DARPA model to the new energy technology sector.[43] Currently funded

41 Van Atta, et al. (2003). *Transformation and Transition. Volume 1*, 64.

42 The following discussion on ARPA-E derives from ongoing discussions since ARPA-E's initial formation between author Bonvillian and ARPA- E's director, its deputy director for operations, and one of its program managers; author Van Atta had similar discussions with ARPA-E officials during this period. Both authors had an extended discussion session with four ARPA-E program managers about the ARPA-E model on April 5, 2011, which was particularly helpful in developing this paper. Both authors have long been observers of the ARPA-E formulation process; both testified before the House Science and Technology Committee on the ARPA-E authorizing legislation, HR-364 (2007); Bonvillian, W. B. (2007). Testimony Before House Science and Technology Committee on ARPA-E Authorizing Legislation, HR-364, April 26, https://www.govinfo.gov/content/pkg/CHRG-110hhrg34719/html/CHRG-110hhrg34719.htm; Van Atta, R. (2007). Testimony before House Science and Technology Committee on ARPA-E Authorizing legislation, HR-364, April 26, https://www.govinfo.gov/content/pkg/CHRG-110hhrg34719/html/CHRG-110hhrg34719.htm. Bonvillian, in addition, wrote about the proposal (Bonvillian. (2006). "Power-Play") which was reviewed by the Committee, and (together with former DARPA Deputy Director Jane Alexander and former DARPA General Counsel Richard L. Dunn) reviewed ARPA-E concepts for Department of Energy Chief Financial Officer (CFO) Steven Isakowitz and his office, over several weeks in February and March 2009, as the CFO's office led the DOE effort to form and stand up ARPA-E within DOE.

43 See, H. R. 364 (2007). Establishing the Advanced Research Projects Agency-Energy, reported by the House Committee on Science and Technology on May 23, 2007,

at $180 million for FY2011, it is about the size of a DARPA program office. It has emphasized speed—particularly, the rapid moving of research breakthroughs into technologies, through a process it labels "Envision-Engage-Evaluate-Establish-Execute".

With $400 million received in the 2009 stimulus legislation cited above, it has awarded funding in six energy technology areas through spring 2011, which are briefly summarized below.[44] These follow a "challenge-based, focused-program" approach modeled on DARPA (this was formed after an initial wide open "early harvest" funding opportunity noted below).

- The "Innovative Materials and Processes for Advanced Carbon Capture Technologies" program (IMPACCT) aims to develop technologies to capture 90 percent of CO2 from coal power plants at much higher efficiency and lower cost; research approaches include advanced new technologies for capturing and converting CO2 at power plants through a range of approaches, from catalysis to membrane sorption.
- ARPA-E's "Electrofuels" initiative aims to synthesize biofuels using micro-organisms to convert CO2 and water into liquid fuels, seeking a tenfold increase in efficiency over current biofuel production processes.

110th Congress, 1st Sess. (as introduced in the House on Jan. 10, 2007, https://www.govtrack.us/congress/bills/110/hr364/text); and America COMPETES Act. (2007). P.L. 110–69, 42 USC 16538, 110th Cong., 1st sess. (as amended, and signed into law 9 August 2007), Sec. 5012. See, also, America COMPETES Act Reauthorization. (2010). P.L. 111–358, HR 5116, 111th Cong., 2nd Sess. (signed by President 4 January 2011), Sec. 904, https://www.govtrack.us/congress/bills/111/hr5116/text, with accompanying report, House Comm. Rep. 111–478, House Committee on Science and Technology, Subtitle B (re: ARPA-E, Other Transactions Authority). The conceptual origin for ARPA-E as a DARPA model for energy stems from the National Academies' report, *Rising above the Gathering Storm*, at 152–58. For a discussion of some of the issues under consideration in the initial Congressional design of ARPA-E, see Bonvillian. (2007). "Will the Search for New Energy Technologies". For an early description of ARPA-E's mission and role from its first director, see Majumdar, A. (2010). Testimony before the House Committee on Science and Technology on the Advanced Research Projects Agency (ARPA-E), https://www.energy.gov/sites/prod/files/ciprod/documents/1-27-10_Final_Testimony_%28Majumdar%29.pdf

44 See ARPA-E. (2010). *Program Awards (Six Areas) Through 2010.* A further award offering was announced by Secretary Chu on 20 April 2011 for rare earths, biofuels, thermal storage, grid controls, and solar power electronics (Department of Energy. (2011). "Secretary Chu Announces $130m for Advanced Research Projects", 20 April).

- The "Batteries for Electrical Energy Storage in Transportation" program (BEEST) seeks ultra-high-density, low cost battery technologies for long range, plug-in electric vehicles, aiming at doubling vehicle ranges and enabling a four-fold reduction in costs from current battery technologies. Technology approaches extend from advanced lithium-ion concepts, to over-the-horizon new battery concepts like lithium-air batteries and an "all-electron battery".
- The "Agile Delivery of Electrical Power Technology" program (ADEPT) seeks to develop materials for advances in magnetics, switches and, high- density storage to improve the efficiency of power electronics to reduce electricity consumption by up to 30 percent. This is an area where the U.S. lead in advanced materials such as SiC and GaN-on-Si could serve as platforms for success in next generation (beyond Si-based) power electronics.
- ARPA-E's "Grid Scale Rampable Intermittent Dispatchable Storage" program (GRIDS) proposes to develop new technologies that create widespread cost-effective grid-scale storage, helping to balance renewables and power supply fluctuations with demand. The program is aiming for new storage systems with efficiency and cost comparable to pumped-hydro. Because the energy storage R&D/technology community has traditionally focused on energy density, new constraints largely on the cost side have brought out numerous new ideas.
- The "Building Energy Efficiency Through Innovative Thermodevices" program (BEETIT) seeks to develop cooling technologies for new and retrofitted buildings to significantly increase energy efficiency. The program has goals to increase air conditioning efficiency by 50 percent and sharply cut refrigerant global warming impacts.

ARPA-E includes two DARPA veterans among its eight PMs, one a former DARPA PM, the other an experienced DARPA performer and advisor. Consistent with its legislative history, it has worked to replicate the DARPA approach. The discussion below lists well-known elements in the DARPA ruleset and reviews how ARPA-E reflects and has adapted that model.

ARPA-E is a ***flat, non-hierarchical*** organization, effectively with only two levels—eight program managers (PMs) and its director,[45] Arun Majumdar, formerly a Berkeley professor with senior administrative experience in DOE's Lawrence Berkeley Lab who has also worked on forming companies.[46] Like DARPA, the ***program managers are "empowered"***, each with strong authority and discretion to administer a portfolio of projects in a related energy field, from storage to biofuels to carbon capture and sequestration. Like DARPA, the ***project approval process is streamlined***—the PMs evaluate and conceive of the research directions for their portfolios, then go through a critique of that approach with the director (and discussions with colleagues); they go through a similar discussion process over proposed contract awards. Essentially, there is only one approval box to check—the director — who retains approval authority before the contract is awarded, which generally goes very quickly. ARPA-E emulates DARPA's reputation for fleet-footed decision-making. Like DARPA, ARPA-E is not bound by the traditional research selection processes, such as peer review or hierarchical bureaucratic lines of authority. It operates through a strong PM selection process outside of peer review. Although ARPA-E uses strong expert reviews to guide PM decisions, there is no "peer review"

45 ARPA-E also has a PM who is deputy director for technology, and several additional teams: an "operations" group supervising its contracting process, including a counsel (who implements ARPA-E's unique personnel and contracting authority despite the very different procedures of DOE's management bureaucracy) and deputy director for operations; a commercialization team (discussed below); and a group of fellows (typically outstanding recent university PhDs, who support the PMs, discussed below). But the R&D operating core of ARPA-E is very flat: its director and its group of PMs. Regarding this term, the original enabling statute uses "Project Managers", see, America COMPETES Act. (2007). Sec. (f)(1)); the term "Program Directors" was substituted by America COMPETES Act Reauthorization. (2010). Sec. 904(f)(1)(C)(i), which amended ARPA-E's enabling statute in 2010. ARPA-E Director Majumdar decided to use the term "Program Directors" to emphasize how empowered its portfolio managers are. However, the term program managers is used here to parallel the term used in DARPA because the functions are similar and it is a functional title which is widely understood in the technology community.

46 Arun Majumdar served as ARPA-E's first director, from 2009 to 2012. After he left ARPA-E, he accepted positions at first Google and then Stanford University, where he is on the faculty at the time of this book's publication. A biography is available on a Stanford University website: https://profiles.stanford.edu/arun-majumdar.

where outside researcher peers make the actual final decisions on what gets funded. As at DARPA, this PM selection process generally avoids the conservatism and caution that often afflicts peer review, which tends to reject higher risk research awards if there are more than four applicants per grant award.

Like DARPA, the PMs use a ***"right-left" research model***—they contemplate the technology breakthroughs they are seeking to emerge from the right end of the pipeline, then go back to the left side of the pipeline to look for proposals for the breakthrough research that will get them there. In other words, like DARPA, ARPA-E uses a ***challenge-based*** research model—it seeks research advances that will meet significant technology challenges. Like DARPA, ARPA-E tends to look for ***revolutionary breakthroughs*** that could be transformative of a sector. Thus far, it has had a penchant for high-risk but potentially high-reward projects. ARPA-E's design is metrics-driven and "challenge-based" for funding opportunities. Metrics are defined in terms of what will be required for cost-effective market adoption in the energy industry. PMs propose to the research community what will be required in terms of technology cost and performance for adoption and then ask this community to pursue this with transformative new ideas.[47]

Like DARPA, ARPA-E's PMs are a highly-respected, technically-talented group,[48] carefully selected by a director who has asserted that there is no substitute for ***world-class talent***. Typically, the PMs have business experience, usually in startups, so they generally know from ***experience in both academic research and in industry*** the journey from research to commercialization. Recognizing that the ability to hire strong talent quickly was a key DARPA enabler, the House Science

47 This approach was used in some specific programs at DARPA. One specific example was the Global Hawk HALE UAV, which had a "firm requirement" of a fly away cost of $10 million per unit. Importantly, while Global Hawk is generally viewed as having had a major impact on U.S. military capabilities, this cost per unit was not met with the initial systems developed under DARPA and subsequently the Air Force has re-designed the Global Hawk to be a much larger and much costlier system (Van Atta, et al. (2003). *Transformation and Transition. Volume 1*, 45–49). On Global Hawk cost and schedule difficulties, see, also, Porter, G., et al. (2009). *The Major Causes of Cost Growth in Defense Acquisition, P-4531, Volume II*. Alexandria, VA: Institute for Defense Analyses, 49–50, https://apps.dtic.mil/dtic/tr/fulltext/u2/a519884.pdf

48 ARPA-E PMs' biographies are available at: http://arpa-e.energy.gov/About/Team.aspx. The PM group includes a deputy director for technology.

and Technology Committee, which initiated the ARPA-E authorization, gave ARPA-E, like DARPA, the ability to supersede the glacial civil service hiring process and rigid pay categories. In fact, ARPA-E's broad ***waiver of civil service hiring authority***[49] may be without precedent in the federal government.

Like DARPA, ARPA-E's research program is organized around the three- to five-year lifetime of its PMs. By statute, ARPA-E's PMs are limited to three years of service (although this can be extended);[50] as with DARPA, this means they must work to get their projects into prototype and implementation stages in the three or so years they are at ARPA-E. Thus, ***the project duration yardstick is the life of the PM***. This means that ARPA-E must forego much long-term research; it must build its project portfolio by seeking breakthroughs that can move to prototype in—for science—a relatively short period. It will aim, therefore, like DARPA, at ***innovation acceleration*** projects that can move from idea to prototype in the program life of its program managers. The House Science and Technology Committee, mirroring DARPA, also emphasized the availability to ARPA-E of highly flexible contracting authority, so-called ***"other transactions authority"***, which enables ARPA-E to emulate DARPA's ability to quickly transact research contracts outside of the slow-moving federal procurement system.[51] Although this authority

49 America COMPETES Act. (2007). Sec. 5012(f)(2)(A), as amended by American COMPETES Act Reauthorization. (2010). Sec. 904(g)(3)(2)(A)(i) (ARPA-E civil service waiver). In contrast, DARPA uses Intergovernmental Personnel Act (IPA) authority to hire PMs promptly (hiring can be completed in a day) from academia or industry, with the employee still paid through his or her former employer at the former salary level.

50 America COMPETES Act. (2007). Sec. 5012(f)(1)(C), as amended by American COMPETES Act Reauthorization. (2010). Sec. 904(g)(2)(C).

51 For DARPA's Other Transactions Authority (OTA), see, P.L. 101–89, 10 U.S.C. 2389 (enacted 1989); P.L. 103–60, Sec. 845. For a discussion of DARPA's OTA authority, see Kaminski, P. G. (1996). Secretary of Defense Memorandum, 14 December, re: 10 U.S.C. 2371, Section 845, Authority to Carry Out Certain Prototype Projects. Arlington, VA: DOD; Dunn, R. L. (2007). *Acquisition Reform, the DARPA Approach*, http://www.authorstream.com/Presentation/Burnell-34588-appe-Why-Business-Everybody-Else-Evolution-Defense-Industry-as-Entertainment-ppt-powerpoint/; Dunn, R. L. (1996). "DARPA Turns to Other Transactions", *Aerospace America* 34/10, 33–37; Dunn, R. L. (1996). DARPA General Counsel Memorandum of Law, Scope of Section 845 Prototype Authority, 24 October; Dunn. R. L. (1995). Testimony of General Counsel of DARPA before the Committee on Science, U.S. House of Representatives on Innovations in Government Contracting Using the Authority to Enter into "Other Transactions" with Industry, 8 November. The Department of

has not yet been fully utilized, it remains promising as ARPA-E moves into new areas, such as prize authority, discussed below.

Like DARPA, ARPA-E is also instituting the ***"hybrid" model,*** providing funding support for both academic researchers and small companies and the "skunk works" operations of larger corporate R&D shops. DARPA has often tied these diverse entities into the same challenge portfolio and worked to convene them together periodically for ongoing exchanges.

This has tended to improve the handoff from research to development, by combining entities from each space, easing technology transition. Like DARPA, ARPA-E has worked from an ***island/bridge model*** for connecting to its federal agency bureaucracy. For innovation entities in the business of setting up new technologies,[52] the best model historically has been to put them on a protected "island" free to experiment, and away from contending bureaucracies—away from "the suits".[53]

ARPA-E, as it was set up within DOE, has required both isolation and protection from rival DOE R&D agencies and the notorious bureaucratic culture at DOE that may battle it for funding and the independence it requires. From the outset, therefore, it needed a bridge back to top DOE leadership to assure it a place in DOE's R&D sun—it received this from Energy Secretary Steven Chu, who was one of the original proponents of ARPA-E while serving on the National Academies' *Gathering Storm* report, and later testified in support of ARPA-E before the House Science and Technology Committee in 2006.[54] It helps, too, as cited above, that

---

Energy received "Other Transactions Authority" in the Energy Policy Act (2005), Sec. 1007 (https://www.govinfo.gov/content/pkg/PLAW-109publ58/pdf/PLAW-109publ58.pdf). However, it was only utilized once (GAO 2008) until the advent of ARPA- E, when ARPA-E used it three times in 2009 in making its initial grant awards. See, America COMPETES Act Reauthorization Act. (2010). House Comm. Rep. 111–478, of House Committee on Science and Technology, Subtitle B, ("To attract non-traditional performers and negotiate intellectual property agreements ARPA-E also uses flexible contracting mechanisms called Technology Investment Agreements authorized for the Department as 'Other Transactions Authority' in the Energy Policy Act of 2005").

52 For analyses of private corporation and DOD defense S&T programs illustrating how successful radical innovation programs are constructed and managed in this manner, see Van Atta, et al. (2003). *Science and Technology*.

53 Bennis, W., and Biederman, P. W. (1997). *Organizing Genius, The Secrets of Creative Collaboration*. New York, NY: Basic Books, 206–07.

54 Secretary Chu also personally selected ARPA-E's director, a Berkeley colleague and friend, who was his former deputy director at Lawrence Berkeley National

ARPA-E's first director was a trusted technical peer and colleague from Secretary Chu's Lawrence Berkeley National Lab days. On a day-to-day basis, DOE Chief Financial Officer Steven Isakowitz oversaw ARPA-E's initial stand-up in its embryonic days immediately after its Stimulus funding[55] was passed to bring ARPA-E to life, serving as an early godfather. Thus, ARPA-E had a critical bridge back to leaders who could protect its independence and funding. It was located by DOE's CFO one block from DOE's Forrestal building in Washington, on the floor of an adjacent non-DOE building—this gave it a handy bridge back to its DOE godparents, but assured that it would have its own island for its own team separated from the DOE bureaucracy.

Well aware of this evolving ruleset, Energy Secretary Chu has remarked—consistent with DARPA's history—that if just one in twenty ARPA-E projects is commercialized, the energy technology landscape could be transformed.[56]

New Elements at ARPA-E

Thus far, we have described ARPA-E as though it were a clone of DARPA. However, ARPA-E faces a very different technology landscape than DARPA. DARPA has been able to launch its technologies into two territories that simplified its tasks. First, it has often been able to place its technologies into the procurement programs of the military services. In this approach, the military is able to serve as the testbed and initial "first" market for new technologies emerging from DARPA. As discussed above, this isn't automatic; it required creative work and senior allies, for example, for DARPA to persuade the military services to adopt stealth and UAV technologies. But, when effective, it greatly eases

Laboratory. This assured a very close connection between ARPA-E and the top agency leadership, somewhat comparable to such noteworthy technology "bridge" relationships as that between, for example, Radiation Lab founder Alfred Loomis and Secretary of War Henry Stimson, and Vannevar Bush, President Franklin Roosevelt's World War II science czar and Harry Hopkins, Roosevelt's chief personal aide. On Loomis, see Conant, J. (2002). *Tuxedo Park: A Wall Street Tycoon and the Secret Palace of Science that Changed the Course of World War II*. New York, NY: Simon & Shuster, 178–289); on Bush, see Zachary, G. P. (1999). *Endless Frontier, Vannevar Bush, Engineer of the American Century*. Cambridge, MA: The MIT Press.

55 American Recovery and Reinvestment Act. (2009). P.L. No: 111–15.

56 Secretary Chu has long been an ARPA-E proponent: see Chu, S. (2006). "The Case for ARPA-E", *Innovation* 4/3, http://www.innovation-america.org/case-arpa-e

technology transition. Second, as discussed at length above, DARPA also launches its technologies into civilian sectors—its keystone role in the IT sector is the most famous example; the Internet, VLSI computing and desktop computing features are among the many noteworthy IT technologies it supported. However, the IT revolution DARPA nurtured was a technology frontier, an example of "open space" technology launch.[57]

In contrast the energy sector that ARPA-E must launch into is occupied territory not open space—energy is already a complex, established "legacy" sector (CELS).[58] New energy technologies have to perform the technology equivalent of parachuting into the Normandy battlefield; there is already a technology-economic-political paradigm that dominates the energy beachhead that must be overcome. Because it faces a very different launch landscape than DARPA, ARPA-E is learning to vary its organizational model. In addition, ARPA-E has assembled what is by all accounts a talented team; they have put in place their own ideas on how to operate their new agency, as well. Thus, ARPA-E is not simply replicating DARPA, it is finding and adding its own elements appropriate to the complex energy sector, where it concentrates, and to its own staff. Some of these constitute new innovation lessons potentially applicable to other agencies and projects. A number of these new elements and variations from the DARPA model, as well as organizational features ARPA-E is focused on, are discussed below.

a. Sharpening the Research Visioning, Selection, and Support Process

Every strong innovation organization, from research groups, to startups and firms, to federal research agencies, must build a strong innovation culture.[59] Organizational cultures in the innovation space tend to lock-in

57 As discussed in Part I of this chapter ("The DARPA Model"), the frontier was not entirely open; DARPA did take on the IBM mainframe model in supporting disruptive technologies to achieve personalized computing and the Internet. But it also took on a set of technologies that were not of primary interest to IBM and in fact were of very low priority to it compared to its mainframe computing.

58 Bonvillian and Weiss. (2009). "Taking Covered Wagons East".

59 A working ruleset for optimal innovation organization cultures is set forth in, Bennis and Biederman. (1997). *Organizing Genius*, 196–218. A number of these Bennis/Biederman "rules" are (as of 2011) painted on the walls near the DARPA Director's office in DARPA's building in Arlington, VA.

quite early in the organizational history, and, once set, patterns of interactions and performance tend to become engrained into the entity's culture. ARPA-E, led by its director and PMs, all of whom have had experience in a range of innovation organizational cultures, including DARPA, have worked to build their own innovation culture within ARPA-E. While it shares many features with DARPA, as noted above, it has its own areas of emphasis.

ARPA-E's director and PMs emphasize that they are working in what they call **"the white space" of technology opportunities**. Starting with their first research award offering,[60] they assert that they have consciously attempted to fund higher risk projects that have the potential to be breakthroughs and be transformational in energy areas where little work previously has been undertaken. This means that their research awards are purposely made seeking transformations, not incremental advance. Comparable to the DARPA model, this approach has placed **technology visioning** at the very front of the ARPA-E's research nurturing process.[61]

ARPA-E has implemented an interesting **two-stage selection process,** offering applicants a chance to offer feedback to the initial round of

60 This perspective evolved from ARPA-E's first award offering for $150 million, issued on 27 April 2009, which was entirely open-ended, simply seeking innovative new ideas for energy technologies from academic energy researchers and firms. While the small ARPA-E staff—the organization was just being assembled—anticipated that they would receive only some 400 applications (assuming, as one ARPA-E official put it later, "Whoever heard of ARPA-E?"); instead, they received over 3500 applications. See, Kosinski, S. (2009). *Advanced Research Projects Agency-Energy (ARPA-E)* (presentation), 8. Because it faced an overwhelming application volume, this forced ARPA-E to assemble a major review effort relying on scientists throughout DOE to assist (which aided in their subsequent DOE community building effort, discussed below). Because they had far too many quality applications for their limited initial award funding (they only made thirty-seven initial awards), they developed their "white space only" approach described in the text above, which has since become a basic agency policy approach. Realizing that the energy tech sector was eager for a DARPA model in energy, and that there was already a major "tech buzz" around ARPA-E, the agency subsequently limited its award offerings to particular technology sectors, discussed above in the text in this section, to control the number of applications and make the review process manageable. However, to avoid disappointing and frustrating the initial wave of applicants, ARPA-E created its innovative energy technology "summit", described below. Thus, two of ARPA-E's more innovative approaches—"white space", and its now annual "summit"—were lessons that came out of the near-nightmare of managing its first open-ended initial award process.

61 Carleton, T. L. (2010). "The Value of Vision in Radical Technological Innovation", PhD thesis, Stanford University, Palo Alto, http://purl.stanford.edu/mk388mb2729.

reviews. Because ARPA-E's director, like many researchers, had been personally frustrated by peer review processes—where the reviewers in their responses to his proposals showed limited understanding of the science and technology advances behind his applications—he implemented a unique review process where his PMs allowed applicants to respond to their application reviews, followed by a further evaluation step. This "second shot" and "feedback loop" in the review process has several upshots: it has improved evaluations, because the PMs know that their conclusions will be critiqued; it has helped educate PMs in new technology developments; and it has resulted in a number of reconsiderations of applications and thus improving the overall ARPA-E research portfolio.

The Empowered program manager Culture—there are eight PMs at ARPA-E at the time of writing; there are no office directors, who serve as an intermediate stage at DARPA between PMs and the director. Because ARPA-E is roughly the size of a large DARPA office, it simply does not need them yet. Each PM picks his or her own inquiry areas; there is no overall technology plan.[62] However, PMs do form macro challenges within the sectors they initiate with the director—for example, seeking a zero emission, long range electric car. PMs therefore retain the flexibility of not being tied to a fixed ARPA-E-wide technology strategy. PMs also retain a great deal of control over their research portfolios, so are "empowered" like DARPA PMs, although they still have to persuade the director to support their program decisions. Director Arun Majumdar has a reputation as a shrewd and intellectually adept judge and analyst. PMs state that he insists on "complete technical and intellectual honesty"; and, as one put it, he is "a cricket batsman—he knows all the pitches". Thus, before a PM can select a technology project, he or she has to "sell" it to the director; the proposal often also has to survive rounds of brainstorming and vetting with PM colleagues. PMs have to have what they refer to as "**religion**"—they have a vision of where they want to take their portfolios, performing as vision champions, in order to sell their projects both inside and outside ARPA-E. Part of "religion", then, is that they must work on being vision implementers. ARPA-E PMs

62 However, ARPA-E's enabling statute does require preparation of a technology "Strategic Vision Roadmap". America COMPETES Act. (2007). Sec. 5012(g)(2), as amended by America COMPETES Act Reauthorization. (2010). Sec. 904(b)(2).

expressed the view that "religion" is the single most critical PM quality, aside from technical excellence. To summarize, ARPA-E uses DARPA's "strong program manager" model for research award selections and calls on its PMs to exert religious zeal in advancing selected technologies through the implementation stage. ARPA-E has purposely not created a formal personnel evaluation process for its PMs—as with DARPA, PMs say they are expected to "manage to results" and they are judged by the director and their colleagues—that is peer pressure-based on the outcomes, impact and results from the portfolios they select.

Additional mechanisms for talent support—ARPA-E has a **fellows program**, of five outstanding recent PhD's who help staff each PM and fill out the capability of each team. This institutional mechanism apparently may be creating a creative process of intergenerational contact and mentoring within ARPA-E, further ensuring that it becomes continuous education environment—a key feature for creative R&D organizations. The new fellows also have been meeting together as a group to attempt to jell their own on new ideas. DARPA currently has no comparable group to help augment internal intellectual ferment.[63] ARPA-E is also considering creating its own team of **senior advisors**—"technology wisemen", in short, who spend time at ARPA-E through frequent visits and so contribute to the PM teams. The group would be somewhat analogous to DOD's "Jasons", a group of experienced technical experts brought in to advise on major technical issues and problems,[64] except that ARPA-E's Jasons would serve a similar function not for DOE in general but within an operational research agency, ARPA-E. This could provide a way to enable technology thought leaders from a range of fields to contribute to energy technology advance, pulling in new perspectives and new ideas.

Portfolio Approach—All ARPA-E projects are selected, as discussed above, to be game changers—to initiate energy breakthroughs. However, within that broad requirement, as PMs assemble their portfolios around a particular challenge area, PMs say they have found they need a "risk mix". They generally include some "out there" projects that may or may not materialize, that are very high risk, but where the technology is so

63 When DARPA was first stood up it had a scientific advisory board and at times some of its Office Directors have empaneled such groups.

64 Finkbinder, A. (2006). *The Jasons.* New York, NY: Viking Books.

potentially important that, although far from implementation, these are well worth pursuing. But for most other portfolio technologies, the PMs want to see that they could be implementable in a reasonable period—that they could reach a cost range that would facilitate entry and commercialization. Some PMs find they need to emphasize more early stage science in their portfolios than other PMs because their portfolio sectors require more frontier advances—so there is a mix, too, of portfolio balance between frontier and applied, science and technology emphasis. The grant approval rate varies between technology sectors, but (following the initial 2009 open ended offering discussed above), PMs indicate the rate ranges from 5 percent to 10 percent. That rate is likely too low for robust portfolios and will discourage some creative applicants; ARPA-E understands this, but is constrained by Congressional budget limits.

As with DARPA, ARPA-E PMs have adopted a "hands-on" relationship with award recipients, with whom they maintain a dialogue, meet at frequent intervals to support their progress and help surmount barriers, and, when ready, promote contacts with venture and commercial funding. In most research agencies, the job of the PM focuses on the award selection process; in ARPA-E, this is only the beginning. PMs view their jobs as technology enablers, helping their tech clients with implementation barriers.

b. Building a Community of Support

While Congress, in designing new science and technology agencies, may get either the substantive design or the political design right, it does not often get both right.[65] In other words, the creation of an agency that is, from a public policy and substantive prospective, sound and effective as well as politically strong enough to survive, is a challenging policy design problem. ARPA-E was founded on a well-tested substantive model, the DARPA model; so as long as its leadership struggled to fulfill that complex design there was some assurance of success from a

65 Bonvillian, W. B. (2011). "The Problem of Political Design in Federal Innovation Organization", in *The Science of Science Policy: A Handbook*, ed. K. Fealing, J. Lane, J. Marburger III, and S. Shipp. Stanford, CA: Stanford University Press. 302–26, https://doi.org/10.1111/j.1541-1338.2011.00523.x

policy perspective. Although the history of DARPA clones is not always a positive one,[66] ARPA-E's leadership has made the ARPA-E clone a widely acknowledged, successful substantive one to date. However, ARPA-E's political design has been a more complex problem; from the outset it has faced a political survival challenge. In part, this is because Congress, on a budget cutback tear, has not fully embraced the need for an energy transformation. In part, it is because it is a small new agency fish in a cabinet agency filled with large agency sharks constantly on the prowl against funding competitors and turf incursions. These include such longstanding major entities as the Office of Science, the applied agencies and the seventeen national energy laboratories. To increase its chances of survival, ARPA-E needed not simply to avoid conflict with its large neighbors but to affirmatively turn them into bureaucratic allies and supporters.[67] Internal allies were not its only need—it also needed to build support outside DOE, from the energy research community it serves and from industry. All this had to be translated into Congressional support.

ARPA-E therefore has worked from the outset on **building internal connections within DOE.** The Department's R&D is organized into stovepipes. The Office of Science, a traditional fundamental science-only agency organized on Vannevar Bush basic research lines,[68] funds its own nest of national labs as well as university research and reports to its own Undersecretary. DOE's applied agencies, including EERE, and fossil, electrical and nuclear offices, fund development work primarily through companies and report to their own Undersecretary. DOE's organization thus severs research from development stages, and historically very few technologies cross over the walls of the two sides of the DOE organizational equation, between basic and applied. In theory, ARPA-E could serve both sides by drawing on basic ideas coming out of the Office of Science that could be accelerated, pushing them to prototypes, then building ties with EERE and the applied agencies to undertake handoffs for late stage development and demonstration stages. ARPA-E could thus serve both sides by working to be a technology connector

66 Bonvillian. (2007). Testimony, 5–6.

67 DARPA over time has attempted to achieve internal support from other defense R&D agencies. See Bonvillian. (2009). "The Connected Science Model", 220.

68 Stokes, D. E. (1997). *Pasteur's Quadrant, Basic Science and Technological Innovation.* Washington, DC: Brookings Institution Press.

within DOE. There are potential downsides to playing the connector role—in some cases at DARPA it has been seen as inconsistent with performing the role of transformation instigator. However, ARPA-E has attempted this task, and met with success in forging a working alliance with EERE, a much larger $2 billion a year applied agency. ARPA-E has EERE experts on its review teams and draws on their expertise; it has received strong support as well from EERE's leadership, who are working with ARPA-E on the handoff process described above (see further discussion below).

Integration with the Office of Science (SC) is still a work in progress. SC very much views itself as a basic research agency, and rejects work on applied research, assuming it is the job of other parts of DOE manage those efforts. It funds a wide variety of basic physical science fields, aside from basic energy-related research. Managers at SC generally view themselves not as technology initiators but as supporters for the actual researchers located in SC's national labs and in academia. This represents a genuine culture clash with the energy breakthrough mission orientation of ARPA-E PMs.

However, some attempts have been made to connect with the forty-six new Energy Frontier Research Centers (EFRCs) formed by SC to focus on energy research in promising areas;[69] two of ARPA-E's PMs report that they have selected one project each from EFRCs located at research universities. Collaboration with the national energy labs has also proven a challenge. Because the labs are large employers, they have tended to become independent political power bases.[70] However, ARPA-E has worked to include Department of Energy national laboratories in its research consortia,[71] hoping the laboratories will view ARPA-E not simply as a funding competitor, but also as a funding supporter.

Summit—ARPA-E has worked at building relations with venture capital firms and large and small companies, and with awardees and non-awardees, through two widely attended annual multi-day forums in the spring of 2010 and 2011.[72] These two energy innovation summits

69 Bonvillian. (2011). "The Problem of Political Design", 315–16.

70 *Ibid.*, 304–05.

71 America COMPETES Act. (2007). Sec. 904(e)(3) authorizes ARPA-E to fund "consortia… which may include federally-funded research and development centers" (FFRDC's—including energy laboratories).

72 See programs for ARPA-E Energy Innovation Summits (2010, 2011).

have become major technology showcase events in Washington, attracting large attendance and featuring prominent business, executive branch and bi-partisan Congressional leaders in speaking roles. ARPA-E featured its awardees at these summits as well as other strong applicants who did not receive awards but deserved attention. VCs and companies have swarmed around their technologies, building good will among attendees, whether they won awards or not. This has helped the growing field of energy technology highlight emerging technologies to potential private sector funders. The summits became, almost overnight, one of the biggest energy annual conference events in the nation and have played a major role in putting ARPA-E on the map as an innovative agency. Importantly, by highlighting new energy technologies of interest to many sectors and firms, the summits have helped in building an advanced energy technology "**community**" around ARPA-E.

Support Community—ARPA-E faced a major funding challenge in FY2011, where a change in political control of the House of Representatives and growing concerns over spiraling federal deficits led to cutbacks in federal agency funding. As noted, because ARPA-E received no funding in FY2010 (it received two years of initial funding in FY2009 through stimulus legislation), it needed affirmative legislation to survive. As a result of the goodwill that had been built in its first two years of operation, a community of support began to collect around ARPA-E to independently advocate for the agency's future with Congressional committees, including venture capital firms, large and small firms that worked with ARPA-E, and universities, all enamored of its research model. Thus, a political support system is growing, separate and apart from ARPA-E (which can't lobby under federal law) to back its efforts and continuation. It has reached a point where ARPA-E has received public support from some very prominent business leaders, including venture capital leader John Doerr of Kleiner Perkins, GE CEO Jeff Immelt, Microsoft's Bill Gates, and FedEx founder Fred Smith. The continued growth of such a political support community could help assure ARPA-E's political future.

In summary, not only has ARPA-E proven a strong substantive success to date from a public policy perspective, a political support base appears to be emerging that could help sustain it over time. ARPA-E could be in a position to achieve that rare combination, an integrated political design model, marrying political support with sound substance.

c. Technology Implementation

ARPA-E's director and PMs are acutely aware of their difficult task in launching technology into the complex, established "legacy" sector of energy. DOE has a four-decade history, as noted, of transitioning some technologies into commercial energy sectors, but comparatively few at a scale where they would make a real difference in U.S. energy consumption.[73] ARPA-E has therefore taken a number of steps to assist in taking its technology to implementation, commercialization and deployment: ARPA-E PMs **consider the implementation** process for technologies they are considering; before they fund a project, they evaluate the technology stand-up process and how that might evolve.

Their focus is not simply on new technology; they also seek to fund projects where they can see a plausible pathway to implementation. This is aided by the fact that ARPA-E PMs generally have both academic and commercial sector experience. On the commercial side, this experience ranges from work in venture capital firms and companies, to participating in technology-based startup firms. This range of background in both academic and private sectors assists in understanding possible commercialization paths. However, in the future, it is likely that ARPA-E will need to explicitly consider, within its R&D program awards, efforts to drive down the costs of technologies it supports to a cost level where they could reach commercial entry.

"In-reach" within DOE—ARPA-E is working on building ties, as suggested above, with applied programs in DOE, so that these agencies can be ready to pick up ARPA-E projects and move them into the applied, later stage implementation programs they run. ARPA-E's PMs have found that building relationships between PMs and both applied line scientists and technologists in the applied entities (particularly EERE, the Fossil Energy Office, and the Electricity Office) is key to this DOE "in-reach". This is a bottom-up connection process.

---

73 National Research Council. (2001). *Energy Research at DOE: Was it Worth It? Energy Efficiency and Fossil Energy Research 1978–2000*. Washington, DC: National Academy Press, http://www.nap.edu/catalog.php?record_id=10165#toc.

Meanwhile, from a top-down perspective, the ARPA-E Director has worked in parallel at building ties between his office and the leadership of the applied agencies at DOE. Nonetheless, the PMs believe "bottom-up" connections are the key to "in-reach" success—without support deep in the applied bureaucracies, transfers simply won't happen, whatever the leadership levels agree to. For example, one ARPA-E PM went to DOE's Fossil office with his leading Carbon Capture and Sequestration funding projects, placing three Fossil experts on his review panel for the selection process and involving them in oversight work and progress meetings. He points out that in-reach is "all relationships and people". There are similar bottom up approaches to build collaborative relations between ARPA-E and EERE on wind and other efforts, and with the Office of Electricity. ARPA-E's Director is also giving consideration to working top down with DOE applied agencies to create more formalized interagency groups around particular technology strands for collaboration across DOE stovepipes. The hope is that the groups can serve as "lead customers", because the resources in applied agencies can promote later development stages.

However, the applied agencies can only take ARPA-E technologies so far. DARPA learned how to work with a "customer" as it tried to collaborate with and encourage the military services to adopt its technologies in their procurement programs. While, as discussed above, this isn't necessarily easy, DOD has acted in many cases as the initial market for DARPA technologies. DOE doesn't offer comparable internal "customers" for ARPA-E technology advances. The efforts to undertake in-reach within DOE are an attempt to improve this situation and can assist in moving into the proof-of-concept, prototype and demonstration stages. The applied offices, which largely fund development work at companies, can also assist ARPA-E with follow-on company support for continued engineering advances for ARPA-E technologies. Learning what it actually means to have and work with a "customer", as DOD does in multiple ways, may prove a vital skill set for an effective R&D agency like ARPA-E, given its concern about affecting technology outcomes. A customer-driven approach, even in the stage of breakthrough research, can be an important driver in technology advance. ARPA-E's leadership and PMs understand that necessity in the energy technology sector. This is shaping up as one of the central questions for ARPA-E's future success, explored in this chapter's Section III, below. While DOE "in-reach", discussed above, is part of the answer, another logical step for ARPA-E is to connect with DOD agencies potentially interested in ARPA-E technologies for DOD needs,[74] given the latter's depth in testbed capabilities and first market opportunities, which remain gaps in DOE's innovation system.

ARPA-E is in fact working on building **ties to DOD for testbeds and initial markets**. DOE has executed a Memorandum of Understanding with DOD, but implementation is still largely at the discussion stage and results are still "in progress". DOD and ARPA-E have recently partnered on two projects, however, in battery storage and power electronics, for a modular energy storage system that can rapidly charge and recharge, and for new ways to combine onsite renewable generation with microgrids for use in military installations.[75]

DOD's own efforts on energy technology are just now coming into effect, but it is pursuing energy technology advances to meet its tactical and strategic needs, as well as to cut energy costs at its 500 installations and 300,000 buildings.[76] As an indication of its serious intent, ARPA-E has on staff a technologist with significant defense contractor experience, as part of the "Commercialization Team", working full time on collaboration with DOD. Since the offices in DOD working on energy technology are in the process of connecting with each other, ARPA-E is helping in convening these groups across the services. The potential role of DOD to test and validate and to offer an initial market for new energy technologies is well-understood at ARPA-E, offsetting the fact that its home organization, DOE, generally does not engage in the innovation process beyond late stage development and prototyping support.

---

74 Alic, J., Sarewitz, D., Weiss, C., and Bonvillian, W. B. (2010). "A New Strategy for Energy Innovation", *Nature* 466, 316–17, https://www.nature.com/articles/466316a; Bonvillian, W. B. (2011). "Time for Plan B for Climate", *Issues in Science and Technology* 27/2: 51–58, http://www.issues.org/27.2/bonvillian.html

75 Hourihan M., and Stepp, M. (2001). "Lean, Mean and Clean: Energy Innovation and the Department of Defense", *Information Technology and Innovation Foundation* (March): 1–26, http://www.itif.org/files/2011-lean-mean-clean.pdf

76 See, for example, Hourihan and Stepp. (2001). "Lean, Mean and Clean"; and testimony of DOD Deputy Under Secretary for Facilities and Environment Dorothy Robyn (Robyn, D. (2010). Testimony before the Senate Homeland Security and Government Affairs Committee, Financial Management, Government Information and Federal Services Subcommittee.

Commercialization Team—ARPA-E has assembled on staff a separate team working full time to promote implementation and commercial advances for ARPA-E technologies. These team members work with particular PMs on the most promising technologies emerging from their portfolios. ARPA-E, in effect, has added a variation to DARPA's famous "Heilmeier Catechism"[77] by requiring PMs and their Commercial Teammates to "tell me how your story will end and how will you get there?" The tactics this team develops in implementing technologies can include creating follow-on approaches for ARPA-E funded technologies through in-reach with DOE applied programs, connections to DOD testbeds and procurement, as well as connections to VCs and interested company collaborators, or combinations of these. Their work includes identifying first markets and market niches for ARPA-E technologies.

"**Halo Effect**"—ARPA-E is consciously taking advantage of the "halo effect", where VCs and commercial firms pick up and move toward commercialization the technologies that are selected by ARPA-E as promising. In other words, the private sector views the ARPA-E project selection process as rigorous and sound enough that it is prepared to fund projects emerging from that process. ARPA-E recently announced, for example, that six of its early projects, which it funded at $23 million, subsequently received over $100 million in private sector financing.[78] This effect has been seen before at DARPA and at the Department of Commerce's Advanced Technology Program (revised in 2007 as the Technology Investment Program). The VC or financing firm will perform its "due diligence" regardless, but ARPA-E's selection helps in identifying and, in effect, validating, a candidate pool.

Connecting to the industry "stage-gate" process[79]—the stage-gate process is used by most major companies in some form in the management of their R&D and technology development. In this approach, candidate

77 George H. Heilmeier was Director of DARPA from 1975–79. See Chapters 1, 8, and 10 for discussions of "The Heilmeier Catechism," a set of questions to ask about proposed research projects. See, generally, Heilmeier, G. H. (1991). *Oral History Interview (by Arthur Norberg)*. Minneapolis, MN: Charles Babbage Institute, https://conservancy.umn.edu/handle/11299/107352

78 Department of Energy. (2011). "Six ARPA-E Projects Illustrate Private Investors Excited About Clean Energy Innovation", 3 February.

79 Cooper, R. G., Edgett, S. J., and Kleinschmidt, E. J. (2002). "Optimizing the Stage-gate Process", *Research Technology Management (Industrial Research Institute, Inc.)* 45/5, https://doi.org/10.1080/08956308.2002.11671532

technology projects are reevaluated at each stage of development, weeded-out and only what appear to be the most promising from a commercial success perspective move to the next stage. This is not a process ARPA-E employs; like DARPA (as discussed above), it places technology visioning up front in its process,[80] and adopts a high-risk/ high-payoff approach to meet the technology vision. Although ARPA-E's is a more fluid and less rigid, vision-based approach, it has recently started to work with its researchers to get their technologies into a format and condition to survive in the industry stage-gate process. For academic researchers in particular, this is not a familiar process. Because most early generation energy technologies are component technologies, and will have to fit into existing systems and platforms controlled by existing companies,[81] ARPA-E PMs are recognizing that many of the technologies it nurtures must slot into the stage-gate industry practice if they are going to link with industry. Therefore, ARPA-E is considering how to prepare its technologies (and technologists) to withstand this process.

Consortia encouragement—aside from stage-gate connections to industry, in a different kind of outreach effort, ARPA-E is building an additional industry connection step between the firms and academics that it works with and the industries they must land in: consortia promotion. ARPA-E tries to pave the way for acceptance of its new technologies at firms by working to encourage companies that work in similar areas to talk to each other on common problems, including on technology solutions that ARPA-E's current or prospective projects could present. This is another facet of its community building efforts referenced above. For example, its PMs are working on this approach with groups of companies potentially interested in ARPA-E's carbon capture and sequestration (CCS) and battery project portfolios. The kinds of problems discussed are not the researcher's "secret sauce", but common issues of organization and general technology advance, including technology needs and standards relevant to all participants—both the researchers and the firms that may be interested in their emerging technologies. This approach helps prepares the ground for technology implementation and acceptance.

80 Carleton. (2010). "The Value of Vision".

81 Weiss, C. and Bonvillian, W. B. (2009). *Structuring an Energy Technology Revolution.* Cambridge, MA: The MIT Press, 185–90.

Prize authority—following in DARPA's footsteps,[82] ARPA-E has authority (America COMPETES Act Reauthorization 2010, Sec. 904(f)) to offer cash prizes for meeting technology challenges and is considering how to use it. This could be an additional creative tool for technology acceleration and implementation but may require unique adaptations to fit the legacy energy sector.

To briefly summarize, then, ARPA-E has not only worked to replicate elements at DARPA, but it has attempted to build new elements in its innovation ruleset as it confronts unique features of the energy sector where its technologies must land, and of the DOE bureaucracy it must work with. These new elements can be grouped into three broad areas, as detailed above: in sharpening the research visioning, selection, and support process; in building a politically survivable support community; and in the implementing and deployment process for its technology advances. Organizational tools in these categories being developed at ARPA-E present lessons that could be relevant and useful to other innovation agencies.

Relevance of the Additional DARPA Features (cited above) for Applicability to ARPA-E

In the discussion of DARPA, above, a number of DARPA capabilities not generally noted in the literature to date have potential relevance to ARPA-E in strengthening its operations and enhancing its future capabilities. These are organizational options not necessarily relevant to ARPA-E's current startup phase, but that it could consider as it continues to evolve. They may also serve as guideposts to help ARPA-E fill gaps and improve its performance. A series of the additional DARPA capabilities discussed above in this chapter's Section I, are reviewed below for relevance to ARPA-E.

Multigenerational Technology Thrust—as noted, DARPA has not only been able to undertake individual technology projects, but to work over an extended period to create enduring "motifs"—generations

82 See, for example, DARPA. (2009). *Network Challenge*, https://www.darpa.mil/program/darpa-robotics-challenge; Wikipedia contributors. (2019). "DARPA Grand Challenge", *Wikipedia*, 17 September, http://en.wikipedia.org/wiki/DARPA_Grand_Challenge

of new applications within a technology thrust that have changed technology landscapes over an extended period. Examples, as noted, include its work in IT, stealth, and precision strike. The approach ARPA-E is now implementing of projects with a three- to five-year duration based on the expected "life" of its PMs, will likely require supplementing with a multigenerational model, because many energy technologies will require ongoing advances before they reach maturity and optimal efficiency. For example, ARPA-E understands lithium ion generation battery advances likely will be displaced by further generations, yet the three- to five-year project approach will not get ARPA-E to the subsequent generational battery advances without further work on its technology organization. ARPA-E has settled on a series of program portfolios which could provide a basis over time for thrust areas, as summarized above. However, it has avoided a technology strategy to date, viewing it as a limiting factor on its PMs' ability to respond to technology opportunities; it may have to consider such an approach to manage the handoffs in the technology sectors it is pursuing as PMs succeed each other. Otherwise, it may not be able to field a multigenerational technology thrust capability to meet the inherently long-term challenges of most energy technologies.

Strategic Relations between Technologies—DARPA has launched related technologies that complement each other, which help build support for the commercialization or implementation of each. For example, its stealth technology advances complemented its precision strike advances, with both serving as mutual enablers. Launching bundles of related technologies could similarly alter the energy landscape. For example, new batteries coupled to biofuel advances could significantly enhance the energy consumption effects of hybrid vehicles, and storage advances are crucial enablers for enhanced renewables technologies. As ARPA-E builds out its technology portfolios, it could work to envision linked and crossover technology advances, supporting complementary efforts.

Confluence with an Advocate Community—DARPA created a broad and sizable community over time from its PM "graduates" and numerous award recipients in both universities and industry. This community was trained in the DARPA model and technology approaches, and in turn constituted a sizable group of change agents that invaded and altered numerous technology sectors. In other words, DARPA has become far larger than simply its onboard staff. ARPA-E began on a much smaller scale than DARPA, but needs to consciously work to build its community to make them not only supporters for its continuation (see subsection II.b. ("Building a Community of Support"), above, for a discussion on its support community) but an allied group of change agents. Its technology task, because it is innovating in a legacy sector, may prove considerably more daunting than DARPA's, so it will need over time to field an army. Its summit, discussed above in subsection II.b. is a useful initial organizing mechanism in this regard, although ARPA-E will need additional mechanisms to achieve this.

Connection to Larger Innovation Elements—DARPA has spawned new technologies that arose and converged with venture capital and entrepreneurial support and led to new economic sectors, particularly in IT fields. Thus, DARPA has been able to play an intermediary role with industry, able to make its advances reinforce sectors that support them, creating a mutual synergy. ARPA-E will need to consider this approach with the firms and sectors it collaborates with, including those providing capital support, as its technologies advance. It is already moving in this direction, as the discussion of the new elements in model suggest, becoming an actor connected with larger innovation efforts. It can play an instrumental role in these larger innovation systems, seldom as a sole actor, but instead as a team creator and player. The DARPA approach where its technologies spawned numerous IT firms which help effectuate its overall vision, and are linked to other supporting elements in DOD, offers lessons for ARPA-E. As its technologies progress, it will need to consider the appropriate models for this kind of confluence in the complex energy sector.

Takes on Incumbents—DARPA historically invaded territory occupied by companies or bureaucracies when it needed to foster technology advances. Perhaps its most famous example, as noted, is how, in an effort to develop new command and control systems, it drove desktop personal computing and the Internet to displace the IBM mainframe model, in a classic example of disruptive technology launch.[83] Because energy is a CELS—a complex, established "legacy" sector—conflict with legacy firms with established technologies will

83 Waldrop. (2001). *The Dream Machine.*

be frequent and inevitable for APRA-E. The energy legacy sectors probably see this coming. The only opposition, for example, to the ARPA-E concept when it was proposed in the National Academies' *Gathering Storm* report of 2006 was from the CEO of a major oil company.[84] While DARPA faced internal bureaucratic battles to launch its technologies, it only occasionally faced industry conflict because it tended to stand up technologies in new territories rather than in existing legacy sectors.[85] ARPA-E, however, will likely face incumbent technologies and firms across its technology portfolios. Accordingly, it will need to further build its support communities if it is to be successful in launching its technologies (see discussion on community building in subsection II.b. above). In addition, it will need to continue to enhance its technology implementation capabilities (subsection II.c. ("Technology Implementation"), above, and Section III in this chapter ("The Remaining Technology Implementation Challenge for DARPA and ARPA-E"), below).

First Adopter/Initial Market Creation Role—DARPA has frequently undertaken a technology insertion role; in coordination with other parts of DOD it has been able to create initial markets for its new technologies, allowing the Department to serve as first technology adopter. As discussed above (subsection II.c.), DOE offers no comparable first market for ARPA-E technologies.

Given DOD's interest in energy technology advances, it could serve as an initial market. ARPA-E will need to develop further strategies to find first adopters and initial markets because the lack of track records on costs and efficiencies constitutes a serious barrier to commercializing and scaling new energy technologies.

Ties to Technology Leadership—DARPA has been particularly effective when it is tied to senior leaders that can effectuate its technologies through DOD or elsewhere. ARPA-E has been effective to date, as discussed above, in securing a network of leaders in the Department, in the White House and on Capitol Hill, to support it, but will need to continually work to bolster its ties to energy decisionmakers in key places throughout the government who can help it fulfill its mission.

84 National Academy of Science. (2007). *Rising above the Gathering Storm*, 152–53.

85 Bonvillian and Weiss. (2009). "Taking Covered Wagons East"; and Weiss and Bonvillian. (2011). "Complex, Established 'Legacy' Sectors".

Doesn't Necessarily Launch into a Free Market—DARPA has embedded itself in a connected innovation system, taking advantage of DOD's ability, as noted above, to operate at all stages of innovation, from research, to development, to prototype, to demonstration, to testbed, to initial market creation. Therefore, it often has been able to launch technology into an integrated system—it doesn't have to toss its prototype technologies over a wall hoping they will be picked up in the private sector, it can ready them for scaling in the private sector, or simply stay in military markets. While, as discussed in Section I, this often is not easy, DARPA has nonetheless made this work. ARPA-E recognizes that because it will be launching its technologies into a CELS, it may be able to use DOD testbed and procurement roles, as discussed above, to further its advances. It can also fund creative companies that have capability to commercialize its technologies into products, and it can otherwise guide its technologies into commercialization, building portfolios of technologies for in depth technology thrusts into emerging markets. It can, in addition, leverage its technologies against regulatory mechanisms, such as fuel economy and appliance standards, or state renewable portfolio standards. These, along with additional tools, will need to be sharpened.

Ahead of the Game—just as DARPA has tried to stay in front of Congressional interference and micro-management, ARPA-E has worked to develop coherent narratives about its technology approaches that show its projects to be "ahead of the game". Like DARPA, it has worked to avoid a situation where Congress captures the narrative and forces ARPA-E involvement. In the next several years, however, particularly as some of its projects approach implementation stages, ARPA-E will need to demonstrate success and further refine its story.

In conclusion, ARPA-E presents an exciting and innovative emerging agency model. It has successfully incorporated the basic operating rules from DARPA into its own ruleset. In addition, because it must operate in the demanding energy sector, which is different from the sectors where DARPA operates, it has evolved a group of its own new rules. There are also useful future lessons for ARPA-E for its organization and strategy from a series of DARPA approaches that have not been covered in depth in the literature on DARPA. Finally, while ARPA-E has taken important steps in the back end of the energy innovation system to implement technologies it nurtures, additional implementation efforts will be needed. This problem is discussed in detail below.

III. The Remaining Technology Implementation Challenge for DARPA and ARPA-E

Both DARPA and ARPA-E face a profound challenge in technology implementation. For DARPA, the Cold War era of major defense acquisition budgets is long gone, and defense "recapitalization"—the replacement of existing generation of aircraft, ships and land vehicles with new defense platforms—is evolving at a glacial pace. Finding homes for its evolving technologies, therefore, has increasingly become a difficult task for DARPA. Because technology transition was once a difficult, but comparatively straightforward, task for DARPA, it has not yet fully faced up to the implications of how complex it has now become. ARPA-E faces a technology transfer problem of the first magnitude: the U.S. has a failed history of moving technology advances into CELS (complex, established "legacy" sectors), including in energy. U.S. Presidents have been calling for energy independence for four decades; the situation has only gotten worse, probably an unparalleled U.S. record for technology failure.

The innovation system used by the defense sector has led most major innovation waves of the twentieth century: aviation, electronics, nuclear power, space, computing and the Internet.[86] In the process, DOD has built a systems approach to its technology advances—it operates, as noted. at each stage of the innovation process: R&D, prototypes, demonstrations and testbeds, engineering and incremental advances, and initial market creation. At each stage, it has created institutions and functions that enable this systems approach.

There are essentially four of these sets of institutions and related functions that match the innovation stages: (1) at the breakthrough R&D stage, DOD uses DARPA, which supplements more traditional service R&D agencies; (2) at the prototype/demonstration/testbed stage, it uses the services, including their system of service labs (FFRDC's); (3) at the engineering and incremental advances stage, for its technologies and

86 Ruttan. (2006). *Is War Necessary.*

the platforms that use them, DOD uses the services' development and procurement programs, based on DOD's "requirements" system;[87] (4) for initial market creation, DOD uses its services-based procurement programs. While the handoffs between these DOD institutions and functions are rarely smooth, it is nonetheless a comparatively integrated system. DOD periodically supplements this system with efforts to launch technologies through civilian markets—DARPA, as noted below, has played a particularly important role in this approach.

An energy technology transformation is going to require a systems approach comparable to DOD's. DOE now has a DARPA-equivalent for the breakthrough R&D stage, complementing other DOE research entities. How will it handle the other three innovation system stages it must put into place to implement its technologies? The discussion below first reviews in detail the challenges DARPA faces in making the DOD innovation system work to implement technologies it originates. This provides lessons for ARPA-E's implementation challenges as well.

A. The Implementation Problem: Launch Pathways for DARPA and ARPA-E Through Military Procurement, Established Industry and the Entrepreneur/VC Model

Can ARPA-E succeed with its focus on transitioning the results of its research to commercial industry, in comparison to DARPA's main mission of developing technology for defense capabilities? The information technology examples discussed in this paper demonstrate that DARPA, using a technology push approach, did develop a successful university-private sector approach for supporting these technologies. While DARPA faces a problem of the "valley of death" between its research and late stage development, there is an equally, if not more, daunting problem of "market launch" lurking behind it.[88] DARPA has

87 See discussion of this engineering stage at DOD in Alic, J. (2011). *Defense Department Energy Innovation: Three Cases*. Presentation at Forum on Leveraging DOD's Energy Innovation Capacity, at the Bipartisan Policy Center, Washington, DC, 25 May; and Gholz, E. (2011). *How Military Innovation Works and the Role of Industry*. Presentation at forum on Leveraging DOD's Energy Innovation Capacity at the Bipartisan Policy Center, Washington, DC, 25 May (see summary at https://bipartisanpolicy.org/wp-content/uploads/2019/03/Energy-Innovation-at-DoD.pdf).

88 The concept of "market launch" is developed in Weiss and Bonvillian. (2009). *Structuring an Energy Technology Revolution*, 14, 20, 34.

five decades of history in attempting to launch its technology using primarily three pathways: military procurement, established industry, and a comparatively new entrepreneur/VC model that DARPA itself helped enable. The launch difficulties for DARPA for each of these pathways will be explored in detail below. Since ARPA-E is too young to have its technologies reach the implementation stage, there is less clarity over the difficulty of its market launch problems. Therefore, the discussion below will focus on DARPA. At the close of this discussion, future technology implementation issues for DARPA will be considered based on these launch pathway issues, followed by a discussion of how DARPA's lessons can provide guidance for ARPA-E.

1. *Market Launch in the Military Sector*

DARPA's main mission implementations, such as unmanned aerial vehicles, large-scale radars for missile defense, and standoff precision strike systems, were effected in a very different environment than commercial industry. Military procurement has enabled implementation of its military technologies as well providing initial markets for a number of technologies implemented primarily through the commercial sector.

Importantly, even with the strong defense imperative and backing from high- levels in DOD for these programs, their implementation, as noted in this chapter's Section II, generally was difficult, costly and time consuming. With only the very exceptional implementation of stealth technology as the F-117A, which was, as noted above, personally overseen by Under Secretary of Defense William Perry, most DARPA-developed military capabilities faced difficult transitions into military service acquisitions. Notably, these transition problems can be attributed to differentiating factors in military systems compared to commercial products, as follows.

Major DOD systems differ substantially from most commercial products:

- Major DOD platforms and systems are massive undertakings compared to almost any other industry endeavor[89]

89 There are very few industry tech developments at a scale comparable to those of major defense systems. One recent example is the Boeing 787 Dreamliner passenger aircraft. Notably, that development has experienced major problems in cost and time

- DOD's recapitalization rate through its procurement programs has been in sharp decline since the end of the Cold War. It builds ever few major systems and each system is likely to be fielded over decades and thus have to meet or respond to projected requirements that are difficult to ascertain and are likely to change in unforeseeable ways.

Commercial and DOD product development processes are substantially different:

- DOD systems are contracted efforts implemented by third parties through program offices. The program is funded based on front-end decision processes based on "needs" criteria and some assessment of feasibility prospects—but these are often at a high level and often with limited means to assess them within the contracted phase.
- In contrast, industry generally makes decisions concerning its own money and investments directly related to developing and implementing the product itself (in conjunction with suppliers and, potentially, outside investors).
- However, this may have changed significantly for commercial industry in recent years, as more of the development is based on outsourced subsystems and components, and even the development itself may have been outsourced. The distributed manufacturing model may have changed the connectivity within the firm between decision and performance and perhaps changed the motivations concerning resource decisions.
- At the same time, DOD contractors have also become more distributed and diffuse, with the concept of lead system

---

to produce (it is three years late and facing on the order of $12 billion in overruns), some of which can be attributed to problems of transitioning and implementing new technologies. See, Gates, D. (2010). "Dreamliner Woes Pile Up", *Seattle Times*, 18 December, http://old.seattletimes.com/html/businesstechnology/2013713745_dreamliner19.html It also appears that Boeing, taking a chapter from the IT sector and its distributed global manufacturing model, hoped to become a global systems integrator to spread and reduce its aircraft development risk, since the capital costs mean each new plane launch is usually a "bet the company" experience. However, complex aero technologies have not yet proven as susceptible as IT to global production distribution. The complexity of managing a global sourcing network for the 787 has been relentlessly problematic.

integrator and a dispersed supply chain, so there may be interesting lessons learned between defense and commercial firms as they both negotiate this new approach to enterprise management.

2. Market Launch Through Established Industry

Although DOD has launched many incremental advances through its service procurements with large firms, for the reasons discussed below, DARPA has had limited success launching its breakthrough advances into established industry.

Commercial and DOD/DARPA technology risk profiles are substantially different: DOD through DARPA has implemented technology push systems that are new and unprecedented, compared to industry, to achieve technological superiority with limited prior knowledge or experience with the proposed technology or its use.

- Several such systems have been developed by DARPA, as a technology push organization, with minimum to little explicit interest or involvement of the services, the "recipient" developer. Often the push for development is a "top-down" mandate from OSD, as noted above, or even Congress. Such systems—e.g., precision strike—are often developed by DARPA as the "innovation hub" explicitly to be disruptive or transformational, but their very nature makes them far riskier, not just from a technological perspective, but from the standpoint of transition and operational risk.
- However, such developments still must be implemented within the existing service acquisition processes, which are relatively cautious about taking on new capabilities beyond their internally developed systems.
- Often the recipient service has very different perspectives and interests from DARPA, the technology developer, on the priorities and value of the technology and its potential application. There are many instances in which the recipient service actively has opposed the technology before being mandated by those at higher levels in OSD to accept it, as was the case, as noted above, with stealth, tactical UAVs,

HALE UAVs, and tactical satellites. In some cases, the service has actually successfully fought the technology or so poorly implemented it that the technological capability eventually succumbed; examples include Discoverer II, Arsenal Ship, and the Aquila UAV.[90]

- The recipient service organization is loath to spend the additional resources required to "de-risk" the "revolutionary" concept particularly if it disrupts or counters its accepted operations and capabilities, and distracts resources that it sees as needed for these.
- Technology push systems usually offer little information to guide their actual use and deployment; what levels of capability are necessary to achieve different levels of performance are difficult to define or assess. Thus, "knowledge risk" might be a major impediment to the adoption of such radical technologies.
- While World War II and the Cold War forced DOD to innovate in an atmosphere of crisis, with the end of the Cold War and short term symmetric threats, there is a diminished sense of crisis, and a corresponding decline in impetus for the services to adopt DARPA's transformative advances, exacerbating its technology implementation problems.[91] This is not simply a military problem; the "knowledge risk" and limited sense of crisis are reasons why industry is usually adverse to radical as opposed to incremental innovation; other reasons are discussed below.

Industry is almost always highly constrained on investments into new endeavors. Within the firm there is constant competition for resources, thus industry generally entertains low technical risk. Firms actively assess risk and value in a series of spaces:

---

90 For a thorough case analysis of the difficulties in the U.S. Army implementation of tactical UAVs, see Knox, W. D. (1999). *Of Gladiators and Spectators: Aquila, the Case for Army Acquisition Reform*. Carlisle, PA: Army War College.

91 Vernon Ruttan has raised the concern that with the post-Cold War decline in impetus in defense innovation, the U.S. innovation system may not now be strong enough to launch new breakthrough technologies in either the public or the private sector (Ruttan, V. W. (2006). "Will Government Programs Spur the Next Breakthrough?", *Issues in Science and Technology* 22/2: 55–61).

- Assessment occurs between the current product, process development and the new endeavor. For example, Sun Microsystems faced a major resource crunch to maintain its current product competitiveness that almost prevented the development of the next generation SPARC work-station.[92] Current product demands dominate production investment decisions, with new risky products requiring external corporate support, which often isolates their development.
- Assessment also occurs between current product divisions and new divisions or enterprises needed to foster the new products. In an example of failed assessment, DEC, after stunning success in microcomputers, was not able to adequately access and manage the transition to desktops and ever more personal computing through a crisis of imagination and innovation organization.[93]
- These risk and value criteria are constantly being evaluated with strong prospects that a system development in industry will be cancelled or severely scaled back if it starts to go off track in any key risk dimension. This is often undertaken through the widely-adopted industry "stage-gate" process for R&D management, which constantly screens and weeds out potential innovations.[94]
- For industry, time-to-product is crucial. The space between market entry vs. the competition is a major criterion; there is a difference in being first and not being in the game. On the other side of this coin, there are also sometimes cost and learning advantages to launching improved products as first follower to market.
- The space between production costs and market price is also crucial. If new underlying technologies for products show

92 Bertrand, H., and Van Atta, R. (1993). *Technology Transfer in the Private Sector: Expert Interviews on Issues, Methodologies, and Problems*, IDA Document D-1407. Alexandria, VA: Institute for Defense Analyses.

93 Schein, E. (2004). *DEC is Dead, Long Live DEC—Lessons on Innovation, Technology and the Business Gene.* San Francisco, CA: BK Berrett-Kohler Publishers.

94 Cooper, Edgett, and Kleinschmidt. (2002). "Optimizing the Stage-Gate Process".

signs of cost escalation and seriously erode profitability, the product is highly likely to be cancelled.

Industry development of technology push systems is rare and difficult given the above issues, as well as problems of business constraints on finance, time-to-product, and internal and external competition. Technology push developments within existing firms are usually incremental, with one new element or component introduced, rarely as a major new integrated system, and market-entry is staged carefully relative to "creating" demand. As Ruttan has explored, incremental advances usually respond to calculated market niches and opportunities,[95] while breakthrough innovation can rely on few such market calculations because the transformative product is unanticipated and disruptive to markets. There are exceptions, but even Apple's properly vaunted latest products fit this model. The iPod was a breakthrough combination of a good MP3 player with a new music access system; its iPad was handheld notebook coupled with broader media access, communication and computing capability. Both were breakthrough products because of the way they combined previously unmixed technologies, but they integrated mostly available and comparatively mature components into new forms and combinations, rather than introducing multiple new technologies. In addition, Apple also relied on distributed global manufacturing to further cut its production risks and to share development costs, allowing it to move rapidly from design to production to quickly capture market share.

Technology push developments within existing firms are usually run differently from the rest of the firm, with different a management structure and oversight, such as the "Skunk Works" approach (at Lockheed and for IBM's PC desktop line), the R&D centers at such large firms such as IBM, GE and P&G, or through the innovation hub notion. Often, such developments report to headquarters and are separate from existing product divisions. Their transition efforts are usually accomplished with strong involvement from central management; while this may be a necessity,[96] it tends to disconnect production and the new products they may have to produce, reducing the potential for "learning by doing".

95 Ruttan, V. W. (2001). *Technology, Growth and Development: An Induced Innovation Perspective.* New York, NY: Oxford University Press.

96 Bennis and Biederman. (1997). *Organizing Genius*, 206–07.

3. Launching Through the Entrepreneur/VC Model: The Role of Entrepreneurs, Startups and VCs

Technology push developments in industry are often undertaken outside of existing firms by startups. Frequently, these are led by entrepreneurs who left existing firms or obtained technologies from existing firms or government research, forming startups and using risk capital from outside investors, usually angel investors, and then venture capitalists (and sometimes as corporate-sponsored "spin-outs"). This Entrepreneurial/VC system was a U.S. model dating from the 1960's and 1970's. Through this system, the U.S. successfully launched the IT innovation wave (then the biotech wave), giving it a significant world competitive advantage and enabling one of the strongest economic growth periods in twentieth century U.S. economic history, in the 1990's. Although many nations have envied this model, few have been able to stand up comparable capability. DARPA played a very significant role its creation because its IT advances coincided with its development, creating strong mutual synergies. This entrepreneur/VC model was the element that enabled DARPA to get around the profound difficulties of trying to introduce its innovations into established industries, with all of the complexities and entry problems listed above. Thus, DARPA did an end run around the established industrial sector in introducing IT.

However, the Entrepreneurial/VC sector has its own limits and requirements, too:

- Such startup ventures are subject to a different set of rules and practices driven by the outside investment community, with well-defined risk assessment and mitigation strategies and practices
- The risk of failure is accepted but because risks are high, the payoffs when successful are required to be very high.
- The investment decision is made early in development, and the endeavor is "given" high-level management support by investors to achieve transition, e.g., Scott McNeely was placed as CEO for Sun Microsystems by venture capital investor Vinod Khosla of Kleiner Perkins. Displacement of the

initial inventor team with more experienced management is frequently the price startups pay for VC support.

- VC investors expect relatively quick payback, so VCs need to be able to move their firms to Initial Public Offerings (IPOs) within a few years of their investments. Thus, VCs won't support technologies more than two or three years from production, and not much longer than that to projected profitability, which are prerequisites to launching an IPO and getting their investors' money back. This worked well in the IT revolution, when new applications could build on an expanding sector as the IT innovation wave gathered momentum and expanded in many directions. Similarly, it worked in biotech, with larger pharmaceuticals ready to produce or buy out biotech companies once their technologies were in range of FDA approval. However, the entrepreneur/VC model, with these relative short timetable requirements for spinning off to the IPO stage, is not readily adaptable to CELS. Even where DARPA sponsors new technologies that can lead to "open territory" innovation not limited by legacy incumbents, the model requires a significant emergence period before it will accept other technology—it took from 1969 to around 1992 before the Internet revolution began to scale.

Therefore, technology push developments in both established industry and through the entrepreneur/VC model are often undertaken using support and subsidies from the government to provide initial buffers against technology and market risks, providing early customer and production learning.

The above discussion has focused on DARPA, which has fifty years of experience trying to move its technologies into implementation. A major embedded point in the above discussion is that DARPA's technology transition is not going to get any easier. Three DARPA launch pathways were identified above: launch through DOD procurement, through established industry, and through the entrepreneur/VC model. Bootstrapping its technology advances onto military procurement may prove more difficult for DARPA over time, as discussed. The services already tend to resist disruptive technologies, and this tendency may accelerate as budgets decline and procurements stretch out. DARPA's ability to launch technologies initially through established industry has never been as strong because, as discussed, their economic constraints make them risk averse, rarely willing to embrace disruptive technologies. DARPA has enlisted established firms through the military procurement system, however, and developed important technologies through such entities as Lockheed's Skunk Works and IBM's research division. It has also supported advances through industry consortia, in semiconductors, for example. Sometimes, in its projects, it will try to tie smaller firms with larger ones, with the larger firms sometimes leading the research management, to facilitate technology scaling. However, as will be explored in more detail in the next section below, DOD faces a challenge to its overall technology leadership due to the decline of the defense manufacturing base, which will affect the production process as DARPA technologies are implemented. This makes this established firm sector a DARPA problem not primarily through initial launch but later implementation of its technologies.

The third launch pathway, through the entrepreneur/VC model, has always provided synergy for DARPA, particularly for the IT advances it has sponsored. While this model adapts well to continued IT advance, its comparatively short timeframe for obtaining capital, through VCs and IPOs, limits its abilities to support the launch of technologies into CELS. In addition, significant pump priming may be required for DARPA to launch new technologies into new unoccupied territories because while the entrepreneurs may be ready, their supporting VC and IPO capital system may take time to sell new ideas to their investors. Thus, DARPA faces serious constraints for implementing its technologies on each of its available launch pathways.

Although ARPA-E is still too young to be pushing products into energy markets, the DARPA launch pathways offer important lessons. Concerning the military procurement pathway, this offers promising implementation opportunities to ARPA-E, as discussed in this chapter's Section II, which it is already starting to pursue. The key will be whether DOD's interest continues or wanes in solving the strategic and tactical operating problems created by its energy dependency, and whether energy efficiency cost reductions and grid security needs can be achieved in reasonable time periods by evolving energy technologies. Concerning launching technologies into established industry, ARPA-E faces all the challenges listed above that DARPA faces, and more. Energy is a classic CELS, and new technologies launched into such legacy sectors generally have to be able to compete on price on day one.[97] Although DARPA could largely ignore established industry because of the inadaptability of breakthrough technologies to commercial constraints, because the established energy sector is there and itself needs to be transformed, ARPA-E can't ignore it, it must confront it. If ARPA-E is to have hopes of achieving this very challenging entry, it must, at a minimum, incorporate into its R&D programs efforts to not only perform research but to drive down the costs to competitive levels to enable entry into the energy CELS. This is one of the most challenging technology tasks any U.S. innovation agency has faced, and there is a long history of problems with such efforts at DOE.

Finally, ARPA-E also faces challenges in utilizing the entrepreneur/VC launch pathway. This is a logical pathway for ARPA-E to emphasize, and DARPA has done so very successfully. However, there are two major difficulties ahead. First, the timeframe for entrepreneurs and their startups, if fueled by VC and IPO capital, is probably on a much longer timetable for successful commercialization than for the IT or biotech sectors. Because energy is such an established sector, and new technologies correspondingly have so many barriers to overcome, technology entry may take a long time, well beyond the three- to five-year timeframe VCs are organized around. Second, most new energy technologies are component technologies, they have to fit into existing systems and platforms—advanced batteries have to fit into cars, fuel cells into homes or commercial buildings, carbon capture technologies into utility systems. The established industries or sectors that control the platforms or systems may all too often be reluctant customers, unwilling to absorb the risk of accepting new technology components until they are fully proven and demonstrated, and costs clear. Without access to a strong testbed system for these demonstrations, the entrepreneur/VC model is unlikely to coalesce around most such technologies emerging from ARPA-E.

97 Bonvillian and Weiss. (2009). "Taking Covered Wagons East".

B. The Problems of Manufacturing and Testbeds

To summarize, the technology implementation challenge faced by both DARPA and ARPA-E is a profound one, although different for each agency. DARPA can still try to use DOD as an initial market, although, as noted above, that is harder than in the past. It has always also tried alternatively to stand up its technologies in the civilian sector, where its success in IT is the leading example. In this "dual-use" approach, the civilian sectors pick up the DARPA technology and fund the ongoing engineering and incremental advances, as well as related applications. Thus, the military leverages from the civilian sector, cutting its own development costs and creating a range of applications that the military itself could never evolve. As long as DARPA is innovating in new as opposed to established sectors, that model, while never easy and longer term, can be made to work. However, DARPA needs to devote new attention to how its innovation can move into both military and civilian markets, given the underlying problem discussed above for its launch pathways.

For example, DOD has long relied on the strength of the U.S. industrial production base, which has been the world's strongest since the late nineteenth century. However, China has now likely passed the U.S. in manufacturing output[98] and the production function for U.S. industry is globalizing. U.S. military superiority has long relied on U.S. technological superiority, and its corresponding ability to implement and mobilize that superiority through on-shore production. That era may be shifting.[99] Accordingly, DARPA, as suggested in this chapter's

98 IHS Global Insight states that in 2010, China accounted for 19.8 percent of world manufacturing output (in current dollars), a fraction ahead of the United States' 19.4 percent; China's manufacturing sector grew 18 percent in 2010 and the U.S. at 12 percent; over 2008–10 China's manufacturing sector grew at a pace of 20.2 percent per year, while the United States grew at 1.8 percent and Japan, the third largest, at 4.25 percent (IHS Global Insight. (2011). "China Passes U.S. in Manufacturing Output", 14 March, http://manubiz.com/china-edges-ahead-of-u-s-in-manufacturing-report-says). See also, Baily, M. N. (2011). "Adjusting to China, A Challenge to the U.S. Manufacturing Sector", *Brookings Policy Brief* 179; and Norris, F. (2011). "As US Exports Soar, It's Not All Soybeans", *New York Times*, 11 February, http://www.nytimes.com/2011/02/12/business/economy/12charts.html?_r=1&src=busl

99 Van Atta, R., Lippitz, M. and Bovey, R. (2005). *DOD Technology Management in a Global Technology Environment*, IDA Paper P-4017, Alexandria, VA: Institute for Defense Analyses.

Section I, above, is now looking hard at whether new manufacturing technologies and processes could improve U.S. production productivity to a point where the U.S. could retain production leadership in critical sectors, a key military capability.[100] This, however, is inherently dual-use technology which would have to be implemented in the private sector. DARPA needs to consider, in parallel to its manufacturing R&D efforts, the implementation tools at DOD it could make use of, including DOD's Mantech program and the Defense Production Act, along with DOD's defense procurement authority.[101] Manufacturing is just one of many technology implementation problems DARPA will face in the future, which may compel it to examine additional implementation models.

As noted above, ARPA-E likewise faces major implementation problems as it launches technologies into the energy CELS. It is working on, as discussed in this chapter's Section II, above, a number of interesting new mechanisms to assist in this task, including: consideration of implementation early in its selection process; selecting PMs with venture or startup experience; an "in-reach" effort within DOE for implementation support from DOE applied agencies; forming its own commercialization team; working with the industry stage-gate process; assisting in forming industry consortia; and connecting with DOD for testbed and initial market creation for its technologies. While both DARPA and ARPA-E move technologies down the innovation pipeline to the prototype or small-scale demonstration stage, neither agency itself has the financing authority to enable initial market commercialization of its potentially breakthrough technologies. Although, as noted, DARPA can work to leverage DOD procurement for product introduction for military or dual-use technologies, DOE has no such capability for ARPA-E to leverage. While DOE has an energy loan guarantee program, it is not structured to finance new technologies without a performance track record.[102]

---

100 MIT Washington Office. (2010). "Survey of Federal Manufacturing Efforts", 4–6, http://web.mit.edu/dc/Policy/MIT%20Survey%20of%20Federal%20Manufacturing%20Efforts.pdf

101 DOD's Office of Manufacturing and Industrial Base Policy is playing a lead role in this area. See Department of Defense Office of Manufacturing and Industrial Base Policy. (2011). "Programs", https://www.businessdefense.gov/Programs/

102 "DOE has had a significant loan guarantee program since 2005 but did not issue loans until 2009. It has a mandate to 'facilitate the introduction of new or

This suggests that there is an earlier stage problem, too, that ARPA-E faces for its energy technologies. Energy technologies are unlikely to be adopted by energy industries, as noted above, until their cost, reliability, performance and efficiency is well-proven. For example, while 40 percent of CO2 emissions come from the building sector, this sector is highly characterized by numerous locally-based firms; it is decentralized, undercapitalized, undertakes little R&D and is risk-adverse. It simply will not adopt technology advances until they are well-proven. Similar problems abound in other energy sectors. A testbed capability has been the remedy for this problem, historically. Such testbeds, then, are increasingly important in energy technology implementation to create the prerequisite demonstrations that will allow commercialization to proceed. This is likely not only going to be ARPA-E's challenge. As entry gets more difficult for DARPA's technologies, a similar testbed capability could become significant. While DOD has long developed testbed capacity, this is not readily connected to DARPA's breakthrough technology model.

DOD has been working on exactly this problem of a connected handoff from R&D to testbed in two interesting energy-related programs. The Strategic Environmental Research and Development Program (SERDP), formed in 1990, is a DOD R&D program housed within the Office of the Deputy Undersecretary of Defense for Installations and Environment in the Office of the Secretary of Defense. DOD, EPA and DOE share an oversight role over the program. It is coupled to the Environmental Security Technology Certification Program (ESTCP), formed in 1995, which tests environmental and energy technologies emerging from SERDP and elsewhere, which are near deployment but require demonstration and validation through a testbed. The programs' online mission statement states that SERDP and ESTCP are DOD's environmental research programs,

---

significantly improved energy technologies with a high probability of commercial success in the marketplace.' Although the program is aimed at helping move technologies past the initial commercialization barrier, the mandate's language builds in potential contradictions. It is limited to deployment-ready projects, so it excludes demonstrations, and the 'high probability of commercial success' clause, perhaps due to the legacy of failed 1980s synfuels projects, significantly limits the risks that the program can take with innovative technologies" (Bonvillian. (2011). "Time for Plan B", 58).

> ... harnessing the latest science and technology to improve DOD's environmental performance, reduce costs, and enhance and sustain mission capabilities. The Programs respond to environmental technology requirements that are common to all of the military Services, complementing the Services' research programs. SERDP and ESTCP promote partnerships and collaboration among academia, industry, the military Services, and other Federal agencies. They are independent programs managed from a joint office to coordinate the full spectrum of efforts, from basic and applied research to field demonstration and validation.[103]

In turn, once through the demonstration process, ESTCP works with DOD's installations programs to enable initial deployment of successfully tested technologies. The director and architect of SERDP and ESTCP, Dr. Jeff Marqusee, has stated regarding ESTCP that, "We can serve as a test bed to get these technologies over the valley of death, and then we can be an early market. The calculation is pretty straightforward. If we test ten technologies, and one is highly successful, we can deploy that in a hundred places [through DOD] and make it profitable".[104] These entities amount to an interesting new model for the energy/environment field from DOD, explicitly and closely linking R&D, testbeds and initial deployment.

A review of pending energy projects by the two connected programs indicates work by United Technologies on methodology and tools for building systems on DOD installations with a 50 percent efficiency improvement, an air source cold climate heat pump with Purdue researchers, and demonstration of high gain solar for distributed energy needs at DOD facilities with Skyline Solar.[105] ESTCP demonstration work in microgrids, storage, building efficiency controls is ongoing. Funding for a new energy testbed capability at DOD installations received $30 million in funding in DOD's FY12 budget.[106] A significant

103 Department of Defense. (2011). "SERDP and ESTCP Program Information", http://www.serdp.org/About-SERDP-and-ESTCP

104 Comment cited in Hourihan and Stepp. (2001). "Lean, Mean and Clean", 17. See also Marqusee, J. (2011). *SERDP and ESTCP*. Presentation at Forum on Leveraging DOD's Energy Innovation Capacity at the Bipartisan Policy Center, Washington, DC, 25 May.

105 See program list in Department of Defense. (2011). "SERDP and ESTCP Energy and Water Projects." An updated description of this effort is available at: https://www.serdp-estcp.org/Program-Areas/Installation-Energy-and-Water

106 Robyn, D. (2011). Testimony before the Senate Armed Services Committee Readiness and Management Support Subcommittee, 17 March, 10–11.

expansion of linked R&D and testbeds as well as a connection to initial market capability may be one answer to the implementation challenge ARPA-E faces.

In the long list of challenges ARPA-E faces, the problem of technology implementation is perhaps the most profound. This is because, to reemphasize the point, of the difficulty new energy technologies face not only with the problem of the "valley of death" in moving from research to late stage development, but the problem endemic to CELS of "market launch"—implementing technology at scale. ARPA-E has worked imaginatively to structure new elements into its model to address this problem. The approach of SERDP and ESTCP provides an interesting new model in the energy area for ARPA-E to consider as it focuses on technology implementation. Collaboration with these programs, which ARPA-E is actively working on, may provide a crucial new toolset.

ARPA-E is not alone in facing this implementation problem; the applied agencies at DOE, led by EERE, face a similar problem and the SERDP/ESTCP combined model of R&D-testbed-deployment offers an interesting new approach. DARPA, too, despite remarkable past successes, is not immune, as discussed above, from the implementation problem. With the budgetary constraints facing DOD for new systems development and the weakening posture of risk investments and venture capital in the commercial U.S. markets with the drive toward outsourcing and offshoring, implementation of DARPA programs appears to be a growing problem. DARPA also might learn lessons and make further uses of the SERDP/ESTCP approach. In addition, DARPA could consider tools such as Mantech and the Defense Production Act as the source of demonstration and initial deployment particularly for its manufacturing initiatives.

In summary, implementation presents a major challenge for both agencies. DARPA needs to consider its existing portfolio of implementation support, including ties to SERDP/ESTCP, building its manufacturing research efforts, and linking with Mantech and the Defense Production Act authority. ARPA-E has worked imaginatively on its implementation capabilities, but the complexity of its task requires it to consider additional mechanisms, including further collaboration with DOD, connecting to the SERDP/ESTCP model, and designing within its research projects efforts to drive down costs. These could be coupled to expansion of the interesting new features it is working on to spur development, as well as adapting DARPA concepts of multi-generational technology thrust, connecting to larger innovation elements, strategic connections between technologies, further building of its support community, and expanding its first adopter/initial market role.

IV. Conclusion—Brief Summary of Key Points

ARPA-E offers a highly interesting new innovation institution to meet the profound energy technology challenge discussed in the introduction. Because it is explicitly modeled on DARPA, this paper has reviewed the noted DARPA approach in detail. Briefly citing well-known features of DARPA, it has explored in detail a number of important features that have not been well discussed in the policy literature on DARPA to date. These included DARPA's ability to undertake multigenerational technology thrusts, the technology synergies it has been able to create through complementary strategic technologies, its ability to build an advocate community, and connections it has built to larger innovation elements downstream from DARPA. In addition, DARPA has been willing to take on incumbent technologies both within DOD and in the private sector. It has used ties to DOD leadership to press its advances, and does not necessarily launch to a free market by playing roles as first adopter and in initial market creation.

The paper subsequently reviewed the new ARPA-E model in detail. It first commented on just how ARPA-E has adopted the key elements of the DARPA approach. It then discussed new features ARPA-E has been moving toward in a series of areas, largely driven by its need to confront the unique and difficult demands of the complex, established energy sector where it operates. In the area of sharpening the research visioning, selection and support process, ARPA-E focuses on: the "white space", where tech opportunities are not being advanced in other parts of the innovation system; an interesting two-stage feedback system for selecting technologies; encouraging its PMs to "get religion" about their technologies to become vision enablers; using a new fellows program to get access intergenerational contact and additional ideas; a portfolio approach that mixes ranges of technology risk; and encouraging a very hands-on relationship between PMs and researchers.

ARPA-E has also been making progress in building a community of support, important to its political survival. This includes building internal connections with other DOE agencies, holding a highly successful community-building energy technology summit, and fostering a broad support community. On the battlefront of technology implementation, ARPA-E has: encouraged consideration of the implementation process in the selection of technology projects; worked on "in-reach" within DOE to move its technologies into the applied side; created ties to DOD for possible test bed and initial market capability; formed an internal commercialization team to work with PMs to move their technologies into implementation; connected technologies to the industry "stage-gate" process; encouraged industry consortia around its projects; and is planning to use prize authority.

In addition, the further DARPA features enumerated above provide potentially useful guideposts to ARPA-E as it continues to support innovation in the energy sector. These include DARPA's multigenerational thrust, strategic relations between technologies, an advocate community, connections to larger innovation elements, coping with incumbents, seeking initial markets for its technologies, and further ties to leadership.

Finally, the paper closed with a discussion of the profound technology implementation problems on the "back end" of the innovation system—including demonstration, test beds, initial markets. The authors believe both agencies must explicitly, imaginatively and actively address the implementation issue. When new, radical, transformational technologies are seen as needed, either for national security, energy security, or economic security, there are sufficient impediments within the existing governmental organizations and within the existing markets that creative partnering between the government and private sector is required to address the downstream risks, while recognizing that the best means to mediate risk is through innovation, not stasis. We believe that the agencies need to expand their innovation efforts in technology implementation by developing further approaches for fostering downstream partnerships between the government and private industry.

References

Alic, J. (2011). *Defense Department Energy Innovation: Three Cases*. Presentation at Forum on Leveraging DOD's Energy Innovation Capacity, at the Bipartisan Policy Center, Washington, DC, May 25.

Alic, J., Sarewitz, D., Weiss, C., and Bonvillian, W. B. (2010). "A New Strategy for Energy Innovation", *Nature* 466: 316–17, http://www.nature.com/nature/journal/v466/n7304/full/466316a.html

America COMPETES Act. (2007). P.L. 110–69, 42 USC 16538, 110th Cong., 1st sess. (as amended, and signed into law 9 August 2007), Sec. 5012.

America COMPETES Act Reauthorization. (2010). P.L. 111–358, HR 5116, 111th Cong., 2nd Sess. (signed by President 4 January 2011), Sec. 904, http://www.govtrack.us/congress/bill.xpd?bill=h111-5116, with accompanying report, House Comm. Rep. 111–478, House Committee on Science and Technology, Subtitle B (re: ARPA-E, Other Transactions Authority), https://www.congress.gov/bill/110th-congress/house-bill/2272?q=%7B%22search%22%3A%5B%22America+COMPETES+Act+of+2007%22%5D%7D&s=6&r=1

American Recovery and Reinvestment Act (ARRA). (2009). P.L. No: 111–15 (signed by President 17 February 2009).

ARPA-E. (2011). *Staff Directory*.

ARPA-E. (2010). *Program Awards (Six Areas) Through 2010*.

ARPA-E. (2010–11). *Energy Innovation Summits*.

Baily, M. N. (2011). "Adjusting to China, A Challenge to the U.S. Manufacturing Sector", *Brookings Policy Brief* 179, https://www.brookings.edu/wp-content/uploads/2016/06/01_china_challenge_baily.pdf

Bennis, W., and Biederman, P. W. (1997). *Organizing Genius, The Secrets of Creative Collaboration*. New York, NY: Basic Books.

Bertrand, H., and Van Atta, R. (1993). *Technology Transfer in the Private Sector: Expert Interviews on Issues, Methodologies, and Problems*, IDA Document D-1407. Alexandria, VA: Institute for Defense Analyses.

Bonvillian, W. B. (2011). "The Problem of Political Design in Federal Innovation Organization", in *The Science of Science Policy: A Handbook*, ed. K. Fealing, J. Lane, J. Marburger III, and S. Shipp. Stanford, CA: Stanford University Press. 302–26, https://doi.org/10.1111/j.1541-1338.2011.00523.x

Bonvillian, W. B. (2011). "Time for Plan B for Climate", *Issues in Science and Technology* 27/2: 51–58, http://www.issues.org/27.2/bonvillian.html

Bonvillian, W. B. (2009). "The Connected Science Model for Innovation—The DARPA Model", in *21st Century Innovation Systems for the U.S. and Japan*, ed. S. Nagaoka, M. Kondo, K. Flamm, and C. Wessner. Washington, DC: National

Academies Press. 206–37, https://doi.org/10.17226/12194, http://books.nap.edu/openbook.php?record_id=12194&page=206 (Chapter 4 in this volume).

Bonvillian, W. B. (2009). "The Innovation State", *The American Interest* 4/6: 69–78, https://www.the-american-interest.com/2009/07/01/the-innovation-state/

Bonvillian, W. B. (2007). "Will the Search for New Energy Technologies Require a New R&D Mission Agency?—The ARPA-E Debate", *Bridges* 14, 12 July, https://ostaustria.org/bridges-magazine/volume-14-july-12-2007/item/2297-will-the-search-for-new-energy-technologies-require-a-new-r-d-mission-agency-the-arpa-e-debate

Bonvillian, W. B. (2007). *Testimony Before House Science and Technology Committee on ARPA-E Authorizing Legislation, HR-364*, Washington, DC, U.S. House of Representatives, April 26, https://www.govinfo.gov/content/pkg/CHRG-110hhrg34719/html/CHRG-110hhrg34719.htm

Bonvillian, W. B. (2006). "Power Play, The DARPA Model and U.S. Energy Policy", *The American Interest* 2/2, November/December, 39–48, https://www.the-american-interest.com/2006/11/01/power-play/

Bonvillian, W. B., and Weiss, C. (2009). "Taking Covered Wagons East, A New Innovation Theory for Energy and Other Established Sectors", *Innovations* 4/4: 289–94, http://www.mitpressjournals.org/userimages/ContentEditor/1259694503297/Bonvillianinov.pdf

Carleton, T. L. (2010). "The Value of Vision in Radical Technological Innovation", PhD thesis, Stanford University, Palo Alto, https://purl.stanford.edu/mk388mb2729

Christiansen, C. (1997). *The Innovator's Dilemma: When New Technologies Cause Great Firms to Fail*. Boston, MA: Harvard Business School Press.

Chu, S. (2006). "The Case for ARPA-E", *Innovation* 4/3, http://www.innovation-america.org/case-arpa-e

Conant, J. (2002). *Tuxedo Park: A Wall Street Tycoon and the Secret Palace of Science that Changed the Course of World War II*. New York, NY: Simon & Shuster.

Continuing Resolution. (FY 2011). HR 1463 (signed by President 14 April 2011).

Cooper, R. G., Edgett, S. J., and Kleinschmidt, E. J. (2002). "Optimizing the Stage-Gate Process", *Research Technology Management (Industrial Research Institute, Inc.)* 45/5, https://doi.org/10.1080/08956308.2002.11671532

DARPA. (1989). Other Transactions Authority (OTA), P.L. 101–89, 10 U.S.C. 2389 (enacted 1989); P.L. 103–60, Sec. 845.

DARPA. (2009). *Network Challenge*.

Department of Defense. (2011). "SERDP and ESTCP Program Information," http://www.serdp.org/About-SERDP-and-ESTCP

Department of Defense. (2011). "SERDP and ESTCP Energy and Water Projects" (an updated description of this effort is available at: https://www.serdp-estcp.org/Program-Areas/Installation-Energy-and-Water).

Department of Defense Office of Manufacturing and Industrial Base Policy. (2011). "Programs", https://www.businessdefense.gov/Programs/

Department of Energy. (2011). "Six ARPA-E Projects Illustrate Private Investors Excited About Clean Energy Innovation", 3 February.

Department of Energy. (2011). "Secretary Chu Announces $130m for Advanced Research Projects", 20 April.

Dunn, R. L. (2007). *Acquisition Reform, the DARPA Approach*, http://www.authorstream.com/Presentation/Burnell-34588-appe-Why-Business-Everybody-Else-Evolution-Defense-Industry-as-Entertainment-ppt-powerpoint/

Dunn, R. L. (1996). "DARPA Turns to Other Transactions", *Aerospace America* 34/10: 33–37.

Dunn, R. L. (1996). DARPA General Counsel Memorandum of Law, Scope of Section 845 Prototype Authority, 24 October.

Dunn. R. L. (1995). Testimony of General Counsel of DARPA before the Committee on Science, U.S. House of Representatives on Innovations in Government Contracting Using the Authority to Enter into "Other Transactions" with Industry, 8 November, https://acquisitioninnovation.darpa.mil/docs/Articles/Dunn%20testimony%20on%20OTs%20Nov%201995.pdf

Energy Policy Act of 2005. (2005). P.L. 109–58 (signed by President August 8, 2005), Other Transactions Authority, Sec. 1007, https://www.govinfo.gov/content/pkg/PLAW-109publ58/pdf/PLAW-109publ58.pdf

Finkbinder, A. (2006). *The Jasons*. New York, NY: Viking Books.

Fong, G. R. (2001). "ARPA Does Windows; the Defense Underpinning of the PC Revolution", *Business and Politics* 3/3: 213–37, https://doi.org/10.2202/1469-3569.1025 (Chapter 6 in this volume).

Fuchs, E. R. H. (2011). "DARPA Does Moore's Law: The Case of DARPA and Optoelectronic Interconnects", in *The State of Innovation: The US Government's Role in Technology Development*, ed. F. Bloch, and W. Keller. Boulder, CO: Paradigm Publishers. 133–48, https://doi.org/10.4324/9781315631905.

Gates, D. (2010). "Dreamliner Woes Pile Up", *Seattle Times*, 18 December, http://old.seattletimes.com/html/businesstechnology/2013713745_dreamliner19.html

General Accounting Office (GAO). (2008). DOE Implementation and Use of Other Transactions Authority, 6 June, http://www.gao.gov/new.items/d08798r.pdf

Gholz, E. (2011). *How Military Innovation Works and the Role of Industry*. Presentation at Forum on Leveraging DOD's Energy Innovation Capacity at the Bipartisan Policy Center, Washington, DC, 25 May (see summary https://bipartisanpolicy.org/wp-content/uploads/2019/03/Energy-Innovation-at-DoD.pdf)

Gupta, U., ed. (2000). *Done Deals, Venture Capitalists Tell Their Stories*. Cambridge, MA: Harvard Business School Press.

Heilmeier, G. H. (1991). *Oral History Interview (by Arthur Norberg)*. Minneapolis, MN: Charles Babbage Institute, https://conservancy.umn.edu/handle/11299/107352

Hourihan M., and Stepp, M. (2001). "Lean, Mean and Clean: Energy Innovation and the Department of Defense", *Information Technology and Innovation Foundation* (March): 1–26, https://www.itif.org/files/2011-lean-mean-clean.pdf

H. R. 364 (2007). Establishing the Advanced Research Projects Agency-Energy, reported by the House Committee on Science and Technology on May 23, 2007, 110th Congress, 1st Sess., https://www.govtrack.us/congress/bills/110/hr364/text

IHS Global Insight. (2011). "China Passes U.S. in Manufacturing Output", 14 March.

Jorgenson, D. (2001). "U.S. Economic Growth in the Information Age", *Issues in Science and Technology* 18/1: 42–50, http://www.issues.org/18.1/jorgenson.html

Kaminski, P. G. (1996). Secretary of Defense Memorandum, 14 December, re: 10 U.S.C. 2371, Section 845, Authority to Carry Out Certain Prototype Projects. Arlington, VA: DOD.

Knox, W. D. (1999). *Of Gladiators and Spectators: Aquila, the Case for Army Acquisition Reform*. Carlisle, PA: Army War College.

Kosinski, S. (2009). *Advanced Research Projects Agency-Energy (ARPA-E)* (presentation).

Majumdar. A. (2010). Testimony before the House Committee on Science and Technology on the Advanced Research Projects Agency (ARPA-E), https://www.energy.gov/sites/prod/files/ciprod/documents/1-27-10_Final_Testimony_%28Majumdar%29.pdf

Marqusee, J. (2011). *SERDP and ESTCP*. Presentation at Forum on Leveraging DOD's Energy Innovation Capacity at the Bipartisan Policy Center, Washington, DC, 25 May.

National Academy of Sciences. (2007). *Rising above the Gathering Storm: Energizing and Employing America for a Brighter Economic Future*. Washington, DC: The National Academies Press, https://doi.org/10.17226/12537, https://www.

nap.edu/catalog/11463/rising-above-the-gathering-storm-energizing-and-employing-america-for#toc

MIT Washington Office. (2010). "Survey of Federal Manufacturing Efforts", http://dc.mit.edu/sites/default/files/pdf/MIT%20Survey%20of%20Federal%20Manufacturing%20Efforts.pdf

National Research Council. (1999). *Funding a Revolution, Government Support for Computing Research*. Washington, DC: National Academies Press, https://doi.org/10.17226/6323

National Research Council. (2001). *Energy Research at DOE: Was it Worth It? Energy Efficiency and Fossil Energy Research 1978–2000*. Washington, DC: National Academy Press, http://www.nap.edu/catalog.php?record_id=10165#toc

Norris, F. (2011). "As US Exports Soar, It's Not All Soybeans", *New York Times*, 11 February, http://www.nytimes.com/2011/02/12/business/economy/12charts.html?_r=1&src=busl

Piore, M. (2008). *Learning on the Fly: Reviving Active Governmental Policy in an Economic Crisis*. Presentation at How Will a New Administration and Congress Support Innovation in an Economic Crisis? Sponsored by the Economic Policy Institute, ITIF, Breakthrough Institute, University of Calif., Washington Center and Ford Foundation, Washington, DC, 1 December, https://www.longviewinstitute.org/blockvideo/view/index.html

Pisano, G. and Shih, W. (2009). "Restoring American Competitiveness", *Harvard Business Review* 87/7–8: 114–25, https://hbr.org/2009/07/restoring-american-competitiveness

Porter, G., et al. (2009). *The Major Causes of Cost Growth in Defense Acquisition, P-4531*. Volume 2. Alexandria, VA: Institute for Defense Analyses, http://www.dtic.mil/cgi-bin/GetTRDoc?Location=U2&doc=GetTRDoc.pdf&AD=ADA519884

Robyn, D. (2011). Testimony for hearing on "Department of Defense Authorization for Appropriations for Fiscal Year 2012 and the Future Years Defense Program," Committee on Armed Services, United States Senate, 17 March, https://www.govinfo.gov/content/pkg/CHRG-112shrg68086/pdf/CHRG-112shrg68086.pdf

Robyn, D. (2010). Testimony for hearing on "Cutting the Federal Government's Energy Bill: An Examination of the Sustainable Federal Government Executive Order," Federal Financial Management, Government, Information, Federal Services, and International Security Subcommittee of the Committee on Homeland Security and Governmental Affairs, United States Senate, 27 January, https://www.govinfo.gov/content/pkg/CHRG-111shrg56839/pdf/CHRG-111shrg56839.pdf

Romer, P. (1990). "Endogenous Technological Change", *Journal of Political Economy* 98: 72–102, https://doi.org/10.1086/261725

Ruttan, V. W. (2006). *Is War Necessary for Economic Growth? Military Procurement and Technology Development*. New York, NY: Oxford University Press.

Ruttan, V. W. (2006). "Will Government Programs Spur the Next Breakthrough?", *Issues in Science and Technology* 22/2: 55–61, https://issues.org/ruttan/

Ruttan, V. W. (2001). *Technology, Growth and Development: An Induced Innovation Perspective*. New York, NY: Oxford University Press.

Schein, E. (2004). *DEC is Dead, Long Live DEC—Lessons on Innovation, Technology and the Business Gene*. San Francisco, CA: BK Berrett-Kohler Publishers.

Schumpeter, J. (1942). *Capitalism, Socialism and Democracy*. New York, NY: Harper & Row.

Solow, R. M. (2000). *Growth Theory, An Exposition*. 2nd ed. Oxford: Oxford University Press,

http://nobelprize.org/nobel_prizes/economics/laureates/1987/solow-lecture.html

Stokes, D. E. (1997). *Pasteur's Quadrant, Basic Science and Technological Innovation*. Washington, DC: Brookings Institution Press.

Tassey, G. (2010). "Rationales and Mechanisms for Revitalizing U.S. Manufacturing R&D Strategies", *Journal of Technology Transfer* 35/3: 283–333, https://doi.org/10.1007/s10961-009-9150-52.

Van Atta, R. (2008). "Fifty Years of Innovation and Discovery", in *DARPA, 50 Years of Bridging the Gap*, ed. C. Oldham, A. E. Lopez, R. Carpenter, I. Kalhikina, and M. J. Tully. Arlington, VA: DARPA. 20–29, https://issuu.com/faircountmedia/docs/darpa50 (Chapter 2 in this volume).

Van Atta, R. (2007). Testimony before House Science and Technology Committee on ARPA-E Authorizing Legislation, HR-364, April 26, https://www.govinfo.gov/content/pkg/CHRG-110hhrg34719/html/CHRG-110hhrg34719.htm

Van Atta, R., Deitchman, S., and Reed, R. (1991). *DARPA Technical Accomplishments. Volume II*. Alexandria, VA: Institute for Defense Analyses, https://apps.dtic.mil/dtic/tr/fulltext/u2/a241725.pdf

Van Atta, R., Deitchman, S., and Reed, S. (1991). *DARPA Technical Accomplishments. Volume III*. Alexandria, VA: Institute for Defense Analyses, https://apps.dtic.mil/dtic/tr/fulltext/u2/a241680.pdf

Van Atta, R., Deitchman, S., and Reed, S. (1990). *DARPA Technical Accomplishments. Volume I*. Alexandria, VA: Institute for Defense Analyses, https://apps.dtic.mil/dtic/tr/fulltext/u2/a239925.pdf

Van Atta, R., Bovey, R., et al. (2003). *Science and Technology in Development Organizations*. Alexandria, VA: Institute for Defense Analyses.

Van Atta, R., Lippitz, M., et al. (2003). *Transformation and Transition, DARPA's Role in Fostering a Revolution in Military Affairs. Volume 1*. Alexandria, VA:

Institute for Defense Analyses, https://doi.org/10.21236/ada422835, https://fas.org/irp/agency/dod/idarma.pdf

Van Atta, R., Lippitz, M., and Bovey, R. (2005). *DOD Technology Management in a Global Technology Environment,* IDA Paper P-4017, Alexandria, VA: Institute for Defense Analyses.

Waldrop, M. M. (2001). *The Dream Machine: J. C. R. Licklider and the Revolution that Made Computing Personal.* New York, NY: Viking Press.

Weiss, C. and Bonvillian, W. B. (2011). "Complex, Established 'Legacy' Sectors: The Technology Revolutions that Do Not Happen", *Innovations* 6/2: 157–87, https://doi.org/10.1162/inov_a_0007, https://www.mitpressjournals.org/doi/pdf/10.1162/INOV_a_00075

Weiss, C. and Bonvillian, W. B. (2009). *Structuring an Energy Technology Revolution.* Cambridge, MA: The MIT Press, https://doi.org/10.7551/mitpress/8161.001.0001

Wikipedia contributors. (2019). "DARPA Grand Challenge", *Wikipedia,* 17 September, http://en.wikipedia.org/wiki/DARPA_Grand_Challenge

Zachary, G. P. (1999). *Endless Frontier, Vannevar Bush, Engineer of the American Century.* Cambridge, MA: The MIT Press.

Zachary, G. P. (1997). "The Godfather", *Wired,* November.

14. IARPA

A Modified DARPA Innovation Model[1]

William B. Bonvillian

The DARPA model for organizing innovation has now been copied in other U.S. agencies. This is in part because DARPA is famous for playing critical roles in the information technology (IT) revolution—from support for personal computing to the Internet, as well as in stealth and drones. As discussed across this volume, DARPA is distinct from other innovation agencies around the world in its rejection of "pipeline" and technology "hand-off' approaches used by most agencies. As an innovation organization, DARPA takes responsibility to bring about technological breakthroughs and nurtures them toward delivering final products. To do this effectively, DARPA has developed a series of specific organizational practices. These have, in turn, been adopted by DARPA clones.

The Advanced Research Project Agency-Energy (ARPA-E) was formed in 2009 to bring a DARPA-like approach to the challenge of advanced energy technologies, and is discussed in Chapter 13. The Intelligence Advance Research Projects Agency (IARPA), reviewed here, began operating in 2007, bringing a DARPA model

1 This paper contains material that originally appeared in 2018 as "DARPA and its ARPA-E and IARPA clones: a unique innovation organization model", *Industrial and Corporate Change* 27/5: 897–914, https://doi.org/10.1093/icc/dty026, https://academic.oup.com/icc/article-abstract/27/5/897/5096003

 https://doi.org/10.11647/OBP.0184.14

to development of intelligence-related technologies. A third DARPA clone, the Homeland Security Advanced Research Projects Agency (HSARPA) was authorized in 2002 as a DARPA-like entity in the then newly-formed Department of Homeland Security. However, it was not adequately established at the time, and much of its early staff, many of whom came from DARPA, left in frustration. It was not allowed to be a separate operating unit within the department's science and technology directorate, subsumed within a more traditional budget and policy office. The Department's Undersecretary for Science and Technology from 2009–2013 worked to reestablish HSARPA during the Obama Administration, however, the Trump Administration has since moved away from it. Because of these operational problems, this chapter does not attempt to evaluate it.

Concerning IARPA, like DARPA, it operates as public sector intermediary, pursuing breakthrough research but also actively promoting its implementation. Like DARPA, it is therefore much more activist than the standard American R&D mission agency, acting as a change agent within the often conservative "legacy" sectors it serves. This chapter examines IARPA in more detail, comparing it to DARPA, and concludes by noting two structural challenges in their innovation systems that DARPA, ARPA-E and IARPA all face.

The DARPA Model in the Context of Innovation Policy

DARPA was a Cold War creation, formed in direct response to a technological crisis. Its operating practices began without any significant inspiration from innovation theorists or growth economists. Its early program officers learned by doing. It is only recently—some sixty years later—that innovation theory is catching up, and consideration is being given to where an agency like DARPA might fit within this theory.

The DARPA model, however, can now be understood against an established policy foundation. In recent years it has been seen to occupy a unique place in the context of the U.S. literature on science, technology and innovation policy, which requires a brief explication here. The economic foundation for the innovation policy field is Robert Solow's work positing technological and related innovation as the dominant causative factor in growth.[2] Paul Romer and other New Growth Theorists argued the importance of technological learning as the underpinning for Solow's technological advance theory.[3] These two strands led to an understanding of two basic underlying innovation factors—support for R&D and follow-on technological advance, and support for Romer's concept of human capital engaged in research that lay behind that system.

Richard Nelson in turn argued the importance in understanding comparative innovation systems of assessing the actors in an innovation system and their comparative strengths.[4] We can enlarge this concept to constitute a third direct innovation factor, innovation organization, which can be analyzed as a connected system of innovation institutions and organizations. Against these factors, particularly the organizational factor, the U.S. innovation system took shape. DARPA and its clones exemplify a unique innovation organization model within that innovation system that deserves explication.

In the postwar, Vannevar Bush's highly influential "pipeline model" for the postwar organization of U.S. R&D agencies was a "technology push" or "technology supply" model, with government support for initial research, but with only a very limited role for government in moving resulting advances (particularly radical or breakthrough innovation) toward the marketplace. Development and the later stages of innovation were left to private industry. Donald Stokes (and others) subsequently sharply critiqued the Bush pipeline model as inherently disconnected, separating the government supported research actors from the industry development actors with few means for technology handoffs between them.[5] Lewis Branscomb and Phillip Auerswald

---

2 Solow, R. M. (2000). *Growth Theory, An Exposition*. 2nd ed. Oxford: Oxford University Press, http://nobelprize.org/nobel_prizes/economics/laureates/1987/solow-lecture.html.

3 Romer, P. (1990). "Endogenous Technological Change", *Journal of Political Economy* 98/5: 72–102, http://pages.stern.nyu.edu/~promer/Endogenous.pdf

4 Nelson, R., ed. (1993). *National Systems of Innovation*. New York, NY: Oxford University Press, 3–21, 505–23. This "innovation organization" factor is also elaborated on at length in Bonvillian, W., and Weiss, C. (2015). *Technological Innovation in Legacy Sectors*. New York, NY: Oxford University Press, 25–27, 181–86, 190–92, https://doi.org/10.1093/acprof:oso/9780199374519.001.0001

5 Stokes, D. E. (1997). *Pasteur's Quadrant, Basic Science and Technological Innovation*. Washington, DC: Brookings Institution Press.

articulated the "valley of death" critique: the disconnect in the U.S. system between research and later stage development led to system failures in commercialization of research results.[6] This concern has been the major focus of U.S. science and technology policy literature for the past twenty years, with resulting discussions of bridging solutions across this valley. Of course, the pipeline model is not the only U.S. innovation system model.

As detailed in Chapter 12 of this work, there are five fundamentally different innovation approaches that help us sort out the roles of DARPA and its clones. These drive the dynamics of innovation in different settings: the innovation pipeline, induced innovation, the extended pipeline, manufacturing-led innovation, and innovation organization.[7] These provide a framework for understanding the place in the innovation system occupied by DARPA and IARPA, as well as ARPA-E. It must also be kept in mind that innovation does not happen entirely through an "invisible hand"; innovation introduction generally requires active efforts by change agents. Such agents are particularly critical for innovation in legacy sectors given the significant barriers innovation faces in these sectors. DARPA and its clones are particularly noteworthy as change agents, not simply research organizations.

The "pipeline" model, as noted above, has long dominated U.S. science and technology thinking. It pictures invention and innovation as flowing from investments in research—predominantly from federal basic research support—at the "front end" of the innovation system. Thus, research is dumped into one end of the innovation pipeline, mysterious things occur, industry picks up their development and new products emerge. However, most technology comes from private sector firms that respond to market opportunities. This constitutes a second model, "induced innovation". Vernon Ruttan is the growth economist who discussed this as the dominant way industry innovates, by identifying market opportunities then innovating to fill them.[8]

6 Branscomb, L., and Auerswald, P. (2002). *Between Invention and Innovation, An Analysis of Funding for Early-State Technology Development,* NIST GCR 02–841. Washington, DC: National Institute of Standards and Technology, https://www.nist.gov/sites/default/files/documents/2017/05/09/gcr02-841.pdf

7 These models are discussed at length in, Bonvillian and Weiss. (2015). *Technological Innovation,* 23–30, 181–76, which is drawn from here.

8 Ruttan, V. W. (2001). *Technology Growth and Development: An Induced Innovation Perspective.* New York, NY: Oxford University Press.

Here, typically the originator—the change agent—is a firm that spots a market opportunity or niche that can be filled by a technology advance—typically an incremental not a radical technology advance. It is a "technology demand" or "technology pull" model—the market creates the demand and pull to induce the technology. The third model can be termed the "extended pipeline", where certain U.S. R&D organizations, particularly through the Defense Department (DOD), and including DARPA, support moving innovations through every innovation stage. Because DOD could not tolerate a disconnected model when faced with Cold War technological demands, it developed an extended pipeline.[9] This means support not just for front end research and development (R&D) but also for each successive "back-end" stage, from advanced prototype to demonstration, testbed, and often to initial market creation, where DOD will buy the first products.[10] While the government's support role in the pipeline model is disconnected from the rest of the innovation system, in this model it attempts to be deeply connected. Most of the major innovation waves of the past three-fourths of a century, have evolved from this system: aviation, nuclear power, electronics, space, computing and the Internet.[11] The extended pipeline facilitates the bridging of the "valley of death" between advanced research and implemented technology. In general, U.S. innovation models in recent decades have tended to stretch their capabilities further down this innovation pipeline.[12]

The fourth model of innovation dynamics, "manufacturing-led" innovation, describes innovations in production technologies, processes and products that emerge from expertise informed by experience in manufacturing.[13] This is augmented by applied research

9 Bonvillian and Weiss. (2015). *Technological Innovation*, 181–86.

10 Bonvillian, W. B., and Van Atta, R. (2011). "ARPA-E and DARPA: Applying the DARPA Model to Energy Innovation", *The Journal of Technology Transfer* 36: 469–513, at 469, https://doi.org/10.1007/s10961-011-9223-x, https://link.springer.com/article/10.1007%2Fs10961-011-9223-x

11 Although he did not use the term "extended pipeline", Vernon Ruttan wrote about the Defense role in evolving these technologies, Ruttan, V. W. (2006). *Is War Necessary for Economic Growth? Military Procurement and Technology Development*. New York, NY: Oxford University Press.

12 Bonvillian, W. B. (2013). "The New Model Innovation Agencies: An Overview", *Science and Public Policy* 41/4: 425–37, https://doi.org/10.1093/scipol/sct059, https://academic.oup.com/spp/article-abstract/41/4/425/1607552?redirectedFrom=fulltext

13 Bonvillian and Weiss. (2015). *Technological Innovation*, 25, 181–85.

and development that is integrated with the production process. It is typically industry-led, but with strong governmental industrial support. While countries like Germany, Japan, Taiwan, Korea and now China have organized their economies around "manufacturing-led" innovation systems, the U.S. in the postwar period did not. It is a major gap in the U.S. innovation system. This system gap is now starting to affect the ability of DARPA and its clones to translate their technologies into actual innovation.

The fifth model, "innovation organization", is different from the others.[14] It calls for improving the means, methods and organization of innovation efforts, both on the innovation front and back ends—it is an organizational model. In this innovation organization model, the innovation system supports the full innovation spectrum, each stage in the innovation process. While the pipeline model supports R&D at the front end, and the manufacturing-led model supports the back end, production stage, the innovation organization model contemplates all stages. It goes beyond the extended pipeline model to orchestrate the institutional and policy changes needed to facilitate innovation not just for a government customer.

Innovation policy theorists, as noted above, have long analyzed the gap between the "front end" of the innovation system—the research side, typically supported by government R&D through university research—and the "back end", the late-stage development through implementation phases, typically a private sector domain. To solve this structural problem, numerous bridging mechanisms have evolved, often with government support. As Philip Shapira and Jan Youtie have noted, this requires technology diffusion approaches, and a wide range of institutional intermediaries.[15]

DARPA and its clones are not basic research agencies; they are public sector intermediaries as well. They work to nurture new technologies from breakthrough stages through applied research and initial development, then to pass off the technologies to entities that will move them into implementation. They intermediate between finding

14 Bonvillian and Weiss. (2015). *Technological Innovation*, 25–27, 186.

15 Shapira, P., and Youtie, J. (2016). *The Next Production Revolution and Institutions for Technology Diffusion*. Presentation at the Conference on Smart Industry: Enabling the Next Production Revolution, OECD and Sweden Ministry of Enterprise and Innovation, Stockholm, 18 September.

the breakthrough to technology implementation. As intermediaries, they also operate as change agents.

DARPA and IARPA are clearly mainstays of the extended pipeline model, able to apply acquisition budgets from their overall agencies to implement technologies they research. Therefore, they are reaching toward the unifying "innovation organization" model. This makes them quite different from other R&D agencies. Nonetheless, it is also important to note that DARPA and later IARPA are able to succeed because the U.S. already had a very rich and complex publicly funded science and technology system, including the federal labs, university-based labs, the National Science Foundation, as well at an earlier time a network of quite significant private sector labs, including, of course, Bell Labs.[16] DARPA and later IARPA could cherry pick the most promising technologists because there were many of them out there to choose from. However, when the talent supply was lacking or tight, DARPA helped produce more experts—its support for the early computer science departments, for example, proved of deep benefit to the emergence of the field as well as to DARPA's many IT advances. ARPA-E and I-ARPA have played similar talent-intermediary roles in their fields.

However, DARPA must play its intermediary role in a defense sector that is often profoundly conservative about technology advances. ARPA-E must be an intermediary in an energy sector that is largely averse to the entry of new technologies. And IARPA faces a comparably conservative intelligence world. These sectors are all complex, established, legacy sectors. The challenge of innovation for intermediaries is already difficult; the difficulty can be multiplied when the technology must be stood up in a legacy sector.

The IARPA Model

IARPA's first director, Lisa Porter, named in 2008, was a former DARPA program manager who understood and consciously attempted to replicate DARPA's strengths and "high-risk/high-payoff" approach. Both IARPA and DARPA hire term-limited program managers with

16 Gertner, J. (2012). *The Idea Factory, Bell Labs and the Great Age of American Innovation.* Penguin Publishing Group: New York, N.Y.

outstanding scientific and engineering credentials and experience.[17] Like DARPA, IARPA competitively selects new projects for funding using "The Heilmeier Catechism"—a set of questions to guide program selection.[18] Like DARPA, IARPA has no lab and conducts no research itself, competitively awarding research contracts and grants to leading teams of academic and industry researchers, using strong program managers without peer review systems. Like DARPA, programs have clear goals and definite ends. Program teams are regularly evaluated and teams are often cut before a program ends, depending on progress. There also are significant differences. While DARPA supports defense missions, IARPA supports national intelligence missions, which can involve quite different technologies. Some of IARPA's key organizational mechanisms to promote its innovation role are discussed below.

1) *Technology Implementation—Tournaments and Testing.* According to its current director, Jason Matheny, many of IARPA's programs are organized as tournaments in which multiple teams are funded in parallel to pursue the same technical goals, scored on a common set of metrics. This competitive approach has tended to produce a range of possible solutions and pathways. As a result, IARPA spends a large percentage of its budget (approximately 25 percent) on independent testing and evaluation. This testing stage plays such a central role at IARPA that it has a Chief of Testing and Evaluation, with contractor support, to ensure that these tests follow best practices in experimental design and statistical inference. The tournament approach and strong emphasis on testing constitute a different approach to technology implementation from DARPA and ARPA-E.

2) *Empowered Program Managers.* The strong program manager role is comparable to DARPA's. IARPA has some twenty-five program managers compared to approximately one hundred at DARPA and fifteen at ARPA-E. Program managers must nurture and pitch their proposed programs and the director

17 Much of the IARPA material below is from Jason Matheny, IARPA director, Personal Communication, 11 July 2017.

18 Chapters 1, 8, and 10 in this volume provide more details about "The Heilmeier Catechism".

> and deputy director then move quickly to approve such new programs for funding. Program managers have broad independence to manage their programs within their approved budgets. They write the solicitations for proposals, they lead proposal reviews, and they make the decisions regarding program direction and evaluation. Every six months, each program is reviewed by the IARPA senior staff, by outside technical reviewers, and by transition partners, to re-evaluate whether continued funding is justified for all research teams, and for the program as a whole. Typically, at least one team is cut per program phase. In some cases, programs are discontinued. As with DARPA and ARPA-E, IARPA program managers have a hands-on relationship with their research teams. Program managers have conference calls every two weeks with each team, they review monthly written reports from each team, and have in-person meetings with each team every quarter, at on-site visits and PI Meetings.

According to its director, IARPA has funded research at over 500 organizations in over a dozen countries. About one-third of IARPA's funding goes to universities and colleges, about one-third to small firms, about one-sixth to large firms, and about one-sixth to FFRDCs and Government labs. In this way, its program managers have a full range of innovation actors to select from. The bulk of its R&D funding goes to research in computing, machine learning, human judgment, sensors, and intelligence information technology platforms.[19]

DARPA and ARPA-E have prided themselves on their ability to hire their program managers quickly, outside of traditional civil service hiring procedures, which helps them move fast on technology challenges. IARPA, however, faces a major challenge because of its lengthy timeline for hiring program managers. This is because its program managers must obtain a high-level security clearance before beginning work. This takes several months and, in some cases, can take more than a year.

3) *Ensuring Buy-In from Agency Customers.* This intelligence technology focus results in organizational changes compared

19 For a summary of current agency work, see IARPA. "Research Program, Current Research", https://www.iarpa.gov/index.php/research-programs

> to DARPA, just as ARPA-E's energy focus required changes. IARPA's research tends to focus on key intelligence problems that have limited commercial markets. For example, programs in quantum computing and superconducting computing have few near-term commercial applications. Its work in natural language processing focuses on languages of little commercial interest. As a result, it has few commercial off-ramps for its research and focuses on technology transition directly to intelligence agencies. Thus, while DARPA stood up its computing initiatives in the private sector, and ARPA-E must stand up its energy initiatives in the private sector, IARPA must focus exclusively on government intelligence agencies as customers for its technologies. While this can mean a more assured route to technology implementation, intelligence is also a long-established bureaucratic sector with legacy features.

There are, however, spillover opportunities over time for the private sector, because it has relatively open research processes. Most of IARPA's research is unclassified. IARPA's research is largely open to university researchers, to foreign participation, it has no publication restrictions, and is published in peer-reviewed journals.

IARPA's agency-focused transition does face technology implementation challenges. Seventy percent of IARPA programs beyond their midpoint, according to its director, have achieved at least one technology transition to an intelligence agency. However, the intelligence community lacks DOD's large industrial base and constellation of labs, so IARPA has to make special efforts to support technology transition directly with intelligence agencies. In particular, it has a full-time Chief of Technology Transition with contractor support to work with these potential government customers. This group is analogous to DARPA's tech to market team.

IARPA works directly with the intelligence community to get its technologies implemented. It involves it agency transition partners in the program pitch, in proposal reviews, and in program reviews. Technology transition plans with the interested agency are typically developed during the second or third year of a program. The Chief of Technology Transition directly supports these efforts. IARPA's strong testing and evaluation emphasis also helps enable agency transitions since technologies they may be considering have been subject to, in effect, a validation process. There are significant lessons from these steps to integrate technology development with customer agencies. These conscious transition efforts mark IARPA as a different kind of R&D entity, using the extended pipeline model.

4) *Multigenerational Technology Development.* Both DARPA and ARPA-E have faced challenges when they undertake multigenerational technology development. In other words, with term-limited program managers, once a program manager nurtures an area, how is it sustained after he or she departs, then built on and moved to the next related set of advances? IARPA has to deal with this problem as well. IARPA program managers often recruit their replacements. Contract employees at IARPA who support the program managers often serve as the institutional memory across multiple program managers. In a number of cases, one program may be organized to lay the groundwork for the next. For example, IARPA's work in quantum computing has been organized along a set of sequential technical milestones, which can move from one program manager to the next.

5) *Cross Disciplinary Thinking Communities.* Like DARPA and ARPA-E, IARPA has worked to build a "thinking community" around its research focus areas. However, IARPA has also worked to add an interesting element. Most IARPA programs require the formation of research teams that cross disciplines. In some cases, these research communities have not previously interacted. For example, according to its director, IARPA's work on the social science of cybersecurity has brought together sociologists and cybersecurity experts, and its work in geopolitical forecasting has brought together political scientists and computer scientists. This multidisciplinary thought community, particularly across social and physical sciences, is an interesting IARPA feature.

Because its technologies serve intelligence needs, it is hard to evaluate IARPA's success metrics. However, IARPA-supported quantum computing research was named a *Science* magazine Breakthrough of the Year in 2010.[20] In 2015, IARPA was named to lead foundational research and development in the interagency National Strategic Computing Initiative, in 2014 it was made part of the interagency BRAIN Initiative and in 2016 it was made part of Nanotechnology-Inspired Grand Challenge for Future Computing.[21] These are all external signals of strong technical capability, in addition to its 70 percent rate of transitioning technologies into agencies.[22]

To summarize, IARPA, in addition to replicating the core of the DARPA model brings interesting variations as well. Its "tournament" approach to many of its projects, where multiple teams are funded in parallel to pursue the same technical goals provides an interesting competitive approach to produce a range of possible solutions and pathways. It spends a large percentage of its budget on independent testing and evaluation under a Chief of Testing and Evaluation. This testing regime has tended to validate its technologies and make them more acceptable to its intelligence agency customers. It involves it agency transition partners in the research program pitch, in proposal reviews, and in program reviews, which has produced further customer buy-in, smoothing the path to technology implementation. In addition, its multidisciplinary approach to building a "thinking community" to contribute to its technology capabilities, particularly across social and physical sciences, is an interesting IARPA feature. All are variations from the basic DARPA model that merit consideration.

---

20 Ford, M. (2010), "Science's Breakthrough of 2010: A Visible Quantum Device", *Ars Technica,* 23 December, https://arstechnica.com/science/2010/12/sciences-breakthrough-of-2010-a-macro-scale-quantum-device/

21 See White House. (2015). "Executive Order: Creating a National Strategic Computing Initiative", July 29; White House. (2014). "Fact Sheet: Over $300m in Support of the BRAIN Initiative", 30 September, 5.; Whitman, L., Bryant, R., and Kalil, T. (2015)., "A Nanotechnology-Inspired Grand Challenge for Future Computing", White House, 30 October.

22 For a useful summary of IARPA's technology progress, see IARPA. (2018). *2018 Year in Review*. Washington DC: IARPA https://www.iarpa.gov/index.php/about-iarpa/2018-year-in-review?highlight=WyJ5ZWFyIiwieWVhcidzIiwiaW4iLCJyZXZpZXciLCJ5ZWFyIGluIiwieWVhciBpbiByZXZpZXciLCJpbiByZXZpZXciXQ==; and IARPA. (2016). *2016 Year in Review*. Washington DC: IARPA, https://www.iarpa.gov/index.php/228-about-iarpa/2016-year-in-review/889-2016-year-in-review?highlight=WyJ5ZWFyIiwieWVhcidzIiwiaW4iLCJyZXZpZXciLCJ5ZWFyIGluIiwieWVhciBpbiByZXZpZXciLCJpbiByZXZpZXciXQ==

Two Challenges to DARPA and its Clones—Manufacturing and Scaling up Startups

DARPA and its clones often innovate in the areas of "hard" technologies that must be manufactured, in addition to work in software. They also rely on innovative, entrepreneurial startups to bring their hard technology projects into implementation. Both systems are under challenge, and this could affect the effectiveness of the DARPA, ARPA-E and IARPA models.

Although there is a substantial argument that manufacturing—particularly initial production of new technologies and complex, high value products—is a significant stage of the innovation system, as Suzanne Berger has articulated,[23] U.S. innovation agencies historically have not organized around it. However, as noted in Chapter 12, other nations have developed what can be termed "manufacturing-led" innovation systems, which is the dominant model in Germany, Japan, Korea, and now China.[24] Emblematic of "manufacturing-led" is Japan's quality manufacturing revolution of the 1970s-80s,[25] Germany's system of industrial support through its Fraunhofer institutes and apprenticeship programs,[26] and lately, China's rapid prototyping and scale-up capacity.[27]

The U.S. missed this model. In the immediate postwar period when it was forming most of its R&D agencies, the U.S. had the strongest manufacturing sector in the world, operating at a level of mass production efficiency that no other economies were close to. There was no reason to bring innovation models to production.[28] Both civilian

23 Berger, S., with the MIT Task Force on Production and Innovation. (2013). *Making in America*. Cambridge, MA: The MIT Press.

24 Bonvillian and Weiss. (2015). *Technological Innovation*, 184–86. See also the discussion of China in, Bonvillian, W. B., and Singer, P. (2018). *Advanced Manufacturing—The New American Innovation Policies*. Cambridge, MA: The MIT Press, 8, 45–52, https://doi.org/10.7551/mitpress/9780262037037.001.0001

25 Womack, J. P, Jones, D. T., and Roos, D. (1991). *The Machine that Changed the World: The Story of Lean Production*. New York, NY: Harper Perennial. See also, discussion of Japan in Bonvillian and Singer. (2018). *Advanced Manufacturing*, 37–44.

26 Bonvillian and Singer. (2018). *Advanced Manufacturing*, 178–83.

27 Nahm, J., and Steinfeld, E. (2013). "Scale-Up Nation: China's Specialization in Innovative Manufacturing", *World Development* 54: 288–300, https://doi.org/10.1016/j.worlddev.2013.09.003

28 Bonvillian and Singer. (2018). *Advanced Manufacturing*, 34–35.

and military innovation models—pipeline and extended pipeline—focused on broader technology development, not on technologies and processes for manufacturing innovation. The U.S. therefore missed manufacturing-led innovation, and subsequently paid a significant price in the decline of its manufacturing base in the early 2000s. The one-third manufacturing job decline from 2000–2010 turned out to be symptomatic of a decline in production capability. Widespread offshoring of manufacturing, encouraged by generations of MBAs and a financial sector taught to focus firms on "core competencies" and to go "asset light", was also a critical factor in limiting domestic production capacity.[29] Linda Weiss has noted the problematic future of American economic primacy and national security as its financialized corporations curtailed investment in manufacturing and related innovation.[30] Production, particularly initial production of new technologies, can be highly innovative, involving creative engineering, design, technology advances and production processes. For the DARPA model agencies to be cut off from these innovation system capabilities, and unable to rely on a strong U.S. manufacturing base for rapid prototyping and innovative production, spells a major potential challenge to their ability to develop and implement hard technologies. Although the U.S. is now pursuing an "advanced manufacturing" model through an innovative group of fourteen new advanced manufacturing institutes,[31] this effort is still in early stages, and it is not clear it will have the political support to be sustained over the extended period required.

The second challenge is that U.S. venture capital (VC) has largely withdrawn from support of startup firms with hard technologies that must be manufactured.[32] VC firms are focused on software, biotech and services startups where they can more readily manage the scale-up process and timetable. Hard technologies typically require more time, risk

29 Berger, S. (2014). "How Finance Gutted Manufacturing", *Boston Review*, 1 April, http://bostonreview.net/forum/suzanne-berger-how-finance-gutted-manufacturing; and Bonvillian and Singer. (2018). *Advanced Manufacturing*, 117–18.

30 Weiss, L. (2014). *America Inc.? Innovation and Enterprise in the National Security State*. Ithaca, NY: Cornell University Press, 203–09.

31 Bonvillian and Singer. (2018). *Advanced Manufacturing*, 135–86.

32 Bonvillian and Singer. (2018). *Advanced Manufacturing*,187–215. These developments are reviewed in further detail in, Singer, P., and Bonvillian, W. B. (2017). "Innovation Orchards: Helping Startups Scale", *Information Technology and Innovation Foundation*, Washington, DC, http://www2.itif.org/2017-innovation-orchards.pdf

and capital for scale-up so increasingly fall outside the VC model. Since VCs dominate the scale-up process for its small, innovative companies, the U.S. is increasingly leaving hard technologies by the technology wayside. Because they leverage the private sector for implementation, this will affect the ability, in particular, of DARPA and ARPA-E to use the entrepreneurial approach they have relied on for scaling up their hard technologies. A new approach, termed "innovation orchards", is now evolving to fill this gap. This entails creating shared technology, equipment and know-how rich spaces for scaling-up startups through advanced prototype, production design and pilot production. In effect, this approach attempts to substitute space for capital. However, it is likewise at a very early stage. In the meantime, this creates a serious implementation challenge for the DARPA model.

Conclusion

DARPA, ARPA-E and IARPA share an ambitious innovation organization model, operating as public sector intermediaries that pursue high-risk/high reward, breakthrough research. Importantly, they also actively promote its implementation. They are therefore much more activist than the standard American R&D mission agency, performing as change agents within the often conservative "legacy" sectors they operate within. The chapter has summarized the DARPA model and reviewed its variations in IARPA in detail. It placed these agencies it in the context of the overall U.S. innovation system—DARPA and IARPA are leading examples of the "extended pipeline" model, while ARPA-E is located within a "pipeline" model agency, trying to reach further down the innovation pipeline. All face the types of innovation barriers common to legacy sectors, which further challenge their efforts to implement their innovations. Despite these challenges, the DARPA model has proven quite dynamic; DARPA has an unparalleled record of technological advance, and the other two are rapidly building their own records. ARPA-E and IARPA show that the DARPA model is now a proven one in the innovation space, clearly relevant to other technology sectors. Therefore, the specifics of their innovation organization present important innovation options deserving close examination, as attempted here. However, because all three agencies work in significant part on "hard" technologies that must be manufactured, they face two significant new structural challenges in the U.S. innovation system: in manufacturing and startup scaling. Their ability to achieve innovation implementation in the future in hard technology fields may depend on progress in addressing these two new innovation system challenges.

References

Berger, S. (2014). "How Finance Gutted Manufacturing", *Boston Review*, 1 April, http://bostonreview.net/forum/suzanne-berger-how-finance-gutted-manufacturing.

Berger, S., with the MIT Task Force on Production and Innovation. (2013). *Making in America*. Cambridge, MA: The MIT Press.

Bonvillian, W. B., and Singer, P. (2018). *Advanced Manufacturing—The New American Innovation Policies.* Cambridge, MA: The MIT Press, https://doi.org/10.7551/mitpress/9780262037037.001.0001

Bonvillian, W. B. (2018), "DARPA and its ARPA-E and IARPA clones: A unique innovation organization model", *Journal of Industrial and Corporate Change*, 27/5, https://academic.oup.com/icc/article-abstract/27/5/897/5096003

Bonvillian, W. B. (2013). "The New Model Innovation Agencies: An Overview", *Science and Public Policy* 41/4: 425–37, https://doi.org/10.1093/scipol/sct059, https://academic.oup.com/spp/article-abstract/41/4/425/1607552?redirectedFrom=fulltext

Bonvillian, W. B., and Weiss, C. (2015). *Technological Innovation in Legacy Sectors*. New York, NY: Oxford University Press, https://doi.org/10.1093/acprof:oso/9780199374519.001.0001

Bonvillian, W. B., and Van Atta, R. (2011). "ARPA-E and DARPA: Applying the DARPA Model to Energy Innovation", *The Journal of Technology Transfer*, 36: 469–513, https://doi.org/10.1007/s10961-011-9223-x (Chapter 13 in this volume).

Branscomb, L., and Auerswald, P. (2002). *Between Invention and Innovation, An Analysis of Funding for Early-State Technology Development*. NIST GCR 02–841. Washington, DC: National Institute of Standards and Technology, https://www.nist.gov/sites/default/files/documents/2017/05/09/gcr02-841.pdf

Bush, V. (1945). *Science: The Endless Frontier*. Washington, DC: Government Printing Office, https://www.nsf.gov/od/lpa/nsf50/vbush1945.htm

Ford, M. (2010), "Science's Breakthrough of 2010: A Visible Quantum Device", *Ars Technica*, 23 December, https://arstechnica.com/science/2010/12/sciences-breakthrough-of-2010-a-macro-scale-quantum-device/

Gertner, J. (2012). *The Idea Factory, Bell Labs and the Great Age of American Innovation.* Penguin Publishing Group: New York, N.Y.

IARPA. "Research Program, Current Research", https://www.iarpa.gov/index.php/research-programs

IARPA. (2018). *2018 Year in Review*. Washington DC: IARPA, https://www.iarpa.gov/index.php/about-iarpa/2018-year-in-review?highlight=WyJ5ZWFyIiwieWVhcidzIiwiaW4iLCJyZXZpZXciLCJ5ZWFyIGluIiwieWVhciBpbiByZXZpZXciLCJpbiByZXZpZXciXQ==

IARPA. (2016). *2016 Year in Review*. Washington DC: IARPA, https://www.iarpa.gov/index.php/228-about-iarpa/2016-year-in-review/889-2016-year-in-review?highlight=WyJ5ZWFyIiwieWVhcidzIiwiaW4iLCJyZXZpZXciLCJ5ZWFyIGluIiwieWVhciBpbiByZXZpZXciLCJpbiByZXZpZXciXQ==

Matheny, J. (2017). Personal Communication, 11 July.

Nahm, J., and Steinfeld, E. (2013). "Scale-Up Nation: China's Specialization in Innovative Manufacturing", *World Development* 54: 288–300, https://doi.org/10.1016/j.worlddev.2013.09.003

Nelson, R., ed. (1993). *National Systems of Innovation.* New York, NY: Oxford University Press.

Romer, P. (1990). "Endogenous Technological Change", *Journal of Political Economy* 98/5: 72–102, https://doi.org/10.1086/261725, http://pages.stern.nyu.edu/~promer/Endogenous.pdf

Ruttan, V. W. (2006). *Is War Necessary for Economic Growth? Military Procurement and Technology Development*. New York, NY: Oxford University Press.

Ruttan, V. W. (2001). *Technology Growth and Development: An Induced Innovation Perspective.* New York, NY: Oxford University Press.

Shapira, P., and Youtie, J. (2016). *The Next Production Revolution and Institutions for Technology Diffusion*. Presentation at the Conference on Smart Industry: Enabling the Next Production Revolution, OECD and Sweden Ministry of Enterprise and Innovation, Stockholm, 18 September.

Singer, P., and Bonvillian, W. B. (2017). "Innovation Orchards: Helping Startups Scale", *Information Technology and Innovation Foundation*, Washington, DC, http://www2.itif.org/2017-innovation-orchards.pdf

Solow, R. M. (2000). *Growth Theory, An Exposition.* 2nd ed. Oxford: Oxford University Press, http://nobelprize.org/nobel_prizes/economics/laureates/1987/solow-lecture.html

Stokes, D. E. (1997). *Pasteur's Quadrant, Basic Science and Technological Innovation.* Washington, DC: Brookings Institution Press.

Weiss, L. (2014). *America Inc.? Innovation and Enterprise in the National Security State.* Ithaca, NY: Cornell University Press.

White House. (2015). "Executive Order: Creating a National Strategic Computing Initiative", 29 July, https://obamawhitehouse.archives.gov/the-press-office/2015/07/29/executive-order-creating-national-strategic-computing-initiative

White House. (2014). "Fact Sheet: Over $300m in Support of the BRAIN Initiative", 30 September, https://obamawhitehouse.archives.gov/the-press-office/2013/04/02/fact-sheet-brain-initiative.

Whitman, L., Bryant, R., and Kalil, T. (2015). "A Nanotechnology-Inspired Grand Challenge for Future Computing", *White House*, 30 October, https://obamawhitehouse.archives.gov/blog/2015/10/15/nanotechnology-inspired-grand-challenge-future-computing

Womack, J. P, Jones, D. T., and Roos, D. (1991). *The Machine that Changed the World: The Story of Lean Production*. New York, NY: Harper Perennial.

15. Does NIH need a DARPA?[1]

Robert Cook-Deegan

The National Institutes of Health (NIH) recently celebrated the fiftieth anniversary of its Division of Research Grants with a symposium on peer review. NIH Director Harold Varmus introduced the theme of the day, likening competitive external peer review to democracy by invoking Churchill's quip: "the worst form of government except all the others that have been tried". This analogy expresses a belief in peer review that is widely shared among those who were in the audience. There are, however, a couple of problems with this analogy. First, it is factually incorrect. Some agencies—notably the Defense Advanced Research Projects Agency (DARPA; ARPA during some periods) and the armed services' R&D operations—have demonstrated that other methods work quite well, arguably as well as or better than those used at NIH. Second, comparing peer review to democracy implies a false dichotomy. A country cannot be at once a democracy and a dictatorship, but an agency can simultaneously use both peer review and other mechanisms to support R&D; indeed, several defense R&D agencies do just that.

The chief alternatives to competitive peer review are formula funding methods, based on political, historical, or performance factors, and what might be called the DARPA model, in which staff experts decide how to distribute research funds. Formula funding would

1 Originally published as Cook-Deegan, R. (1997). "Does NIH Need a DARPA?", *Issues in Science and Technology* 13/2, https://issues.org/cookde/

 https://doi.org/10.11647/OBP.0184.15

surely reduce transaction costs and could provide a stable flow of support to good researchers. The price of reducing transaction costs through formula funding, however, is the loss of expert judgment about innovative promise. The desire to invest in such promise, as opposed to past performance alone, is a major reason that agencies have come to rely on outside expert advice. But the DARPA approach is also a way to foster innovation.

DARPA's effectiveness depends on expert staff, clear mission, focused effort, and lean management. DARPA's main function is to quickly exploit new inventions, ideas, and concepts with potential military utility. Its eighty or so program managers distribute between $2 billion and $2.5 billion annually and are supervised by a half-dozen office directors, who in turn report to the DARPA director. Thus, only one management layer exists between the DARPA director and the program managers. The entire DARPA staff is roughly comparable in size to that responsible for administering extramural funds for the National Human Genome Research Institute (NHGRI) or one of the smaller NIH institutes that expends between $100 million and $200 million.

DARPA managers are hired for their expertise, often from industry or academia, and typically serve for four years or less. Each handle from $10 million to $50 million of research funding per year, of which at least 20 percent is intended for new investments. The money for new programs is a direct result of DARPA's ruthless willingness to kill programs that are not meeting expectations. Success results from a long-term strategy pursued by highly expert staff who are given great discretion to manage substantial funding commitments. Those staff members are held accountable for the results produced by the programs they fund, in quarterly reviews and detailed annual assessments by the DARPA director.

In DARPA culture, managers are self-avowed scientific and technological fanatics. Their base skill is recognizing talent that is relevant to defense needs and providing funds for its expression. The institutional ethos is described as "80 decision makers linked by a travel office", which emphasizes its highly interactive (and, at times, intrusive) style. It is ironic that within one of the world's most notorious bureaucracies, the Department of Defense, resides a tribe of rambunctious technological entrepreneurs.

Created by the Eisenhower administration in the wake of the Soviet launch of Sputnik, DARPA played a crucial early role in the development of computer time-sharing, interactive computing, space launch vehicles, satellite surveillance, lasers, stealth technology, and many other technological innovations. Its twenty-five-year-old Information Processing Techniques Office (IPTO) is DARPA's best-known program outside defense technologies. IPTO spawned the first departments of computer science, bolstered an academic base for large-scale integrated chip design at a time when that foundation was eroding perilously, and created the prototype for today's Internet. It is safe to say that many computing activities we take for granted in the 1990s, such as e-mail, computer graphics, interactive computing, alternative chip architectures, and networking, can be traced to DARPA funding decisions made in the 1960s and 1970s.

Biomedical Success

This period has also been a time of remarkable progress in biomedical research, and NIH has played a central role. NIH funding accounts for almost 30 percent of the world's biomedical research literature, compared to about 40 percent from other U.S. sources and about 30 percent from all foreign sources. The volume and excellence of U.S. biomedical research, as well as the innovative power of industries dependent on such research (such as pharmaceuticals, medical devices, and biotechnology), can largely be attributed to NIH and its system of peer review.

But is peer review the only way to achieve success in this field? In materials science, telecommunications, space, lasers, and microelectronics—other fields in which the United States is the world leader—the nation's advantages in R&D arguably derive as much from mission-oriented agency-directed research and technology development as from peer-reviewed science. In many fields of engineering, mathematics, and physical sciences, the National Science Foundation's (NSF) base of peer-reviewed grants is complemented by other agencies' dynamic portfolio of mission-related science and technology, much of which is funded outside of peer review.

Many of these fields do seem more like engineering than pure science, and some people assume that DARPA's funding procedures are suited to technology with definite aims but not to science. Experience suggests otherwise, however. Packet switching for electronic communication, computer time-sharing, integrated large-scale chip design, and networking were as conceptually "basic" when DARPA was funding them as most molecular biological experiments are today. Nothing was there except a notion that computers could be made to do things they had never done before. When NSF and NIH both frowned upon funding work on neural networks, Leon Cooper received funding thanks to the judgment of a program manager at the Office of Naval Research (ONR), which uses a mix of peer review and DARPA-like funding mechanisms. ONR also led the way toward single-atom chemistry, "squeezed" states of light, and acoustics—all fields with a heavy dose of basic science.

Another reason to consider the DARPA approach is its lower transaction costs. Administrative review costs at NIH or NSF rise arithmetically with the number of applications. External costs, however, rise much faster as the percentage of proposals that are funded falls. If half of all proposals result in funding, which was the case at NIH several decades ago, one unfunded grant proposal is prepared for each one funded. When success rates fall to one in five or six, as they have in several areas, four or five proposals are wasted for everyone funded. Preparing a grant proposal is a substantial effort, and the total external costs for all applicants may approach or even exceed the amount awarded to the successful one. Physicist Leo Szilard once noted that, at some point in a competitive grant system, applying for grants would consume all of a scientist's time, leaving none for research. With 15- to 20-percent success rates, a "Szilard point" (where waste exceeds benefit) is no longer a frivolous speculation, but a real possibility. Whereas NIH extramural administrators spend most of their time crafting rules for competition and then selecting among applicants, DARPA staff spend most of their time keeping abreast of their fields and camping in sparsely populated outposts along the technological and scientific frontiers.

Many scientists and engineers fear that grant competition has pushed peer review well past its power to distinguish the truly outstanding from the merely excellent. The least painful solution to this problem, at least for the scientists and engineers seeking funds, is more money for grants, so that more are funded, the success rate rises, the relative external costs fall, and reviewers need only separate the good from the excellent. To relieve the tension in the peer review system would require at least a doubling of federal research support in combination with a "birth control" policy to stem the growth of the applicant pool. Although NIH enjoys stalwart bipartisan support, a budget increase of this magnitude is unlikely; and even if budgets grow, the applicant pool may well grow faster, if history is any guide.

Although important, budget constraints and administrative inefficiency are not the most compelling reasons to experiment with DARPA-like funding mechanisms. The most serious threat to science under the peer review system is conservatism—the safe squeezing out the novel. A look at the history of NIH involvement in DNA sequencing illustrates how a DARPA-like mechanism might prove more effective than external, prospective peer review. In 1981, Leroy Hood and his colleagues at Caltech applied for NIH (and NSF) funding to support their efforts to automate DNA sequencing. They were turned down. Fortunately, the Weingart Institute supported the initial work that became the foundation for what is now the dominant DNA sequencing instrument on the market. By 1984, progress was sufficient to garner NSF funds that led to a prototype instrument two years later. In 1989, the newly created National Human Genome Research Institute (NHGRI) at NIH held a peer-reviewed competition for large-scale DNA sequencing. It took roughly a year to frame and announce this effort and another year to review the proposals and make final funding decisions, which is a long time in a fast-moving field. NHGRI wound up funding a proposal to use decade-old technology and an army of graduate students but rejected proposals by J. Craig Venter and Leroy Hood to do automated sequencing. Venter went on to found the privately funded Institute for Genomic Research, which has successfully sequenced the entire genomes of three microorganisms and has conducted many other successful sequencing efforts; Hood's groups, first at Caltech and then at the University of Washington, went on to sequence the T cell receptor region, which is among the largest contiguously sequenced expanses of human DNA. Meanwhile, the army of graduate students has yet to complete its sequencing of the bacterium Escherichia coli. The point is not that the study section bet wrong—any research funding must be fault-tolerant and take risks-but that it bet on old technology over new.

NIH and NSF have long struggled with the tendency toward conservatism in peer review. NSF has set aside small grants for exploratory research that is subject only to expeditious staff review. With NSF's tradition of grant managers rotating into and out of their fields in academia, this is similar in spirit to DARPA, although the dollar amounts are generally too small to fund more than pilot projects. NSF has a good idea, but there is no reason to believe that innovative projects are always small. Besides, requiring that innovation prove itself early with small grants may lead to premature declarations of failure and force investigators to write a follow-up grant at the same time as they have only a few months' funding to do the pilot work. At NIH, some study sections set aside specific grants or are given the option of selecting one or a few especially novel proposals for special consideration. But this does not avoid the inefficiencies of the group process and of grant proposal preparation, and it ultimately amounts to a few groups doing sporadically what individual experts might do better.

A Small Dose of DARPA

A DARPA-like funding mechanism cannot cover the same breadth of science and technology as NIH or NSF. Even if a pilot test of a DARPA-like program is a success, it still should be considered as an alternative for a few select programs only. Much of the most important work supported by NIH and NSF is conducted through tens of thousands of relatively small grants. Innovation bubbles up in unexpected places thanks to the flexibility of the grant mechanism, which leaves funds largely under the control of investigators. NIH handles 45,000 grant applications per year. It would be folly to adopt DARPA's methods for so many small projects covering enormous areas of science. The DARPA system cannot scale up easily, because its effectiveness depends on a flat bureaucracy and strong direct accountability from manager to agency director. The DARPA process is best suited to force scientific and technical progress in critical areas and to accomplish tasks when a new technology is promising but not yet proven. It is not suited to sustaining the bulk of scientific research.

DARPA-like pilot projects might be tried first by one or a few NIH institutes or center directors working with their respective councils to foster specific fields or to develop needed technical capacities. If NIH were to experiment with a DARPA-like mechanism, it should focus on areas that are ripe for such experimentation, such as:

An emerging technological capacity that would be widely beneficial if successfully developed,

- An advance promising a major leap, not an incremental improvement,
- A capacity whose development requires substantial sustained funding,
- A field or technique that is unlikely to be developed by ongoing academic efforts or within industrial firms,
- An emerging scientific field or technical area that lacks a natural disciplinary base, or
- A promising new field populated by only a few individuals.

NIH has amply demonstrated its agility and excellence, maintaining scientific quality and administering a credible and effective process for allocating funds. That solid base of peer-reviewed science should be not be chipped and fragmented. The edifice could benefit from a new wing, however, that poses little danger to its foundations. One or two institute directors could hire some rising stars and make them responsible for moving their fields ahead rapidly. After four or five years, the results of NIH's "DARPA corps" could be compared to the record of peer review groups in similar areas.

Testing a DARPA mechanism within NIH is not a call to end peer review as we know it, or even a substantial fraction of it. But neither is the generally excellent track record of NIH and NSF any proof that a DARPA-like mechanism can't improve the system. In the 1960s, C. Jackson Grayson wrote a classic work on oil drilling that demonstrated why a long-term diversified strategy is important for success when confronting uncertainty. Peer review is best regarded as a way to contend with moderate uncertainty, but it is not a good way to decide where to wildcat. DARPA's methods seem better suited to that, and some wildcatting is a good idea.

PART IV

CONCLUSIONS

16. Lessons from DARPA's Experience

Richard Van Atta, Patrick Windham and William B. Bonvillian

DARPA has been considered unique in having successfully promoted transformative innovation for more than six decades. Its ability to do so is based on several key features, which have been elaborated in the chapters of this book. It should be noted that these features have varied in emphasis over the years. Not all of these features existed when DARPA was created—indeed they evolved as the agency evolved. DARPA initially began with little explicit structure, organizational architecture, or management processes. It was largely *ad hoc.* Its first programs were large collections of projects aimed to tackle various aspects of the three Presidential issues it was given as its first assignments: (1) get the U.S. into space; (2) missile defense; (3) nuclear test detection. These were large, umbrella tasks for which there was no well-defined path—they were all highly exploratory and required multiple approaches. Soon after it was established, DARPA took on another area of research—Project AGILE, which was to provide technical support to counterinsurgency in Vietnam. This began a decade-long on-going program which in many ways, in retrospect, had features contradictory to almost all of those identified by the chapters in this book. AGILE also was an ignominious failure. This failure could be due to the fact that it was trying to develop technical solutions to intrinsically political problems. But it was also due to the fact

 https://doi.org/10.11647/OBP.0184.16

is it was run without any of the discipline and clear management focus that have become associated with DARPA programs. There is perhaps an important lesson here: just being ambitious and taking on a major problem, such as "counterinsurgency" is not enough—it is necessary to bound that problem and apply well-founded management principles.

Important Features

In this Conclusion, we want to highlight some of the most important of these features:

Ambitious technical goals. DARPA focuses on high-risk, high-potential projects. Moreover, it is not simply a research agency. Its focus and goal is to create and demonstrate new significant technologies and systems. DARPA works on both basic technologies and components *and* on prototype systems that use these advanced technologies to demonstrate new and valuable equipment, processes, and other systems. It therefore both helps to create new technologies and applies them in useful and novel ways.

Organizational independence. DARPA takes on problems that are beyond or outside those of other defense organizations. One way this has been phrased is that if there is a defined "requirement" for something—that is, there is a known approach for accomplishing a specific objective—then it is not a job for DARPA. The Defense Department chartered DARPA to focus on new technologies and approaches, not incrementally improve on what exists.

Freedom from bureaucratic procedures. Moreover, the agency is free of day-to-day bureaucratic interference from other parts of the Defense Department. It can move quickly, without having to check with outside organizations or committees. While the Director and the Office Directors may take overall direction from the Secretary of Defense and some other high-level OSD executives, this is almost entirely focused on stating defense priorities and not how to do the research itself.

Highly-talented managers and a lean management structure. DARPA focuses on some of the hardest and most important technical problems in the U.S., and will succeed only if both DARPA and the R&D performers DARPA funds are among the very best technical people in the country. The agency hires excellent program managers and then lets them propose programs, run competitions, select R&D performers, and work with those performers. The agency has only two layers of management above the program manager: the office director and deputy and the agency director and deputy. Moreover, these managers are themselves technical experts and can make informed technical judgments quickly.

Temporary R&D teams. DARPA does not have its own internal laboratory and instead funds outside R&D performers, usually for projects that last three to five years. As a result, the agency has great flexibility. If progress is strong in a particular program area, DARPA can extend funding over several generations of programs. If a program fails, then it is terminated and funds are used for other work. In addition, once a new technology is created and demonstrated, DARPA can move into other, newer areas.

A technically-sophisticated and well-funded customer. The Defense Department is a sophisticated customer, which makes technology transfer practical. Moreover, DARPA is most successful when senior Defense Department officials not only support the agency's independence, but also work to transfer its new technologies to the military services.

Continuous management, not post hoc evaluation. Because DARPA program managers are themselves technical experts, they can quickly judge whether R&D projects are succeeding or not and, equally important, can work with R&D performers to change projects when surprises and changes inevitably occur. DARPA therefore has a process of continuous learning. And if a project fails even after changes, then it is terminated. DARPA does not perform post hoc assessments of its research program as is done by many other government research organizations. In this sense it is run more like a business where the "evaluation" is in results, recognizing that not all its projects will succeed.

A credible process for accountability. R&D quality and agency accountability are ensured by picking an excellent director and excellent program managers, by this process of continuous evaluation, and by having oversight (but not heavy paperwork) from senior government officials. There is no need for supervision by committees, outside evaluations and audits, or other bureaucratic steps that would actually slow down the agency's work.

An effective political design. DARPA has built strong, enduring political support based on its performance, and therefore has had stable budgets and continuing independence. It has this support because it performs a vital mission (keeping the U.S. military technologically advanced), focuses on long-term challenges and opportunities facing the Defense Department, does not threaten the budgets of other DOD agencies, has won the respect and support of the U.S. technical community, and has credible procedures for tracking program progress and maintaining quality.

The chapters in this book present greater detail on these features and also other aspects of its management and operations that are seen as useful for it to fulfill its special mission. Some of these mechanisms have been introduced to assure that DARPA does not get bogged down in stultifying bureaucratic processes that inhibit its flexibility and adaptability. We have emphasized that a key feature is the ability to bring on technically expert program managers for explicit, short term appointments. Another feature is the ability to quickly undertake, but also if needed, quickly cancel specific projects. To make these feasible, DARPA uses ***flexible hiring and contracting authorities***.

Creating New Technical Communities

By funding multi-disciplinary teams that both compete and cooperate with each other, DARPA often helps create new technical communities and new academic fields. Examples over the years include materials science and engineering, computer science, and, more recently, artificial intelligence, autonomous systems, and synthetic biology/engineering biology. In fact, one can argue that DARPA actually makes two very important contributions: it not only helps create and demonstrate new technologies, but also helps create important new technical communities.

These researchers then can perform additional R&D, teach students, and contribute further ideas to DARPA, as well as commercialize the technologies. Indeed, DARPA-funded communities are a primary means for transitioning the newly developed technologies to the military and to commercial companies.

DARPA and the Future

DARPA has existed for over sixty years and has had massive impacts on many areas of defense capabilities. It has also produced much broader, revolutionary advances in information technologies, microelectronics, materials, and other areas, that have had profound economic and societal impacts. DARPA has garnered a reputation as the innovation icon—often pointed to as the most successful U.S. innovation agent. DARPA has changed over time in response to the changing security, technological and governmental landscapes.

DARPA's higher-risk, longer-term R&D agenda distinguishes it from other defense R&D organizations. Perhaps the most important effect of DARPA's work is to change people's minds as to what is possible. DARPA's sixty-year history reveals an institution driven by a constant imperative to create novel, high-payoff capabilities by pushing the frontiers of knowledge. DARPA has many of the same features as its research. DARPA began as an experiment aimed at overcoming the usual incremental processes of technology development. Like the research it is chartered to develop, DARPA consistently has been purposively "disruptive" and "transformational". Over the decades, there have been various efforts to tone down DARPA, make its research more compatible and integrated into the rest of DOD R&D, and have it focus more heavily on nearer term, more incremental applications—that is, to shift its focus away from disruptive possibilities.

Additionally, there have been efforts to broaden its charter into prototyping systems beyond the proof-of-concept demonstrations DARPA traditionally has carried out. However, with strong internal leadership, both within DARPA and in the OSD, as well as with support from Congress, DARPA has been able to perform a truly unique role for six decades. It has been, and continues to be, DOD's "Chief Innovation Agency", pushing the frontiers of what is possible for the benefit of national security and the nation.

DARPA remains an impressive "opportunity farm". For example, DARPA helped move "artificial intelligence" (AI) from an inchoate notion with almost no technological underpinnings into pervasive capabilities affecting our everyday lives and supporting real-time military operational decision-making. It is now pursuing similar advances in cognitive computing and robotics. It is pursuing fundamental advances in materials such as biomaterials, and accelerated materials development. The Agency has revolutionized the realm of distributed sensing. Among many current DARPA research topics that populate the opportunity farm are heterogeneous electronics, engineering biology, agile access to space, and hypersonic systems.

Looking to the future, the question is not whether DARPA can still pursue new change-state prospects. The question is this: how can DARPA and the Department of Defense identify and focus on what these should be in the changing geopolitical and technological environments? DARPA has been adroit in addressing emerging technological prospects—but, in today's world, it has to be yet more focused on where it can have leverage, as others are investing, often massively, in the very technologies that DARPA initially championed.

With global investments in robotics, AI, synthetic biology, quantum computing, and advanced materials, on what should DARPA focus? From a military applications perspective, what should DARPA do to harness and promote the potential use of such emerging technologies into defense uses? To what needs, as opposed to today's defined requirements, should DARPA seek to employ these technologies? Crucially how should DOD achieve the fruition of these efforts?

Today DARPA faces new challenges that raise a key issue concerning its future success—the ability to draw upon extraordinary technical talent for program managers. The commercial high-tech sector, particularly in such areas as information technology, autonomous systems, advanced biology—areas in which DARPA is focused—is aggressively spending vast sums and hiring the very best. These firms attract this talent with high salaries, relatively unfettered work environments, in locations far from Washington, DC, with foreign nationals making up a growing proportion. Moreover, many leading tech companies are now outside of the U.S., in Asia and Europe. Thus, there is greater competition for technical talent and greater competition worldwide in advanced technologies. Even as DARPA now must confront a tougher recruiting context that it has in the past, it still presents prospective program managers unique opportunities to affect the future that few other organizations can offer. These talent dynamics are crucial to understanding today's DARPA and its ongoing mission to identify, demonstrate and develop the technologies of the future.

These questions emphasize a crucial point—DARPA does not succeed by itself. Its success resides in the opportunities it creates that *others* bring into fruition. Thus, its success must build upon the larger U.S. innovation infrastructure. That innovation ecosystem has changed fundamentally over the past twenty-five years. For DARPA research to be successful, it must eventually culminate in transition, whether in an operational military capability or a new field of technology that expands frontiers for decades. DARPA itself is not responsible for executing transitions, but it depends on effective transition paths being there. These paths need to be better understood, and other stakeholders, beyond DARPA, need to support the measures that foster transition—whether within DOD or within industry. Some worry that military transition mechanisms within the DOD have eroded. In the broader commercial economy, transition paths have become more uncertain and diffuse. For DARPA to continue to have transformative impacts, it must exist within an economic and policy environment that encourages implementation. These are critical technology policy concerns that the U.S. must address to ensure that DARPA can continue to deliver breakthrough technologies in the decades to come.

Further Reading

DARPA Documents

The following DARPA reports provide good summaries of the agency's mission, philosophy, and programs at different points in time:

DARPA. (2015). *Breakthrough Technologies for National Security*. Arlington, VA: Defense Advanced Research Projects Agency, https://www.esd.whs.mil/Portals/54/Documents/FOID/Reading%20Room/DARPA/15-F-1407_BREAKTHROUGH_TECHNOLOGIES_MAR_2015-DARPA.pdf

DARPA. (2013). *Driving Technological Surprise: DARPA's Mission in a Changing World*. Arlington, VA: Defense Advanced Research Projects Agency, https://defenseinnovationmarketplace.dtic.mil/wp-content/uploads/2018/02/DARPAStrategicPlan.pdf

DARPA. (2005). *DARPA—Bridging the Gap, Powered by Ideas*. Arlington, VA: Defense Advanced Research Projects Agency, http://www.dtic.mil/cgi-bin/GetTRDoc?Location=U2&doc=GetTRDoc.pdf&AD=ADA433949

General Histories of DARPA

Barber Associates, R. (1975). *The Advanced Research Projects Agency, 1958–1974*. Report prepared for the Advanced Projects Research Agency. Springfield, VA: Defense Technical Information Center.

Van Atta, R., Deitchman, S., and Reed, S. (1990). *DARPA Technical Accomplishments. Volume I.* Alexandria, VA: Institute for Defense Analyses, https://apps.dtic.mil/dtic/tr/fulltext/u2/a239925.pdf

Van Atta, R., Deitchman, S., and Reed, R. (1991). *DARPA Technical Accomplishments. Volume II.* Alexandria, VA: Institute for Defense Analyses, https://apps.dtic.mil/dtic/tr/fulltext/u2/a241725.pdf

Van Atta, R., Deitchman, S., and Reed, S. (1991). *DARPA Technical Accomplishments. Volume III.* Alexandria, VA: Institute for Defense Analyses, https://apps.dtic.mil/dtic/tr/fulltext/u2/a241680.pdf

Van Atta, R., Lippitz, M., et al. (2003). *Transformation and Transition, DARPA's Role in Fostering a Revolution in Military Affairs. Volume 1.* Alexandria, VA: Institute for Defense Analyses, https://doi.org/10.21236/ada422835, https://fas.org/irp/agency/dod/idarma.pdf

Weinberger, S. (2017). *The Imagineers of War: The Untold Story of DARPA, The Pentagon Agency That Changed the World.* New York, NY: Alfred A. Knopf.

Histories of DARPA's Contributions to Information Technology

Several excellent books describe ARPA's early and very pivotal contributions to computing and computing networking, including its role in the creation of the ARPANET and Internet Protocol. These books also stress how ARPA/DARPA both drew on the overall U.S. technical community and, in turn, contributed to that community:

Flamm, K. (1988). *Creating the Computer: Government, Industry, and High Technology.* Washington, DC: The Brookings Institute.

Hafner, K., and Lyon, M. (1996). *Where Wizards Stay Up Late: The Origins of the Internet.* New York, NY: Simon & Schuster.

Segaller, S. (1998). *Nerds 2.0.1: A Brief History of the Internet.* New York, NY: TV Books.

Waldrop, M. M. (2001). *The Dream Machine: J. C. R. Licklider and the Revolution that Made Computing Personal.* New York, NY: Viking Press.

List of Tables and Illustrations

Chapter 3

Chapter 5

Chapter 6

Chapter 7

Chapter 8

Index

This book need not end here...

At Open Book Publishers, we are changing the nature of the traditional academic book. The title you have just read will not be left on a library shelf, but will be accessed online by hundreds of readers each month across the globe. OBP publishes only the best academic work: each title passes through a rigorous peer-review process. We make all our books free to read online so that students, researchers and members of the public who can't afford a printed edition will have access to the same ideas.

This book and additional content is available at:
https://doi.org/10.11647/OBP.0184

Customise

Personalise your copy of this book or design new books using OBP and third-party material. Take chapters or whole books from our published list and make a special edition, a new anthology or an illuminating coursepack. Each customised edition will be produced as a paperback and a downloadable PDF. Find out more at:

https://www.openbookpublishers.com/section/59/1

Donate

If you enjoyed this book, and believe that research like this should be available to all readers, regardless of their income, please become a member of OBP and support our work with a monthly pledge — it only takes a couple of clicks! We do not operate for profit so your donation will contribute directly to the creation of new Open Access publications like this one.

https://www.openbookpublishers.com/supportus

You may also be interested in:

ANZUS and the Early Cold War
Strategy and Diplomacy Between Australia, New Zealand and the United States, 1945-1956
Andrew Kelly

https://doi.org/10.11647/OBP.0141

History of International Relations
A Non-European Perspective
Erik Ringmar

https://doi.org/10.11647/OBP.0074

Peace and Democratic Society
Amartya Sen (ed.)

https://doi.org/10.11647/OBP.0014

www.ingramcontent.com/pod-product-compliance
Lightning Source LLC
LaVergne TN
LVHW011858200926
848484LV00002B/477